Power Systems

Werner Leonhard, Control of Electrical Drives

Springer

Berlin
Heidelberg
New York
Barcelona
Hong Kong
London
Milano
Paris
Singapore
Tokyo

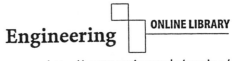

Engineering ONLINE LIBRARY

http://www.springer.de/engine/

Werner Leonhard

Control
of Electrical Drives

Third Edition

With 299 Figures

 Springer

Professor WERNER LEONHARD
Technische Universität Braunschweig
Institut für Regelungstechnik
Hans Sommer Str. 66
38106 Braunschweig
e-mail: W.Leonhard@tu-bs.de
Internet: http://www.ifr.ing.tu-bs.de

ISBN 3-540-41820-2 Springer-Verlag Berlin Heidelberg NewYork

Library of Congress Cataloging-in-Publication Data

Leonhard, Werner:
Control of Electrical Drives. – Third edition. / Werner Leonhard. – Berlin; Heidelberg; New York;
Barcelona; Hong Kong; London; Milan; Paris; Singapore; Tokyo: Springer 2001
(Power Systems)
ISBN 3-540-41820-2

Springer-Verlag is a company in the BertelsmannSpringer publishing group
http://www.springer.de
© Springer-Verlag Berlin Heidelberg New York 2001
Printed in Germany

Typesetting: Camera ready by author
Cover-Design: de´blik, Berlin
Printed on acid free paper   SPIN: 10833196   62/3020/kk - 5 4 3 2 1 0

Preface

Electrical drives play an important role as electromechanical energy converters in transportation, material handling and most production processes. The
ease of controlling electrical drives is an important aspect for meeting the increasing demands by the user with respect to flexibility and precision, caused
by technological progress in industry as well as the need for energy conservation. At the same time, the control of electrical drives has provided strong
incentives to control engineering in general, leading to the development of
new control structures and their introduction to other areas of control. This
is due to the stringent operating conditions and widely varying specifications
— a drive may alternately require control of torque, acceleration, speed or
position — and the fact that most electric drives have — in contrast to chemical or thermal processes — well defined structures and consistent dynamic
characteristics.

During the last years the field of controlled electrical drives has undergone
rapid expansion due mainly to the advances of semiconductors in the form of
power electronics as well as analogue and digital signal electronics, eventually culminating in microelectronics and microprocessors. The introduction
of electronically switched solid-state power converters has renewed the search
for adjustable speed AC motor drives, not subject to the limitations of the
mechanical commutator of DC drives which dominated the field for a century.
This has created new and difficult control problems since the mechanically
simpler AC machine is a much more involved control plant than a DC motor;
on the other hand, the fast response of electronic power switching devices and
their limited overload capacity have made the inclusion of protective control
functions essential. The present phase of evolution is likely to continue for
some years, a new steady-state is not yet in sight.

This book, originally published 1974 in German as "Regelung in der elektrischen Antriebstechnik", was an outcome of lectures the author held for a
number of years at the Technical University Braunschweig. In its updated English version it characterises the present state of the art without laying claim
to complete coverage of the field. Many interesting details had to be omitted,
which is not necessarily a disadvantage since details are often bound for early
obsolescence. In selecting and presenting the material, didactic view points
have also been considered. A prerequisite for the reader is a basic knowledge

of power electronics, electrical machines and control engineering, as taught in most under-graduate electrical engineering courses; for additional facts, recourse is made to special literature. However, the text should be sufficiently self contained to be useful also for non-experts wishing to extend or refresh their knowledge of controlled electrical drives.

They consist of several parts, the electrical machine, the power converter, the control equipment and the mechanical load, all of which are dealt with in varying depths. A brief resume of mechanics and of thermal effects in electrical machines is presented in Chaps. 1–4 which would be skipped by the more experienced reader. Chaps. 5–9 deal with DC drives which have for over a century been the standard solution when adjustable speed was required. This part of the text also contains an introduction to line-commutated converters as used for the supply of DC machines. AC drives are introduced in Chap. 10, beginning with a general dynamic model of the symmetrical AC motor, valid in both the steady-state and transient condition. This is followed in Chap. 11 by an overview of static converters to be employed for AC drives. The control aspects are discussed in Chaps. 12–14 with emphasis on high dynamic performance drives, where microprocessors prove invaluable in disentangling the multivariate interactions present in AC machines. Chapter 15 finally describes some of the problems connected with the industrial application of drives. This cannot by any means cover the wide field of special situations with which the designer is confronted in practice but some frequently encountered features of drive system applications are explained there. It will become sufficiently clear that the design of a controlled drive, in particular at larger power ratings, cannot stop at the motor shaft but often entails an analysis of the whole electro-mechanical system.

In view of the fact that this book is an adaptation and extension of an application-orientated text in another language, there are inevitably problems with regard to symbols, the drawing of circuit diagrams etc. After thorough consultations with competent advisors and the publisher, a compromise solution was adopted, using symbols recommended by IEE wherever possible, but retaining the authors usage where confusion could otherwise arise with his readers at home. A list of the symbols is compiled following the table of contents. The underlying principle employed is that time varying quantities are usually denoted by lower case latin letters, while capital letters are applied to parameters, average quantities, phasors etc; greek letters are used predominantly for angles, angular frequencies etc. A certain amount of overlap is unavoidable, since the number of available symbols is limited. Also the bibliography still exhibits a strong continental bias, eventhough an attempt has been made to balance it with titles in english language. The list is certainly by no means complete but it contains the information readily available to the author. Direct references in the text have been used sparingly, when the origin was to be acknowledged. Hopefully the readers are willing to accept these shortcomings with the necessary patience and understanding.

The author wishes to express his sincere gratitude to two British colleagues, R. M. Davis, formerly of Nottingham University and S. R. Bowes of the University of Bristol who have given help and encouragement to start the work of updating and translating the original German text and who have spent considerable time and effort in reviewing and improving the initial rough translation; without their assistance the work could not have been completed. Anyone who has undertaken the task of smoothing the translation of a foreign text can appreciate how tedious and time-consuming this can be. Thanks are also due to the editors of this Springer Series, Prof. J. G. Kassakian and Prof. D. H. Naunin, and the publisher for their cooperation and continued encouragement.

Braunschweig, October 1984   *Werner Leonhard*

Preface to the 2nd edition

During the past 10 years the book on Control of Electrical Drives has found its way onto many desks in industry and universities all over the world, as the author has noticed on numerous occasions. After a reprinting in 1990 and 1992, where errors had been corrected and a few technical updates made, the book is now appearing in a second revised edition, again with the aim of offering to the readers perhaps not the latest little details but an updated general view at the field of controlled electrical drives, which are maintaining and extending their position as the most flexible source of controlled mechanical energy.

The bibliography has been considerably extended but in view of the continuous stream of high quality publications, particularly in the field of controlled AC drives, the list is still far from complete. As those familiar with word processing will recognise, the text and figures are now produced as a data set on the computer. This would not have been possible without the expert help by Dipl.-Ing. Hendrik Klaassen, Dipl.-Math. Petra Heinrich, as well as Dr.-Ing. Rüdiger Reichow, Dipl.-Ing. Marcus Heller, Mrs. Jutta Stich and Mr. Stefan Brix, to whom the author wishes to express his sincere gratitude. Peter Wilson has been reading some of the chapters, offering valuable suggestions. The final layout remained the task of the publishers, whose patience and helpful cooperation is gratefully appreciated.

Braunschweig, May 1996   *Werner Leonhard*

Preface to the 3rd Edition

The early depletion of the publishers stocks offered an opportunity of again updating the book, making minor corrections and enlarging some subjects, while the main part of the book was left unchanged. Tackling this work required the encouragement by good friends but the author would still have been unable to realise it alone. He expresses his sincere gratitude to the helpful researchers at the Institut für Regelungstechnik, particularly Dipl.-Ing. Jan Becker and Dipl.-Ing. Frithjof Tobaben who were always ready to resolve software-related crises on the computer, as well as Dipl.-Ing. Klaus Jaschke and cand. Wirtsch.-Inform. Danny Wallschläger, our definitive LaTeX-experts whose participation was essential for undertaking the task. Dr.-Ing. Sönke Kock suggested a clearer definition of the magnetic leakage which had gone unnoticed before; finally, I want to thank Prof. Walter Schumacher, now head of the laboratory, for his continued interest.

Braunschweig, Spring 2001   *Werner Leonhard*

Contents

Abbreviations and Symbols

1 Equations

In all equations comprising physical variables, they are described by the product of a unit and a dimensionless number, which depends on the choice of the unit.

Some variables are nondimensional due to their nature or because of normalisation (p.u.).

2 Characterisation by Style of Writing

$i(t)$, $u(t)$, etc.	instantaneous values
$\bar{i}$, I_d, $\bar{u}$, U_d, etc.	average values
I, U, etc.	RMS-values
$\underline{I}$, $\underline{U}$, etc.	complex phasors for sinusoidal variables
$\underline{i}(t)$, $\underline{u}(t)$, etc.	complex time-variable vectors, used with multiphase systems
$\underline{i}^*(t)$, $\underline{u}^*(t)$, $\underline{I}^*$, $\underline{U}^*$, etc.	conjugate complex vectors or phasors
${}^1\underline{i}(t)$, ${}^1\underline{u}(t)$, etc.	vectors in special coordinates
$I(s) = L(i(t))$ etc.	Laplace transforms

3 Symbols

Abbreviation	Variable	Unit
$a(t)$	current distribution	A/m
	linear acceleration	$\mathrm{m/s^2}$
	nondimensional factor	
A	area	$\mathrm{m^2}$
b	nondimensional field factor	
B	magnetic flux density	$\mathrm{T = Vs/m^2}$
C	electrical capacity	$\mathrm{F = As/V}$
	thermal storage capacity	$\mathrm{J/^\circ C = Ws/^\circ C}$
D	damping ratio	

$e(t)$, E, $\underline{E}$	induced voltage, e.m.f.	V
f	frequency	Hz = 1/s
	force	N
$F(s)$	transfer function	
g	gravitational constant	m/s^2
$g(t)$	unit impulse response	
G	weight	N
	gain	
h	airgap	m
$i(t)$, I, $\underline{I}$	current	A
J	inertia	kg m^2
k	nondimensional factor	
K	torsional stiffness	Nm/rad
l	length	m
L	inductance	H = Vs/A
$m(t)$	torque	Nm
M	mass	kg
	mutual inductance	H
n	speed, rev/min	1/min
N	number of turns	
$p(t)$, P	power	W
Q	reactive power	VA
r	radius	m
R	resistance	Ω
$s = \sigma + j\,\omega$	Laplace variable	rad/s
s, x	distance	m
S	slip	
t	time	s
T	time constant	s
$u(t)$, U, $\underline{U}$	voltage	V
$v(t)$	velocity	m/s
	unit ramp response	
V	volume	m^3
$w(t)$	unit step response	
	energy	J = Ws
x	control variable	
y	actuating variable	
z	disturbance variable	
$z = e^{sT}$	discrete Laplace variable	
Y	admittance	$1/\Omega = $ S
Z	impedance	Ω
α	coefficient of heat transfer	W/m^2 $^\circ$C
	firing angle	
	angular acceleration	

$\alpha, \beta, \delta, \zeta, \xi, \lambda, \mu, \varrho$ etc.	angular coordinates	rad
$\gamma = 2\pi/3$		
δ	load angle	rad
Δ	difference operator	
ε	angle of rotation	rad
η	efficiency	
ϑ	temperature	°C
	absolute temperature	K
Θ	magnetomotive force, m.m.f.	A
μ_0	coefficient of permeability	H/m
ν	integer number	
σ	leakage factor	
$\tau = \int \omega\,dt,\ \omega t$	normalised time, angle	rad
φ	phase shift	rad
$\cos\varphi$	power factor	
Φ	magnetic flux	Wb = Vs
ψ	flux linkage	Wb = Vs
ω	angular frequency	rad/s

4 Indices

i_a	armature current
i_e	exciting current
u_F	field voltage
i_S	stator current
i_R	rotor current
$i_{Sd},\ i_{Sq}$	direct and quadrature components of stator current
$i_{Rd},\ i_{Rq}$	direct and quadrature components of rotor current
i_m	magnetising current
i_{mR}	magnetising current representing rotor flux
i_{mS}	magnetising current representing stator flux
m_L	load torque
m_M	motor torque
m_p	pull-out torque
S_p	pull-out slip

5 Graphical Symbols

$$T \frac{dx}{dt} = y$$ integrator

$$T \frac{dx}{dt} = G \left(T \frac{dy}{dt} + y\right)$$ PI controller

$$T \frac{dx}{dt} + x = G\,y$$ first order lag

$$T_1 \frac{dx}{dt} + x = G \left(T_2 \frac{dy}{dt} + y\right)$$ first order lead/lag

$$x(t) = G\,y\,(t - T)$$ delay

$$x = y_1 - y_2$$ summing point

$$x = y_1\,y_2$$ multiplication

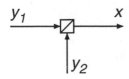

$$x = y_1/y_2$$

division

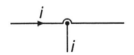

$$x = f(y)$$

nonlinearity

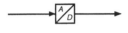

$$x = y \quad \text{for } x_{\min} < y < x_{\max}$$
$$x = x_{\min} \text{ for } y \leq x_{\min}$$
$$x = x_{\max} \text{ for } y \geq x_{\max}$$

limiter

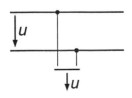

A/D- or D/A-
converter

current sensor

voltage sensor

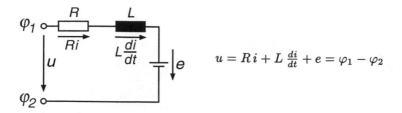

The voltage arrows indicating voltage sources (u, e) or voltage drops $\left(R\,i,\; L\,\frac{di}{dt}\right)$ represent the differences of electrical potential, pointing from the higher to the lower assumed potential. Hence the voltages in any closed mesh have zero sum, $\sum u = 0$.

Introduction

Energy is the basis of any technical and industrial development. As long as only human and animal labour is available, a main prerequisite for social progress and general welfare is lacking. The energy consumption per capita in a country is thus an indicator of its state of technical development, exhibiting differences of more than two orders of magnitude between highly industrialised and not yet developed countries.

In its primary form, energy is widely distributed (fossil and nuclear fuels, hydro and tidal energy, solar and wind energy, geothermal energy etc.), but it must be developed and made available at the point of consumption in suitable form, for instance chemical, mechanical or thermal, and at an acceptable cost. This creates problems of transporting the energy from the place of origin to the point of demand and of converting it into its final physical form. In many cases, these problems are best solved with an electrical intermediate stage, Fig. 0.1, where the bold numbers refer to the European grid, because electricity can be

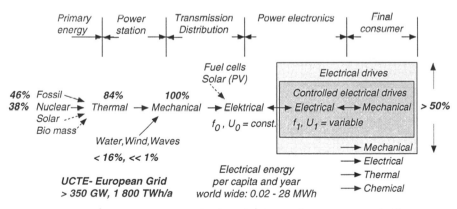

Fig. 0.1. From primary energy to final use, a chain of conversion processes

- generated from primary energy (chemical energy in fossil fuel, potential hydro energy, nuclear energy) in relatively efficient generating stations,
- transported with low losses over long distances and distributed simply and at acceptable cost,
- converted into any final form at the point of destination.

This flexibility is unmatched by any other form of energy.

Of particular importance is the mechanical form of energy which is needed in widely varying power ratings wherever physical activities take place, involving the transportation of goods and people or industrial production processes. For this final conversion at the point of utilisation, electromechanical devices in the form of electrical drives are well suited; it is estimated, that about half the electricity generated in an industrial country is eventually converted to mechanical energy. Most electrical motors are used in constant- speed drives that do not need to be controlled except for starting, stopping or protection, but there is a smaller portion, where torque and speed must be matched to the need of the mechanical load; this is the topic of this book. Due to the progress of automation and with a view to energy conservation, the need for control is likely to become more important in future. As an example, Fig. 0.2 shows the mechanical power needed by a centrifugal pump, when the flow is controlled by a variable speed drive or, still in frequent use, by throttle and bypass valves.

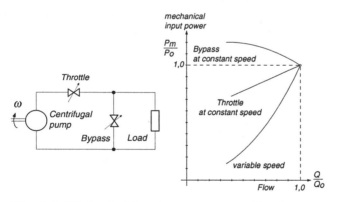

Fig. 0.2. Mechanical input power to a centrifugal pump using different methods of flow control

The predominance of electrical drives is caused by several aspects:

- Electric drives are available for any power, from 10^{-6} W in electronic watches to $> 10^8$ W for driving pumps in hydro storage plants.
- They cover a wide range of torque and speed, $> 10^7$ Nm, for an ore mill motor, $> 10^5$ 1/min, for a centrifuge drive.

- Electric drives are adaptable to almost any operating conditions such as forced air ventilation or totally enclosed, submerged in liquids, exposed to explosive or radioactive environments. Since electric motors do not require hazardous fuels and do not emit exhaust fumes, electrical drives have no detrimental effect on their immediate environment. The noise level is low compared, for instance, with combustion engines.
- Electric drives are operable at a moment's notice and can be fully loaded immediately. There is no need to refuel, nor warm-up the motor. The service requirements are very modest, as compared with other drives.
- Electrical motors have low no-load losses and exhibit high efficiency; they normally have a considerable short-time overload capacity.
- Electrical drives are easily controllable. The steady state characteristics can be reshaped almost at will, so that railway traction motors do not require speed-changing gears. High dynamic performance is achieved by electronic control.
- Electrical drives can be designed to operate indefinitely in all four quadrants of the torque-speed-plane without requiring a special reversing gear, Fig. 0.3. During braking, i.e. when operating in quadrants 2 or 4, the drive is normally regenerating, feeding power back to the line. A comparison with combustion engines or turbines makes this feature look particularly attractive.

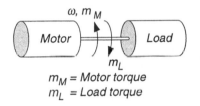

m_M = Motor torque
m_L = Load torque

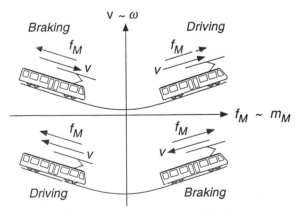

Fig. 0.3. Operating modes of an electric drive in all quadrants of the torque/speed plane

- The rotational symmetry of electrical machines and (with most motors) the smooth torque results in quiet operation with little vibrations. Since there are no elevated temperatures causing material fatigue, long operating life can be expected.
- Electrical motors are built in a variety of designs to make them compatible with the load; they may be foot- or flange-mounted, or the motor may have an outer rotor etc. Machine-tools which formerly had a single drive shaft and complicated mechanical internal gearing can now be driven by a multitude of individually controlled motors producing the mechanical power exactly where, when and in what form it is needed. This has removed constraints from machine tool designers.

  In special cases, such as machine-tools or the propulsion of tracked vehicles, linear electric drives are also available.

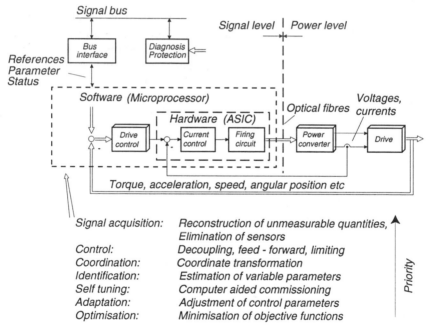

Fig. 0.4. Digital control structure of an electrical drive

As would be expected, this long list of remarkable characteristics is to be supplemented by disadvantages of electric drives which limit or preclude their use:

- The dependance on a continuous power supply causes problems with vehicle propulsion. If no power rail or catenary is available, an electric energy source must be carried on-board, which is usually bulky, heavy and expensive (storage battery, rotating generator with internal combustion engine or turbine, fuel- or solar cells). The lack of a suitable storage battery has so far prevented the wide-spread use of electric vehicles. The weight of a present day lead-acid battery is about 50 times that of a liquid fuel tank storing equal energy, even when taking the low efficiency of the combustion engine into account.

- Due to the magnetic saturation of iron and cooling problems, electric motors are likely to have a lower power-to-weight ratio than, for instance, high pressure hydraulic drives that utilise normal instead of tangential forces. This is of importance with servo drives on-board vehicles, e.g. for positioning the control surfaces of aircraft.

The electromechanical energy conversion in controlled drives is subject to the terminal quantities of the electrical machine which can be changed with low losses by controllable power electronic converters consisting of semiconductor switches; they can produce voltages and currents of almost any waveform as prescribed by control which today is executed by microelectronic components. Thus semiconductor technology, combining power conversion and high speed signal processing at reasonable cost, has been the essential force behind the development of todays high performance drives; this is part of the general transition from analogue to digital control systems using microcomputers and signalprocessors. Fig. 0.4 gives an impression of how the mechanical, power electronic and control functions, combining hardware and software, are interleaved in a modern drive system.

1. Some Elementary Principles of Mechanics

Since electrical drives are linking mechanical and electrical engineering, let us recall some basic laws of mechanics.

1.1 Newtons Law

A mass M is assumed, moving on a straight horizontal track in the direction of the s–axis, Fig. 1.1 a. Let $f_M(t)$ be the driving force of the motor in the direction of the velocity v and $f_L(t)$ the load force opposing the motion, then Newtons law holds

$$f_M - f_L = \frac{d}{dt}(Mv) = M\frac{dv}{dt} + v\frac{dM}{dt} \, , \qquad (1.1)$$

where Mv is the mechanical momentum.

Usually the forces are dependent on velocity v and position s, such as gravitational or frictional forces.

If the mass is constant, $M = M_0 = $ const., Eq. (1.1) is simplified,

$$f_M - f_L = M_0\frac{dv}{dt} \, ; \qquad (1.2)$$

with the definition of velocity $v = ds/dt$, this results in a second order differential equation for the displacement,

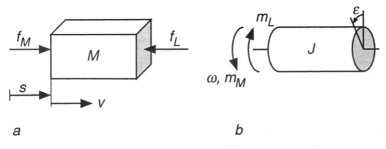

a   b

Fig. 1.1. Translational and rotational motion of lumped masses

$$f_M - f_L = M_0 \frac{d^2 s}{dt^2} \, , \tag{1.3}$$

where

$$a = \frac{dv}{dt} = \frac{d^2 s}{dt^2} \tag{1.4}$$

is the acceleration. If the motion is rotational, which is usually the case with electrical drives, there are analogous equations, Fig. 1.1 b,

$$m_M - m_L = \frac{d}{dt}(J\omega) = J\frac{d\omega}{dt} + \omega\frac{dJ}{dt} \, , \tag{1.5}$$

with m_M being the driving- and m_L the load torque. $\omega = 2\pi n$ is the angular velocity, in the following called speed. J is the moment of inertia of the rotating mass about the axis of rotation, $J\omega$ is the angular momentum. The term $\omega\,(dJ/dt)$ is of significance with variable inertia drives such as centrifuges or reeling drives, where the geometry of the load depends on speed or time, or industrial robots with changing geometry. In most cases however, the inertia can be assumed to be constant, $J = J_0 = \text{const.}$, hence

$$m_M - m_L = J_0 \frac{d\omega}{dt} \, . \tag{1.6}$$

With ε the angle of rotation and $\omega = d\varepsilon/dt$ the angular velocity, we have

$$m_M - m_L = J_0 \frac{d^2\varepsilon}{dt^2} \, , \tag{1.7}$$

where

$$\alpha = \frac{d\omega}{dt} = \frac{d^2\varepsilon}{dt^2} \, , \tag{1.8}$$

is the angular acceleration.

It should be noted that m_M is the internal or electrical motor torque, not identical with the torque available at the motor shaft. The difference between internal torque and shaft torque is the torque required for accelerating the inertia of the motor itself and overcoming the internal friction torque of the motor.

Translational and rotational motions are often combined, for example in vehicle propulsion, elevator- or rolling mill-drives. Fig. 1.2 shows a mechanical model, where a constant mass M is moved with a rope and pulley; when neglecting the mass of the pulley and with

$$m_M = r\,f_M \, , \qquad m_L = r\,f_L \qquad \text{und} \qquad v = r\,\omega$$

we find with $M = \text{const.}$

$$m_M - m_L = r\frac{d}{dt}(Mv) = M\,r^2\frac{d\omega}{dt} \, . \tag{1.9}$$

$J_e = M\,r^2$ represents the equivalent moment of inertia of the linearly moving mass, referred to the axis of the pulley. Apparently the mass M can be thought of as being distributed along the circumference of the wheels.

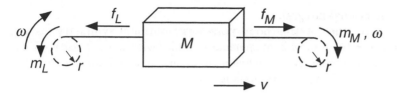

Fig. 1.2. Linking linear and rotational motion

1.2 Moment of Inertia

The moment of inertia, introduced in the preceding section, may be derived as follows:

A rigid body of arbitrary shape, having the mass M, rotates freely about a vertical axis orientated in the direction of gravity, Fig. 1.3. An element of the mass dM is accelerated in tangential direction by the force element df_a, which corresponds to an element dm_a of the accelerating torque

$$dm_a = r\, df_a = r\, dM\, \frac{dv}{dt} = r^2\, dM\, \frac{d\omega}{dt}\ .$$

The total accelerating torque follows by integration

$$m_a = \int_0^{m_a} dm_a = \int_0^M r^2\, \frac{d\omega}{dt}\, dM\ . \tag{1.10}$$

Due to the assumed rigidity of the body, all its mass elements move with the same angular velocity; hence

$$m_a = \frac{d\omega}{dt} \int_0^M r^2\, dM = J\, \frac{d\omega}{dt}\ . \tag{1.11}$$

The moment of inertia, referred to the axis of rotation,

$$J = \int_0^M r^2\, dM \tag{1.12}$$

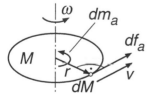

Fig. 1.3. Moment of inertia

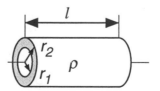

Fig. 1.4. Moment of inertia of concentric cylinder

is a three-dimensional integral.

In many cases the rotating body possesses rotational symmetry; as an example, consider the hollow homogeneous cylinder with mass density ϱ,Fig. 1.4. As volume increment dV we define a thin concentric cylinder having the radius r and the thickness dr; its mass is

$$dM = \varrho \, dV = \varrho \, 2 \, \pi \, r \, l \, dr \, .$$

This reduces the volume integral to a simple integration along the radius,

$$J = \int_0^M r^2 \, dM = \varrho \, 2 \, \pi \, l \int_{r_1}^{r_2} r^3 \, dr = \frac{\pi}{2} \varrho \, l \, (r_2^4 - r_1^4) \, . \tag{1.13}$$

Hence the moment of inertia increases with the 4th power of the outer radius. Introducing the weight of the cylinder

$$G = \varrho \, g \, l \, \pi \, (r_2^2 - r_1^2) \tag{1.14}$$

results in

$$J = \frac{G}{g} \frac{r_2^2 + r_1^2}{2} = \frac{G}{g} r_i^2 \, , \tag{1.15}$$

where g is the gravitational acceleration. The quadratic mean of the radii

$$r_i = \sqrt{\frac{1}{2} \, (r_1^2 + r_2^2)} \tag{1.16}$$

is called the radius of gyration; it defines the radius of a thin concentric cylinder with length l and mass M, that has the same moment of inertia as the original cylinder.

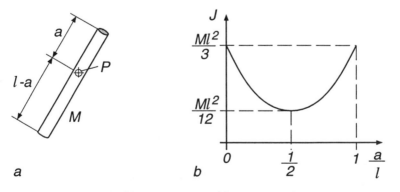

Fig. 1.5. Moment of inertia of a rod, pivoted out of centre

Another example is seen in Fig. 1.5 a, where a homogeneous thin rod of length l and mass M is pivoted around a point P, the distance of which from one end of the rod is a. With the mass element $dM = (M/l)\, dr$ we find for the moment of inertia

$$J = \int_0^M r^2\, dM = \frac{M}{l}\left[\int_0^a r^2\, dr + \int_0^{l-a} r^2\, dr\right]$$

$$= \frac{M\, l^2}{12}\left[1 + 3\left(1 - 2\frac{a}{l}\right)^2\right]. \tag{1.17}$$

The minimum inertia is obtained, when the rod is pivoted at the centre, Fig. 1.5 b.

1.3 Effect of Gearing

Many applications of electrical drives call for relatively slow motion and high torque, for instance in traction or when positioning robots. Since the tangential force per rotor surface, i.e. the specific torque of the motor, is limited to some N/cm^2 by iron saturation and heat losses in the conductors, the direct coupling of a low speed motor with the load may result in an unnecessarily large motor. It is then often preferable to employ gears, operating the motor at a higher speed and thus increasing its power density. This also affects the inertia of the coupled rotating masses.

In Fig. 1.6 an ideal gear is shown, where two wheels are engaged at the point P without friction, backlash or slip. From Newtons law it follows for the left hand wheel, assumed to be the driving wheel,

$$m_{M1} - r_1\, f_1 = J_1\, \frac{d\omega_1}{dt}, \tag{1.18}$$

where f_1 is the circumferential contact force exerted by wheel 2. If there is no load torque applied we have, correspondingly, for wheel 2

$$r_2\, f_2 = J_2\, \frac{d\omega_2}{dt}. \tag{1.19}$$

f_2 is the force driving wheel 2.

Since the forces at the point of contact are in balance and the two wheels move synchronously,

$$f_1 = f_2, \qquad r_1\, \omega_1 = r_2\, \omega_2, \tag{1.20}$$

elimination of f_1, f_2, ω_2 results in

$$m_{M1} = J_1\, \frac{d\omega_1}{dt} + \frac{r_1}{r_2}\, J_2\, \frac{d\omega_2}{dt} = \left[J_1 + \left(\frac{r_1}{r_2}\right)^2 J_2\right]\frac{d\omega_1}{dt}$$

$$= J_{1e}\, \frac{d\omega_1}{dt}. \tag{1.21}$$

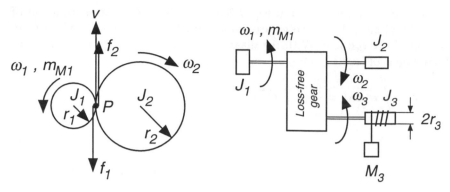

Fig. 1.6. Effect of gearing on inertia   **Fig. 1.7.** Hoist drive with gear

J_{1e} is the moment of inertia effective at the axis of wheel 1; it contains a component reflected from wheel 2. In most cases it is easier to determine the speed ratio rather than the radii,

$$J_{1e} = J_1 + \left(\frac{\omega_2}{\omega_1}\right)^2 J_2 , \tag{1.22}$$

which indicates that a rotating part, moving at higher speed, contributes more strongly to the total moment of inertia.

In Fig. 1.7 an example of a multiple gear for a hoist drive is seen. J_1, J_2, J_3 are the moments of inertia of the different shafts. The total effective inertia referred to shaft 1 is

$$J_{1e} = J_1 + \left(\frac{\omega_2}{\omega_1}\right)^2 J_2 + \left(\frac{\omega_3}{\omega_1}\right)^2 \left[J_3 + M_3\, r_3^2\right] , \tag{1.23}$$

including the equivalent inertia of the mass M_3 being moved in vertical direction. Applying Newtons law, taking the load of the hoist into account, results in

$$m_{M1} = J_{1e}\frac{d\omega_1}{dt} + \frac{\omega_3}{\omega_1}\, r_3\, g\, M_3 . \tag{1.24}$$

1.4 Power and Energy

The rotational motion of the mechanical arrangement shown in Fig. 1.8 is described by a first order differential equation for speed

$$m_M = m_L + J\frac{d\omega}{dt} . \tag{1.25}$$

Multiplication by ω yields the power balance

$$\omega\,m_M = \omega\,m_L + J\,\omega\,\frac{d\omega}{dt}\,,\qquad(1.26)$$

where $p_M = \omega\,m_M$ is the driving power, $p_L = \omega\,m_L$ the load power and $J\,\omega\,(d\omega/dt)$ the change of kinetic energy stored in the rotating masses.

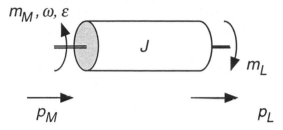

Fig. 1.8. Power flow of drive

By integrating Eq. (1.26) with the initial condition $\omega\,(t = 0) = 0$ we find the energy input

$$w_M(t) = \int_0^t p_M\,d\tau = \int_0^t p_L\,d\tau + \int_0^t J\,\omega\,\frac{d\omega}{d\tau}\,d\tau$$

$$= \int_0^t p_L\,d\tau + J\int_0^\omega \Omega\,d\Omega$$

$$= w_L(t) + \frac{1}{2}\,J\,\omega^2\,.\qquad(1.27)$$

The last term represents the stored kinetic energy; it is analogous to

$$\frac{1}{2}\,M\,v^2\,,\qquad \frac{1}{2}\,L\,i^2\qquad \frac{1}{2}\,C\,u^2\,.$$

of other energy storage devices. Since the energy content of a physical body cannot be changed instantaneously — this would require infinite power — the linear or rotational velocity of a body possessing mass must always be a continuous function of time. This is an important condition of continuity which will frequently be used.

Because of the definitions

$$v = \frac{ds}{dt}\quad\text{and}\quad \omega = \frac{d\varepsilon}{dt}$$

the positional quantities s and ε are also continuous functions of time, due to finite speed. This is also understandable from an energy point of view, since position may be associated with potential energy, as seen in Fig. 1.7, where the mass M_3 is positioned vertically depending on the angle of rotation of the drive shaft.

1.5 Experimental Determination of Inertia

The moment of inertia of a complex inhomogeneous body, such as the rotor of an electrical machine, containing iron, copper and insulating material with complicated shapes can in practice only be determined by approximation. The problem is even more difficult with mechanical loads, the constructional details of which are normally unknown to the user. Sometimes the moment of inertia is not constant but changes periodically about a mean value, as in the case of a piston compressor with crankshaft and connecting rods. Therefore experimental tests are preferable; a very simple one, called the run–out or coasting test, is described in the following. Its main advantage is that it can be conducted with the complete drive in place and operable, requiring no knowledge about details of the plant. The accuracy obtainable is adequate for most applications.

First the input power $p_M(\omega)$ of the drive under steady state conditions is measured at different angular velocities ω and with the load, not contributing to the inertia, being disconnected. From Eq. (1.26),

$$p_M = p_L + J\omega\frac{d\omega}{dt}\,,\tag{1.28}$$

the last term is omitted due to constant speed, so that the input power p_M corresponds to the losses including the remaining load, $p_M = p_L$. This power is corrected by subtracting loss components which are only present during power input, such as armature copper losses in the motor. From this corrected power loss p'_L the steady–state effective load torque $m'_L = p'_L/\omega$ is computed for different speeds; with graphical interpolation, this yields a curve $m'_L(\omega)$ as shown in Fig. 1.9.

For the run–out test, the drive is now accelerated to some initial speed ω_0, where the drive power is switched off, so that the plant is decelerated by the loss torque with the speed measured as a function of time, $\omega(t)$. Solving the equation of motion (1.25) for J results in

$$J \approx \frac{-m'_L(\omega)}{\frac{d\omega}{dt}(\omega)}\,,\qquad m_M = 0\,.\tag{1.29}$$

Hence the inertia can be determined from the slope of the coasting curve, as shown in Fig. 1.9.

Graphical constructions, particularly when a differentiation is involved, are only of moderate accuracy. Therefore the inertia should be computed at different speeds in order to form an average. The accuracy requirements regarding inertia are modest; when designing a drive control system, an error of $\pm$ 10% is usually acceptable without any serious effect.

Two special cases lead to particularly simple interpretations:

a) Assuming the corrected loss torque m'_L to be approximately constant in a limited speed interval,

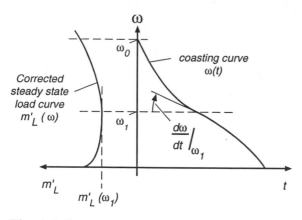

Fig. 1.9. Run - out test

$$m'_L \approx \text{const}, \qquad \text{for} \qquad \omega_1 < \omega < \omega_2 \,,$$

then $\omega(t)$ resembles a straight line; the inertia is determined from the slope of this line.

b) If a section of the loss torque may be approximated by a straight line,

$$m'_L \approx a + b\omega \,, \qquad \text{for} \qquad \omega_1 < \omega < \omega_2 \,.$$

a linear differential equation results,

$$J \frac{d\omega}{dt} + b\omega = -a \,.$$

The solution is, with $\omega(t_2) = \omega_2$,

$$\omega(t) = -\frac{a}{b} + \left(\omega_2 + \frac{a}{b}\right) e^{-b(t-t_2)/J} \,, \qquad t \geq t_2 \,.$$

Plotting this curve on semi-logarithmic paper yields a straight line with the slope $-b/J$, from which an approximation of J is obtained.

2. Dynamics of a Mechanical Drive

2.1 Equations Describing the Motion of a Drive with Lumped Inertia

The equations derived in Chap. 1

$$J \frac{d\omega}{dt} = m_M\left(\omega, \varepsilon, y_M, t\right) - m_L\left(\omega, \varepsilon, y_L, t\right), \qquad (2.1)$$

$$\frac{d\varepsilon}{dt} = \omega, \qquad (2.2)$$

describe the dynamic behaviour of a mechanical drive with constant inertia in steady state condition and during transients. Stiff coupling between the different parts of the drive is assumed so that all partial masses may be lumped into one common inertia. The equations are written as state equations for the continuous state variables ω, ε involving energy storage, e. g. [38,88]). Only mechanical transients are considered; a more detailed description would have to take into account the electrical transients defined by additional state variables and differential equations. The same is true for the load torque m_L which depends on dynamic effects in the load, such as a machine tool or an elevator. Also the control inputs y_M, y_L to the actuators on the motor and load side have to be included. Fig. 2.1 shows a block diagram, representing the interactions of the mechanical system in graphical form. The output variables of the two integrators are the continuous state variables, characterising the energy state of the system at any instant. Linear transfer elements, such as integrators with fixed time constants, are depicted by blocks with single frames containing a figure of the step response. A block with double frame denominates a nonlinear function; if it represents an instantaneous, i. e. static, nonlinearity, its characteristic function is indicated. The nonlinear blocks in Fig. 2.1 may contain additional dynamic states described by differential equations.

Dependence of the driving torque m_M on the angle of rotation is a characteristic feature of synchronous motors. However, the important quantity is not the angle of rotation ε itself but the difference angle δ against the no–load angle ε_0 which is determined by the torque. Under steady state conditions the load angle δ is constant.

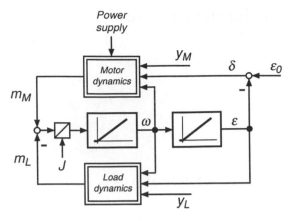

Fig. 2.1. Simplified block diagram of lumped inertia drive

In order to gain a better insight, let us first assume that the electrical transients within the motor and the internal load transients decay considerably faster than the mechanical transients of ω and ε; as a consequence, it follows that the motor and load torques m_M, m_L are algebraic, i.e. instantaneous functions of ω, ε and δ. Hence, by neglecting the dynamics of motor and load, we arrive at a second order system that is completely described by the two state equations (2.1), (2.2).

So far we have assumed that all moving parts of the drive can be combined to form a single effective inertia. However, for a more detailed analysis of dynamic effects it may be necessary to consider the distribution of the masses and the linkages between them. This leads to multi–mass–systems and in the limit to systems with continuously distributed masses, where transients of higher frequency and sometimes insufficient damping may be superimposed on the common motion. The frequency of these free oscillations, describing the relative displacement of the separate masses against each other, increases with the stiffness of the connecting shafts; they are usually outside the frequency range of interest for control transients but must be considered for the mechanical design of the drive.

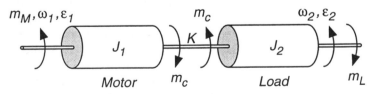

Fig. 2.2. Drive consisting of motor and load coupled by flexible shaft

However, when the partial masses are coupled by flexible linkages, such as with mine hoists, where drive and cage are connected by the long winding rope or in the case of paper mill drives with many gears, drive shafts and large rotating masses particularly in the drying section, a more detailed description becomes necessary. The free oscillations may then have frequencies of a few Hertz which are well within the range of a fast control loop.

In Fig. 2.2 an example is sketched, where the drive motor and the load having the moments of inertia J_1, J_2 are coupled by a flexible shaft with the torsional stiffness K. The ends of the shaft, the mass of which is ignored, have the angles of rotation ε_1, ε_2 and the angular velocities ω_1, ω_2. Assuming a linear torsional law for the coupling torque m_c,

$$m_c = K \left(\varepsilon_1 - \varepsilon_2 \right) , \tag{2.3}$$

and neglecting internal friction effects, the following state equations result

$$J_1 \frac{d\omega_1}{dt} = m_M - m_c = m_M \left(\omega_1, \varepsilon_1, y_M \right) - K \left(\varepsilon_1 - \varepsilon_2 \right) , \tag{2.4}$$

$$J_2 \frac{d\omega_2}{dt} = m_c - m_L = K \left(\varepsilon_1 - \varepsilon_2 \right) - m_L \left(\omega_2, \varepsilon_2, y_L \right) , \tag{2.5}$$

$$\frac{d\varepsilon_1}{dt} = \omega_1 , \tag{2.6}$$

$$\frac{d\varepsilon_2}{dt} = \omega_2 . \tag{2.7}$$

A graphical representation is seen in Fig. 2.3 a. Here too, the torques m_M, m_L are in reality defined by additional differential equations and state variables. If only the speeds are of interest, the block diagram in Fig. 2.3 b may be useful, which, containing three integrators, is described by a third order differential equation.

With increasing stiffness of the shaft, the quantities ε_1, ε_2 and ω_1, ω_2 become tighter coupled; in the limit the case of lumped inertia emerges, where $\varepsilon_1 = \varepsilon_2$, $\omega_2 = \omega_1$.

If the transmission is affected by mechanical backlash, where the two inertias separate when the shaft torque changes sign, the linear torsional branch K is replaced by a nonlinear function, as shown in Fig. 2.3 c. This is of importance with reversible drives, for instance servo drives.

Obviously the subdivision of the inertia may be continued indefinitely; every time a new partial inertia is separated, two more state variables have to be defined, transforming Fig. 2.3 into a chainlike structure. A typical example, where many partial masses must be taken into account for calculating stress and fatigue, is a turbine rotor.

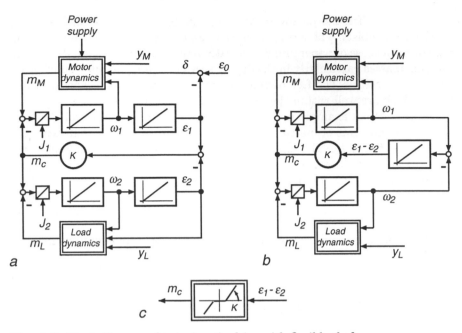

Fig. 2.3. Block diagram of twin–inertia drive with flexible shaft
(**a**) Model of mechanical plant, (**b**) Reduced order model,
c) Mechanical transmission with backlash

2.2 Two Axes Drive in Polar Coordinates

On machine tools or robots there are normally several axes of motion, that
must be independently driven or positioned. An example is seen in Fig. 2.4
a, where an arm, carrying a tool or workpiece, is rotated by an angle $\varepsilon(t)$
around a horizontal axis. The radial distance $r(t)$ from the axis to the center of the mass M_2 represents a second degree of freedom, so that M_2 can
be positioned in polar coordinates in a plane perpendicular to the axis. The
rotary and radial motions are assumed to be driven by servo motors, producing a controlled driving torque m_M and a driving force f_M through a rotary
gear and a rotary to translational mechanical converter, for instance a lead
screw. With fast current control the motors are generating nearly instantaneous impressed torques, serving as control inputs to the mechanical plant.
For simplicity, the masses are assumed to be concentrated in the joints, resulting in the inertias J_1, J_2. The coupling terms of the motion can be derived
by expressing the acceleration of the mass M_2 in complex form.

$$\frac{dr}{dt} = \frac{d}{dt}\left(r\,e^{j\,\varepsilon}\right) = (v + j\,r\,\omega)\,e^{j\,\varepsilon}\,, \tag{2.8}$$

$$\frac{d^2r}{dt^2} = \frac{d^2}{dt^2}\left(r\,e^{j\,\varepsilon}\right) = \left(\frac{dv}{dt} - \omega^2\,r\right)\,e^{j\,\varepsilon} + j\left(2\,\omega\,v + r\,\frac{d\omega}{dt}\right)\,e^{j\,\varepsilon}\,, \tag{2.9}$$

where $\omega = d\varepsilon/dt$ and $v = dr/dt$ are the rotational and radial velocities. After separating the terms of acceleration in tangential and radial direction and superimposing frictional and gravitational components, Newtons law is applied in both directions, resulting in the equations for the mechanical motion of the centre point of M_2

$$\overbrace{\left(J_1 + J_2 + M_2\, r^2\right)}^{J} \frac{d\omega}{dt} =$$

$$m_M - \overbrace{2\, M_2\, r\, \omega\, v}^{Coriolis} - \overbrace{M_2\, g\, r\, \cos\varepsilon}^{Gravitation} -m_F - m_L\ , \quad (2.10)$$

$$\frac{d\varepsilon}{dt} = \omega\ , \qquad\qquad\qquad\qquad\qquad\qquad (2.11)$$

$$M_2\, \frac{dv}{dt} = f_M + \overbrace{M_2\, r\, \omega^2}^{Centrifugal} - \overbrace{M_2\, g\, \sin\varepsilon}^{Gravitation} -f_F - f_L\ , \qquad (2.12)$$

$$\frac{dr}{dt} = v\ . \qquad\qquad\qquad\qquad\qquad\qquad (2.13)$$

The equations (2.10)–(2.13) are depicted in Fig. 2.4 b in the graphical form of a block diagram, containing four integrators for the state variables. Despite the simple mechanics, there are complicated interactions, which become more prominent with increasing rotary and radial velocities. The control of this mechanical structure is dealt with in a later chapter.

Clearly, the two motions are nonlinearly coupled though gravitational, Coriolis– and centrifugal effects; they are described by four nonlinear state equations. m_F, f_F and m_L, f_L are due to friction and external load forces with may exhibit their own complicated geometric or dynamic dependencies. If it is important for the application to express the position of mass M_2 in cartesian coordinates, this is achieved by a polar–cartesian conversion

$$x(t) = r\, \cos\varepsilon\ , \qquad\qquad\qquad\qquad\qquad (2.14)$$
$$y(t) = r\, \sin\varepsilon\ . \qquad\qquad\qquad\qquad\qquad (2.15)$$

Moving the arm also in the direction of the axis of rotation, so that the mass M_2 can be positioned in cylindrical coordinates, would introduce a third decoupled degree of freedom.

The dynamic interactions for a general motion, involving six degrees of freedom (three for the position, three for the orientation of the tool) are exceedingly complicated, they must be dealt with when controlling the motions of multi–axes robots with high dynamic performance [M53, O13, S49].

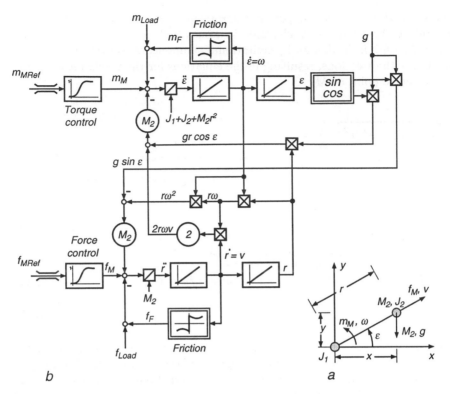

Fig. 2.4. Two axes drive in polar coordinates
(a) Mechanical plant (b) Block diagram

2.3 Steady State Characteristics of Different Types of Motors and Loads

Consider first the steady state condition, when the torque and speed of a single axis lumped inertia drive are constant and the angle changes linearly with time; this condition is reached when $m_M - m_L = 0$. With some motors, such as single phase induction motors, or loads, for example piston compressors or punches, the torque is a periodic function of the angle of rotation; in this case, the steady state condition is reached, when the mean values of both torques are equal, $\bar{m}_M - \bar{m}_L = 0$. The speed then contains periodic oscillations, which must be kept within limits by a sufficiently large inertia.

The steady state characteristics of a motor or load are often functions given in graphical form, connecting main variables, such as speed and torque; the provision is that auxiliary or control inputs, for example supply voltage, field current, firing angle, brush position or feed rate, are maintained constant. In Fig. 2.5 a, three typical characteristics of electric motors are shown. The "synchronous" characteristic is only valid for constant speed, since the variable is the load angle δ, i. e. the displacement of the shaft from its no–load angular position. When the maximum torque is exceeded, the motor falls out of step; asynchronous operation of larger motors is not allowed for extended periods of time because of the high currents and pulsating torque. The electrical transients usually cannot be neglected with synchronous motor drives.

The rigid speed of synchronous motors when supplied by a constant frequency source makes them suitable for only a few applications, for example large slow–speed drives for reciprocating compressors or synchronous generators operating as motors in pumped storage hydro power stations; at the low end of the power scale are electromechanical clocks. The situation is different, when the synchronous motor is fed from a variable frequency inverter because then the speed of the drive can be varied freely (Chap. 14). With the progress of power electronics, these drives are now more widely used.

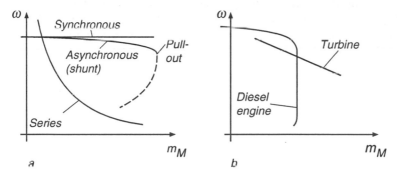

Fig. 2.5. Steady state torque–speed characteristics of (a) electrical and (b) mechanical drives

The "asynchronous" or "shunt-typ" characteristic in Fig. 2.5 a is slightly drooping; often there is also a pronounced maximum torque. The lower portion of the curve is forbidden in steady state in view of the high losses. With three- phase asynchronous motors the rotor angle has no effect on the torque in steady state.

Motors with "series–type" characteristic show considerably larger speed drop under load; with DC or AC commutator motors, this is achieved by a suitable connection of the field winding. The main area of applications at larger ratings are traction drives because the curved characteristic resembling a hyperbola facilitates load sharing on multiple drives and permits nearly constant power operation over a wide speed range without gear change; this is particularly suited to a Diesel–electric or turbo- electric drive, where the full power of the thermal engine must be used.

For comparison, some typical characteristics of a turbine and a Diesel engine at constant fuel injection per stroke are seen in Fig. 2.5 b.

The curves in Fig. 2.5 a are "natural" characteristics which can be modified at will by different control inputs, e. g. through the power supply. With closed loop control a shunt motor could assume the behaviour of asynchronous or of a series type motor. As an example, typical steady state curves of a controlled DC drive are shown in Fig. 2.6; they consist of a constant speed branch (normal operation) which is joined at both ends by constant torque sections activated under overload condition through current limit. Figure 2.7 depicts the steady state curves of the motor for driving a coiler. If the electrical power reference is determined by the feed velocity v of the web or strip to be wound and includes the friction torque, the coiler operates with constant web force f independent of the radius r of the coil, $p_L = vf + p_F$.

The steady state characteristics of mechanical loads are of great variety; however, they are often composed of simple elements. This is seen in Fig. 2.8 with the example of a hoist and a vehicle drive. The gravitational lift torque m_L is independent of speed (Fig. 2.8 c); in the first quadrant the load is lifted, increasing its potential energy, hence the drive must operate in the motoring region. In the fourth quadrant, the power flow is reversed with the load releasing some of its potential energy. Part of that power is flowing back to the line, the remainder is converted to heat losses. The lower half of the winding rope, seen in Fig. 2.8 a, serves to balance the torque caused by the

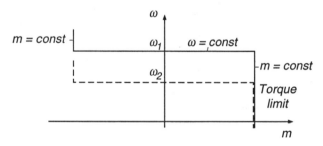

Fig. 2.6. Torque/speed curves of controlled motor with constant speed branch and torque limit

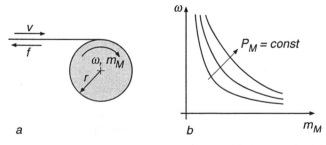

Fig. 2.7. Coiler drive (**a**) Mechanics and (**b**) static load curves

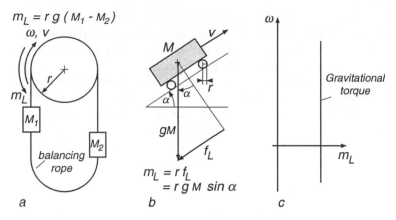

$$m_L = r g \, (M_1 - M_2)$$

$$m_L = r f_L = r g M \sin \alpha$$

a   b   c

Fig. 2.8. Drives involving gravitational forces

weight of the rope; this effect can be substantial on a winder for a deep mine, tending to destabilise the drive.

All mechanical motion is accompanied by frictional forces between the surfaces where relative motion exists. There are several types of friction, some of which are described in Fig. 2.9. In bearings, gears, couplings and brakes one observes dry or Coulomb friction (a), which is nearly independent of speed; however one has to distinguish between sliding and sticking friction, the difference of which may be considerable, depending at the roughness of the surfaces. The forces when cutting or milling material also contain Coulomb type friction.

In well lubricated bearings there is a component of frictional torque which rises proportionally with speed; it is due to laminar flow of the lubricant and is called viscous friction (b). At very low speed and without pressurised lubrication, Coulomb type friction again appears.

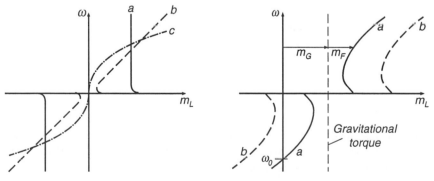

Fig. 2.9. Different types of frictional torque

Fig. 2.10. Torque/speed curves of a hoist drive

With pumps and ventilators, where turbulent flow occurs, the torque rises with the square of speed; the drag force due to air flow around vehicles or the torque required by a centrifugal blower pushing cooling air through an electric motor also have this characteristic (c).

In practical drives, with the motor as well as load, all these types of friction exist simultaneously, with one or the other component dominating. When driving a paper mill, printing presses or machine tools, Coulomb friction is usually the main constituent but with centrifugal pumps and compressors the torque following the square law is most important, representing the useful mechanical output. Note that frictional torques are always opposed to the direction of relative motion.

In Fig. 2.10 various torques, acting on a crane under load, are drawn; m_G is the constant gravitational torque caused by the load of the crane, m_F is the frictional torque, resulting in the total load torque curve (a). The speed ω_0 corresponds to the run–away speed with no external braking torque. When for safety reasons a self-locking transmission, such as a worm gear (b), is employed the crane must be powered even when lowering the load; this is due to the large sticking friction.

2.4 Stable and Unstable Operating Points

By ignoring the dependence of driving and load torques on the angle of rotation, the corresponding interaction seen in Fig. 2.1 vanishes and so does the effect of Eq. (2.2) on the static and transient behaviour of the drive. Equation (2.2) then becomes an indefinite integral having no effect on the drive. If we also neglect the electrical transients and the dynamics of the load, the remaining mechanical system is described by a first order, usually nonlinear, differential equation but, in view of the simplifications introduced, its validity is restricted to relatively slow changes of speed, when the internal transients and interactions in the electrical machine and the load can be neglected.

$$ J \frac{d\omega}{dt} = m_M (\omega, t) - m_L (\omega, t) . \tag{2.16} $$

Apparently, a steady state condition with a constant rotational speed ω_1 is possible, if the characteristics are intersecting at that point, i. e. for

$$ m_M (\omega_1) - m_L (\omega_1) = 0 . $$

In order to test whether this condition is stable, Eq. (2.16) can be linearised at the operating point ω_1, assuming a small displacement $\Delta\omega$. With $\omega = \omega_1 + \Delta\omega$, we find the linearised equation

$$ J \frac{d\Delta\omega}{dt} = \left. \frac{\partial m_M}{\partial \omega} \right|_{\omega_1} \Delta\omega - \left. \frac{\partial m_L}{\partial \omega} \right|_{\omega_1} \Delta\omega , $$

or in normalised form,

$$\frac{J}{k} \frac{d\Delta\omega}{dt} + \Delta\omega = 0, \qquad \text{where} \qquad k = \frac{\partial}{\partial\omega}(m_L - m_M)\,|_{\omega_1}. \qquad (2.17)$$

This is illustrated in Fig. 2.11 with some examples. The assumed steady state is stable for $k > 0$; in this case a small displacement $\Delta\omega$, that may have been caused by a temporary disturbance, is decaying along an exponential function with the time constant $T_m = J/k$. k can be interpreted as the slope of the retarding torque at the operating point, as seen in Fig. 2.11. For $k < 0$ the operating point at ω_1 is unstable, i. e. an assumed deviation of speed increases with time; a new stable operating point may or may not be attained. The case $k = 0$ corresponds to indifferent stability; there is no definite operating point, with the speed fluctuating due to random torque variations.

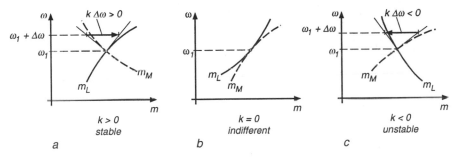

Fig. 2.11. Stable and unstable operating points

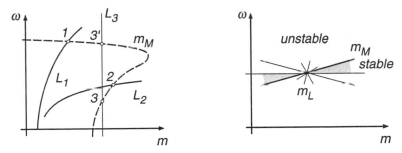

Fig. 2.12. Induction motor with differ- Fig. 2.13. Rising torque/speed curve as
ent types of load   cause of instability

Figure 2.12 depicts the steady state characteristic of an induction motor (m_M) together with some load curves (Sect. 10.2). L_1 could be the characteristic of a ventilating fan; the intersection 1 is stable, roughly corresponding to nominal load. With L_2 there is also a stable operating point, but the motor would be heavily overloaded. With the ideal lift characteristic L_3, there is an unstable (3) and a stable (3') operating point, where the motor would also be overloaded. In addition, the drive would fail to start since the load would pull the motor in the lowering direction when the brakes are released and there is insufficient frictional torque.

A particularly critical case is seen in Fig. 2.13. A slightly rising motor characteristic, which on a DC motor could be caused by armature reaction due to incorrect brush setting, leads to instability with most load curves except those intersecting in the shaded sector.

The stability test based on a linearised differential equation does not fully exclude instability if electrical transients or angle–dependent torques should have been included. The condition $k > 0$ is only to be understood as a necessary condition, even though it is often a sufficient condition as well.

3. Integration
of the Simplified Equation of Motion

With the assumptions introduced in the preceding section the motion of a single axis lumped inertia drive is described by a first order differential equation, Fig. 3.1,

$$J\frac{d\omega}{dt} = m_M\left(\omega, t\right) - m_L\left(\omega, t\right) = m_a\left(\omega, t\right), \tag{3.1}$$

which upon integration yields the mechanical transients. Several options are available for performing the integration.

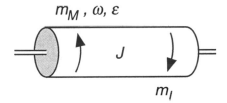

m_M , ω, ε

J

m_l

Fig. 3.1. Drive with concentrated inertia

3.1 Solution of the Linearised Equation

In the linearised homogeneous equation, Eq.(2.17),

$$T_m\frac{d(\Delta\omega)}{dt} + \Delta\omega = 0, \qquad T_m = \frac{J}{k} \tag{3.2}$$

is $\Delta\omega$ a small deviation from the steady state speed ω_1 and

$$k = \frac{\partial}{\partial\omega}\left(m_L - m_M\right)\Big|_{\omega_1} \tag{3.3}$$

a measure for the slope of the retarding torque at the operating point. $T_m = J/k$ has the meaning of a mechanical time constant. The general solution is

$$\Delta\omega(t) = \Delta\omega(0)\, e^{-t/T_m} \tag{3.4}$$

where $\Delta\omega(0)$ is the initial deviation, that may be caused by switching from one motor characteristic to another resulting in a new steady state speed. Because of the stored kinetic energy, the speed is a continuous state variable (the only one because of the simplifications introduced).

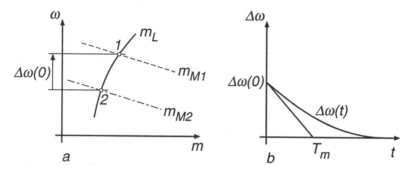

Fig. 3.2. Mechanical transient of linearised drive system

In Fig. 3.2 a it is assumed that the motor, initially in a steady state condition at point 1, is switched to another supply voltage so that a new characteristic for the driving torque m_{M2} is valid. This causes an initial deviation $\Delta\omega(0)$ with respect to the new operating point. The deviation vanishes along an exponential function (Fig. 3.2 b), the time constant of which is determined by the inertia and the slope of the load– and the driving–torques. This is discussed with the help of some simple examples.

3.1.1 Start of a Motor with Shunt–type Characteristic at No–load

A motor which initially is at rest is started at $t = 0$, Fig. 3.3. The torque/speed–curve of the motor is assumed to be linear between no-load speed ω_0 (neglecting friction load) and stalled torque m_0. With normal motors of medium size, the stalled torque may be 8 or 10 times nominal torque; it is determined by extrapolating the drooping shunt-type characteristic to standstill and hence represents only a reference quantity that cannot be measured in practice. However, in the present example we assume that a starting resistor has been inserted, reducing the stalled torque to perhaps twice nominal torque and thus extending the validity of the linear characteristic down to standstill. This, by the way, makes the simplifying assumptions more realistic since the electrical transients become faster while the mechanical transients are delayed.

Under ideal no-load conditions, we have $m_L = 0$; also, due to the assumed linearity of the curves $m_M(\omega)$, $m_L(\omega)$ the deviation $\Delta\omega$ is not restricted to small values.

The slope of the retarding torque is

$$k = \frac{\partial}{\partial \omega} (m_L - m_M) = \frac{m_0}{\omega_0} > 0 \, ;$$

hence the differential equation is stable, it has the form

$$T_m \frac{d(\Delta \omega)}{dt} + \Delta \omega = 0 \, , \tag{3.5}$$

where

$$T_m = \frac{J \, \omega_0}{m_0} \, .$$

is the mechanical time constant. The steady state operating point at the intersection of the two torque curves is $\omega_1 = \omega_0$; due to the initial condition at standstill, we find $\Delta \omega(0) = -\omega_0$, which leads to the particular solution

$$\Delta \omega(t) = -\omega_0 \, e^{-t/T_m}$$

or with $\omega = \omega_0 + \Delta \omega$

$$\omega(t) = \omega_0 \, (1 - e^{-t/T_m}) \, . \tag{3.6}$$

From the torque/speed curve of the motor

$$\frac{\omega}{\omega_0} = 1 - \frac{m_M}{m_0} \tag{3.7}$$

the motor torque during the start–up is obtained (Fig. 3.4),

$$m_M(t) = m_0 \, e^{-t/T_m} \, . \tag{3.8}$$

The discontinuity of the motor torque is caused by the omission of the electrical transients. In reality the torque is also associated with energy states and, hence, is continuous.

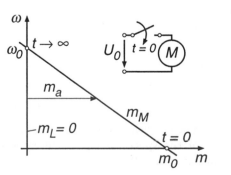

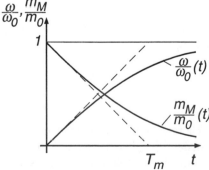

Fig. 3.3. Starting of motor at no–load   Fig. 3.4. Starting transient

3.1.2 Starting the Motor with a Load Torque Proportional to Speed

In Fig. 3.5 the curves of a drive with speed–proportional load torque are displayed. The steady state operating point is now at $\omega_1 < \omega_0$, hence $\Delta\omega(0) = -\omega_1$.

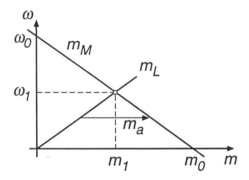

Fig. 3.5. Starting of motor with linearly rising load torque

The slope of the retarding torque is

$$k_1 = \frac{\partial}{\partial\omega}(m_L - m_M) = \frac{m_0}{\omega_1} > 0, \tag{3.9}$$

which leads to a reduced mechanical time constant,

$$T_{m1} = \frac{J\omega_1}{m_0} = \frac{\omega_1}{\omega_0}T_m. \tag{3.10}$$

The speed transient is

$$\omega(t) = \omega_1\left(1 - e^{-t/T_{m1}}\right). \tag{3.11}$$

Elimination of speed from the motor characteristic Eq. (3.7) results in the driving torque

$$m_M(t) = m_0\left(1 - \frac{\omega}{\omega_0}\right) = m_0\left[1 - \frac{\omega_1}{\omega_0}\left(1 - e^{-t/T_{m1}}\right)\right]$$
$$= m_1 + (m_0 - m_1)e^{-t/T_{m1}}. \tag{3.12}$$

Both transients are plotted in Fig. 3.6, together with the no–load starting transient ($m_L = 0$). At $t = 0$, the slope of the speed curves is the same, because the accelerating torques are identical at $\omega = 0$.

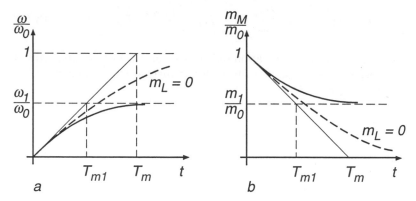

Fig. 3.6. Starting transient with load

3.1.3 Loading Transient of the Motor
Initially Running at No–load Speed

The motor is now supposed to be initially in a no–load condition at the speed ω_0 with the starting resistor short circuited, so that the speed droop is reduced and the extrapolated stalled torque assumes its large nominal value m_{0n}. The torque/speed–curve is then linear only in a narrow speed range around no–load speed.

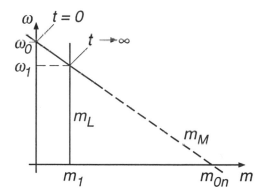

Fig. 3.7. Applying constant load torque to motor running at no–load

At $t = 0$ a constant torque load without inertia is applied to the motor shaft, possibly by a mechanical brake, Fig. 3.7, causing the motor to slow down to the steady state speed ω_1; hence the initial deviation is $\Delta\omega(0) = \omega_0 - \omega_1$.

In view of the constant load torque, we find

$$k = \left.\frac{\partial}{\partial \omega} (m_L - m_M)\right|_{\omega_1} = \frac{m_{0n}}{\omega_0}, \qquad T_m = \frac{J \omega_0}{m_{0n}} = T_{mn} . \qquad (3.13)$$

Since m_{0n} corresponds to the extrapolated stalled torque with short circuited starting resistor, T_{mn} is also called short circuit mechanical time constant.

The solution of Eq. (3.5) is

$$\Delta\omega(t) = (\omega_0 - \omega_1) e^{-t/T_{mn}}$$

or

$$\omega(t) = \omega_1 + (\omega_0 - \omega_1) e^{-t/T_{mn}} . \qquad (3.14)$$

The driving torque is again obtained from Eq. (3.7),

$$m_M(t) = m_{0n} \left(1 - \frac{\omega}{\omega_0}\right) = m_{0n} \left(1 - \frac{\omega_1}{\omega_0}\right) (1 - e^{-t/T_{mn}})$$

$$= m_1 (1 - e^{-t/T_{mn}}) . \qquad (3.15)$$

In Fig. 3.8 the transients are drawn for loading and unloading the motor.

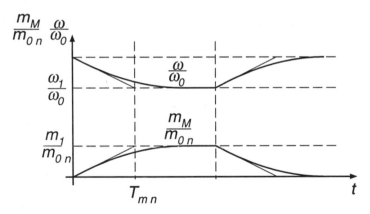

Fig. 3.8. Loading transient

The driving torque apparently follows the applied load torque with a lag determined by T_{mn}. The reason for this is that during the deceleration phase some of the kinetic energy stored in the inertia is released and makes up for part of the load torque; when the drive is accelerated after disconnecting the load, the missing kinetic energy is restored. The mechanical energy storage thus acts as a buffer between the load and the electrical supply feeding the motor. This effect may be accentuated by adding a flywheel to the drive, in order to protect the supply system from high load surges such as those caused by large rolling mill drives; conversely, a sensitive load, for example a computer, can be made immune against short time disturbances on the line side, when it is supplied by a rotating motor–generator set having a flywheel attached.

3.1.4 Starting of a DC Motor
by Sequentially Short-circuiting Starting Resistors

With DC motors fed from a constant voltage bus the starting circuit shown in Fig. 3.9 may be used, where the contactors $C_1, \ldots C_N$ are closed in succession. By suitable choice of the resistors, the droop of the resulting torque/speed characteristic is progressively reduced. In order to achieve fast starting without overloading the motor, the torque is specified to vary between the limits m_1, m_2 during the start, as is shown in Fig. 3.10. The armature current, being proportional to torque, then varies between the limits i_1, i_2. A similar starting procedure exists for AC induction motors having a wound rotor and starting resistors in the rotor circuit.

As the motor is accelerated, the torque m_M decreases until the lower limit m_1 is reached; then the next section of the resistor is short-circuited causing the torque to jump again to the upper limit m_2. This is repeated until all the starting resistors are switched out and the motor operates on its natural torque/speed curve. In steady state the torque is determined by the load. The required number of sections of the starting resistor and their appropriate values can be calculated in closed form [37].

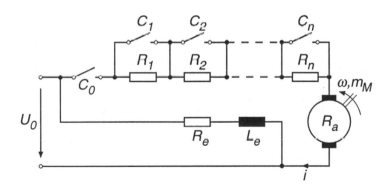

Fig. 3.9. Starting circuit for a DC motor

From Fig. 3.10 the following relation is found

$$\frac{m_2 - m_1}{m_2} = \frac{\omega\,(\nu + 1) - \omega\,(\nu)}{\omega_0 - \omega\,(\nu)}, \tag{3.16}$$

which results, with the abbreviation $m_1/m_2 = a < 1$, in a recursive formula

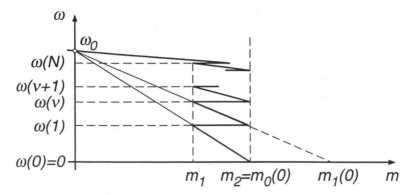

Fig. 3.10. Torque/speed-curves for different values of armature resistor

$$\omega\,(\nu + 1) = (1 - a)\,\omega_0 + a\,\omega\,(\nu) = \omega(1) + a\,\omega\,(\nu)\ .$$

Assuming $\omega(0) = 0$, this leads to a geometric progression

$$\omega(2) = (1 + a)\,\omega(1)$$
$$\omega(3) = (1 + a + a^2)\,\omega(1)$$
$$\vdots$$
$$\omega(\nu) = (1 + a + \ldots + a^{\nu - 1})\,\omega(1)$$

with the sum

$$\omega(\nu) = \frac{1 - a^\nu}{1 - a}\,\omega(1) = (1 - a^\nu)\,\omega_0\ . \tag{3.17}$$

The last of the N switching instants is reached when the natural torque/speed curve yields a value of the torque which is less than the maximum, m_2,

$$\omega(N) = (1 - a^N)\,\omega_0 \geq \left(1 - \frac{m_2}{m_{0n}}\right)\omega_0\ ;$$

solving for N, this results in

$$N \geq \frac{\ln(m_{0n}/m_2)}{\ln(1/a)}\ , \qquad N \text{ integer.} \tag{3.18}$$

From previous results it is known that, when assuming linear torque/speed curves, all variables are exponential functions of time, with the time constants depending on the angles of the curves at the point of intersection.

In Fig. 3.11 a starting transient without load is shown; in the interval $\omega(\nu) \leq \omega \leq \omega(\nu + 1)$

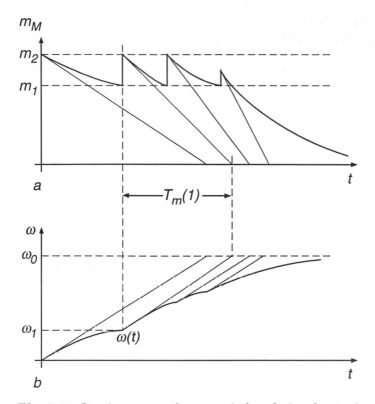

Fig. 3.11. Starting a motor by successively reducing the starting resistor

$$\omega(t) = \omega(\nu) + [\omega_0 - \omega(\nu)]\left[1 - e^{-(t-t_\nu)/T_m(\nu)}\right] , \tag{3.19}$$

where t_ν is the instant, when C_ν closes.

The driving torque in the same interval follows also an exponential function

$$m_M(t) = m_2\, e^{-(t-t_\nu)/T_m(\nu)} ; \tag{3.20}$$

as soon as $m_M(t)$ has dropped to m_1, the next contactor $C_{\nu+1}$ is closed.

The mechanical time constant, valid in the interval, is

$$T_m(\nu) = \frac{J\,\omega_0}{m_0(\nu)} = \frac{J\,\omega_0}{m_2}\,a^\nu = T_{mn}\,\frac{m_{0n}}{m_2}\,a^\nu ; \tag{3.21}$$

it decreases continuously as the starting resistor is reduced.

3.2 Analytical Solution
of Nonlinear Differential Equation

A direct solution of the equation

$$J \frac{d\omega}{dt} = m_M(\omega) - m_L(\omega)$$

by separation and integration,

$$t_2 - t_1 = J \int_{\omega_1}^{\omega_2} \frac{1}{m_M(\omega) - m_L(\omega)} d\omega \qquad (3.22)$$

is rarely possible because there is either no analytical expression of the functions $m_M(\omega)$ and $m_L(\omega)$ or, if there is such a formula, the integral cannot be solved in general terms.

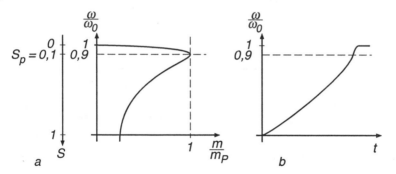

Fig. 3.12. Starting of induction motor
(a) Steady state torque/speed curve **(b)** Starting transient at no-load

An exception is the no–load starting of an induction motor without stator resistance and skin effect in the rotor bars; the normalised steady state torque/speed function is

$$m_M = \frac{2\,m_p}{S/S_p + S_p/S},$$

with $S = 1 - \omega/\omega_0$ being the slip and S_p the pull–out slip at which maximum torque m_p occurs. The function which will be derived in Chap. 10 is seen in Fig. 3.12 a for $S_p = 0.1$.

With no load ($m_L = 0$) and the initial condition $\omega(t_1 = 0) = 0$, $S(0) = 1$ the integration yields

$$t_2 = J \int_0^{\omega_2} \frac{d\omega}{m_M} = \frac{J\omega_0}{2\,m_p} \int_{S_2}^0 \left(\frac{S}{S_p} + \frac{S_p}{S} \right) dS$$

$$= \frac{J\omega_0}{2\,m_p} \left[\frac{1 - S_2^2}{S_p} - S_p \ln S_2 \right] . \qquad (3.23)$$

This function is plotted in Fig. 3.12 b. Since the speed approaches the synchronous speed ω_0 asymptotically, the definition of a starting time requires the choice of a final steady state slip S_2.

Computing the starting transient on the basis of the steady state torque/speed characteristic makes sense only if the starting is sufficiently slow in order to justify ignoring the electrical transients. This condition is met when the load contributes substantial additional inertia or when the start is delayed due to reduced voltage. Otherwise the start of a smaller motor would be completed within a few periods of the line voltages, long before the electrical transients have decayed; this will be discussed in Sect. 10.3.

3.3 Numerical and Graphical Integration

The most versatile method — with regard to accuracy and applicability — of integrating nonlinear differential equations is the stepwise integration with the digital computer.

It is known that by introducing state variables $x_i(t)$, $i = 1, 2, ..., n$, a differential equation of nth order may be written as a system of n first order differential equations

$$\frac{dx_i}{dt} = \varphi_i(x_j, y_k, t), \qquad i, j, k = 1, \dots, n , \qquad (3.24)$$

where $y_k(t)$ are the independent forcing functions and t the time as variable of integration. The functions φ_i may be arbitrary, given as analytical expressions, curves or even discrete samples. Once the state variables $x_i(t)$ are known, the output variables $z_i(t)$ can be calculated from algebraic expressions, i.e. without further integration.

$$z_i(t) = \Phi_i(x_j, y_k, t) . \qquad (3.25)$$

The simultaneous solution of the n state equations (3.24) is performed in steps of length Δt

$$x_i((\nu + 1)\,\Delta t) = x_i(\nu\,\Delta t) + \int_{\nu\,\Delta t}^{(\nu+1)\,\Delta t} \varphi_i(x_j, y_k, t)\,dt . \qquad (3.26)$$

$x_i(\nu\,\Delta t) = x_i(\nu)$ are the results obtained in the preceding step, $x_i(\nu + 1)$ is the new set of values at time $(\nu+1)\,\Delta t$. Based on approximations provided by numerical mathematics and with sufficiently small step length we can write

$$x_i(\nu + 1) \approx x_i(\nu) + F[\varphi_i(\nu), \varphi_i(\nu - 1), \ldots] \,, \tag{3.27}$$

where F is a linear combination of the functions φ_i at previous intervals. There exists a large number of integration formulae which differ in complexity, i.e. computing time, sensitivity to numerical instability and accuracy [22]. The ones best known are square-, trapezoidal- and Simpsons-rules, furthermore the Newton- and Runge-Kutta-algorithms. Because of its favourable properties the last mentioned algorithm is widely used; it is usually available as a complete subroutine so that only the functions φ_i and the initial conditions $x_i(0)$ need to be inserted as well as the integration interval Δt and the time t_2 at which the integration is to be terminated. Often the integration interval Δt is chosen automatically, depending on the functions φ_i and a specified accuracy limit. We are not going to discuss details of numerical integration; there can be subtle problems interrelating step size, accuracy and numerical stability. Since a large number of steps may have to be computed sequentially, even minute systematic errors can accumulate under unfavourable conditions.

Obviously the two differential equations, Eqs. (2.1), (2.2), correspond to this general scheme; when taking the angle of rotation into account, the order is $n = 2$, i. e. a very small system of differential equations results. The state variables are the rotational or linear speed and the angle of rotation or the linear distance. These variables represent storage effects and, hence, are continuous. If a more accurate description is desired, the electrical transients have to be included, thus increasing the number of differential equations and state variables; this is discussed in later sections.

When a timetable for a branch of a railway is to be computed, it is necessary to take into account the mass M of the trains, including the equivalent inertia of the wheel sets, the speed–dependent driving forces of the locomotives, distance–dependent conditions, such as grades, curves or speed restrictions, as well as maximum values for deceleration. The task requires the simultaneous solution of the differential equations for velocity and distance

$$\frac{dv}{dt} = \frac{1}{M} \left[f_M(v, t) - f_L(s, v, t) \right] ,$$
$$\frac{ds}{dt} = v \,,$$

where f_M is the internal driving force of the locomotive and f_L is the total load force, including frictional and braking forces as well as gravitational forces on grades. The task of finding an acceptable timetable is an optimisation problem of considerable complexity, since many constraints and boundary conditions must be taken into account. This includes the use of the same track by trains having different acceleration and velocity limits, such as intercity or heavy goods trains. The problem can only be solved by iteration, repeatedly integrating Eqs. (2.1), (2.2) with different initial conditions. Today this is a typical problem for numerical integration with a digital computer. However, in the past it had to be solved manually using graphical methods.

Since it cannot be ruled out that, even today, this may have to be done occasionally, let us briefly consider the principle of graphical integration, using a simple example. Eq. (3.1), where $m_M(\omega)$ and $m_L(\omega)$ are given as empirical curves, Fig.3.13,

$$J \frac{d\omega}{dt} = m_M(\omega) - m_L(\omega) = m_a(\omega);$$
(3.28)

the equation is first normalised with the help of arbitrary values ω_1, m_1,

$$\frac{J\omega_1}{m_1} \frac{d}{dt}\left(\frac{\omega}{\omega_1}\right) = \frac{m_M}{m_1} - \frac{m_L}{m_1} = \frac{m_a}{m_1},$$
(3.29)

where m_a/m_1 corresponds to the normalised accelerating torque.

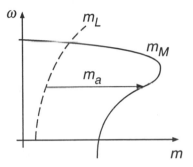

Fig. 3.13. Torque/speed curves of motor and load

With the abbreviations

$$\frac{J\omega_1}{m_1} = T_1, \qquad \frac{t}{T_1} = \tau, \qquad \frac{\omega}{\omega_1} = x, \qquad \frac{m_a}{m_1} = y$$

a nondimensional equation results

$$\frac{dx}{d\tau} = y(x).$$
(3.30)

Normalisation avoids the awkward choice of scale factors; the reference values ω_1, m_1 are of no significance, even though it is recommendable to choose characteristic values such as nominal speed and torque in order to work with handy numbers.

In Fig. 3.14 the normalised curve $y(x)$, obtained from Fig. 3.13 is shown, with the normalised speed plotted against normalised torque. The x-axis is subdivided into several intervals, not necessarily of equal length; it is in fact appropriate to reduce the lengths of the intervals in sections where $y(x)$ is changing rapidly. In each interval y is approximated by a constant, visually

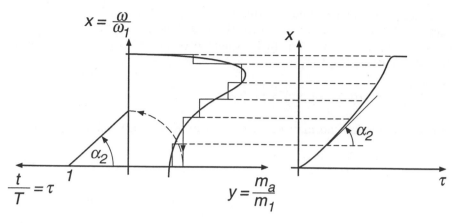

Fig. 3.14. Principle of graphical integration

representing the average of $y(x)$ in that interval. Approximating $y(x)$ by a staircase function has the consequence that $dx/d\tau$ is assumed constant in each interval, so that $x(\tau)$ assumes polygonal form.

By selecting the point $\tau = 1$ at a suitable distance to the left of the origin, radii with the slope $y(x)$ can be drawn which are then joined to form the polygonal curve approximating the exact solution $x(\tau)$. The graphical construction is indicated in Fig. 3.14.

The method described is related to the rectangular rule of integration, however with considerably improved accuracy because of the visual averaging process. Normally it does not pay to choose the x–intervals too small because drafting inaccuracies may then accumulate.

If, besides the speed, the angular rotation or the distance travelled is of interest, a second integration is required; the two graphical constructions must proceed in alternate steps if acceleration depends on distance; this is the case with a train due to grades or curves with speed restrictions and changing frictional forces.

4. Thermal Effects in Electrical Machines

4.1 Power Losses and Temperature Restrictions

So far our considerations were only dealing with mechanical phenomena and the pertinent steady-state and dynamic conditions, but suitable torque/speed-curves and adequate power are not the only criteria for designing electrical drives.

Of equal importance are the thermal transients in the motor caused by the unavoidable power losses during the process of energy conversion and the ensuing heat flow from the points of origin to the cooling medium. The various materials used in an electrical machine naturally have different temperature limits; of particular importance are the insulating materials, which determine the temperature classes of the machine, for example

$$
\begin{array}{lll}
\text{class } A, & \overline{\Delta\vartheta} < 60°\ \text{C}, & \text{cotton, synthetic, paper,} \\
B, & \overline{\Delta\vartheta} < 80°\ \text{C}, & \text{resin, shellac,} \\
F, & \overline{\Delta\vartheta} < 105°\ \text{C}, & \text{mica, epoxy,} \\
H, & \overline{\Delta\vartheta} < 125°\ \text{C}, & \text{glass fibre, silicone rubber.}
\end{array}
$$

$\overline{\Delta\vartheta}$ is the mean temperature rise above an assumed ambient temperature of $\vartheta_0 = 40°$C. All temperatures are measured in degree Celsius (centigrades).

For a given frame size and type of cooling, the output power and the allowable power losses are higher for the elevated temperature insulation classes; at the same time the cost of the machine increases.

The main causes of power losses in an electrical machine are:

a) *Conductor heat loss (copper losses)* in the windings, cables, brushes, slip-rings and commutator. If there is alternating current, the effective resistance of the conductor may be noticeably increased by eddy currents (skin effect); this is accentuated when large conductors are surrounded by magnetic material and may have to be alleviated by the use of stranded conductors.

b) *Iron heat losses* in stationary or moving magnetic material exposed to changing magnetic fields. The losses consist of eddy current- and hysteresis-losses; being of separate nature, they are determined by different properties of the material and follow different laws with regard to amplitude and frequency of the magnetic field.

c) *Friction losses*, including bearing-, brush- and ventilation losses.

The various types of losses depend in complicated fashion on the operational state of the machine. The main factors of influence are speed and load torque, as well as voltages and currents with their RMS–values and waveforms.

In order to induce heat flow from the points of origin to the cooling surfaces, a temperature gradient arises inside the machine determining the direction and intensity of the heat flow. The temperatures are usually highest in the windings because the loss density in the conductors being covered by insulating material is high; also the windings are partly embedded in slots and thus are not directly exposed to the cooling air. The hot–spot temperature with class B insulation could be, for example,

$$\vartheta_{max} = \overline{\Delta\vartheta} + (\Delta\vartheta_{max} - \overline{\Delta\vartheta}) + \vartheta_0$$
$$40°C = 130°C .$$

When operating the machine at elevated ambient temperatures, e.g. in the tropics, the power rating may have to be reduced.

An accurate prediction of the heat flow and the temperature distribution in an electrical machine is exceedingly difficult; this is due to the complex geometrical shapes, the use of heterogeneous and anisotropic materials (laminated iron, insulation), furthermore the complicated laws of heat generation with respect to space and time, as well as the different cooling conditions. Also, the heat conductivity of the various materials does not differ by orders of magnitude, as is the case with electrical or magnetic fields. As a result, the temperature distribution can only be computed in restricted sections and with considerable simplifications.

The user of electrical drives has normally little influence on the temperature distribution within the machine. He must be confident that the designer has chosen sufficiently large conductors and cooling channels so as not to exceed temperature limits under specified operating conditions. With this assumption it is usually adequate to drastically simplify the thermal model of the "motor" by regarding it as a homogeneous body exhibiting thermal storage, whose internal transients are unknown and of no interest. Naturally, such a crude model cannot offer any detailed information on specific internal thermal conditions.

4.2 Heating of a Homogeneous Body

The simplified thermal model is characterised by the following assumptions, Fig. 4.1:

A homogeneous body with the surface A, the thermal capacity C, measured in Ws/°C, and the mean surface temperature ϑ is heated by the input

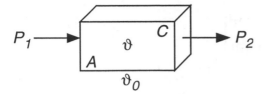

Fig. 4.1. Homogeneous body with thermal storage capacity

power p_1, at the same time emitting heat power p_2 by convection. The ambient temperature is ϑ_0, the coefficient of heat transfer α. The component of radiated heat is assumed to be negligible due to the relatively low operating temperatures (it rises with the fourth power of absolute temperature) and due to back radiation, for example at the cooling fins. Hence the equation, describing the power balance is

$$C\frac{d\vartheta}{dt} = p_1 - p_2 \;. \tag{4.1}$$

The heat transfer by convection is a linear function of the temperature difference

$$p_2 = \alpha\,A\,(\vartheta - \vartheta_0)\;; \tag{4.2}$$

with $\vartheta - \vartheta_0 = \Delta\vartheta$ this results in

$$C\frac{d(\Delta\vartheta)}{dt} + \alpha\,A\,\Delta\vartheta = p_1 \tag{4.3}$$

or

$$T_\vartheta\frac{d\Delta\vartheta}{dt} + \Delta\vartheta = \frac{p_1}{\alpha\,A}\;, \tag{4.4}$$

a linear first order differential equation; $T_\vartheta = C/\alpha A$ is the thermal time constant.

With $p_1 = p_{10} = $ const. and $\Delta\vartheta(0) = 0$ the solution is

$$\Delta\vartheta(t) = \Delta\vartheta(\infty)\,(1 - e^{-t/T_\vartheta})\;, \tag{4.5}$$

where

$$\Delta\vartheta(\infty) = p_{10}/\alpha\,A \tag{4.6}$$

is the steady state end temperature, at which the heat transfer by convection equals the input power.

In Fig. 4.2 a thermal transient is seen. The shaded area between $p_1/\alpha A$ and $\Delta\vartheta(t)$ corresponds to the energy stored as heat. Hence, the storage effect tends to delay the temperature changes, causing the temperature to be a continuous function of time.

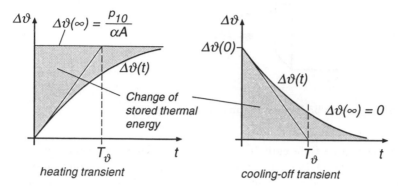

Fig. 4.2. Thermal transients

The approximation incurred with this simple model is mainly due to the assumption of evenly distributed heat sources in the homogeneous body and the neglect of an internal heat flow; hence, transportation delays are not contained in the simplified model.

A rough estimate of a typical thermal lag may be derived from the data sheet of an enclosed 100 kW standard induction motor cooled by an external fan (TEFC):

p_n = 100 kW,   nominal power
M =   800 kg,   mass
η =   0.92,   nominal efficiency
$\Delta\vartheta(\infty) = 50°$C is the steady–state temperature rise due to the losses p_1 at nominal power

$$p_1 = p_n \left(\frac{1}{\eta} - 1\right) ;$$

the convective heat conductivity is

$$\alpha A = \frac{p_1}{\Delta\vartheta(\infty)} .$$

Assuming that the motor consists of solid iron having the specific heat c_{Fe}, the heat capacity is

$$C = c_{Fe}\, M .$$

This yields the thermal time constant

$$T_\vartheta = \frac{C}{\alpha A} = \frac{c_{Fe}\, M\, \Delta\vartheta(\infty)}{\frac{1-\eta}{\eta}\, p_n} = \frac{0.48\,\text{kWs/kg°C}\ 800\ \text{kg}\ 50°\text{C}}{9\ \text{kW}} \approx 34\ \text{min.}$$

Typically, thermal transients are much slower than electrical or mechanical effects. Of course, there are exceptions such as disk motors, where the armature winding is printed directly on an insulating disk and possesses very little thermal storage capacity. The following time scales are usually valid with electrical machines:

electromagnetic	mechanical	thermal transients
0.1 − 100 ms	10 ms − 10 s	10 − 60 min

Travelling wave phenomena in windings are not considered; they are faster yet by several orders of magnitude. The large difference of more than a factor of 100 between the time scales of mechanical and thermal transients usually permits a separate treatment which results in considerable simplification.

The heat transfer by convection from a ventilated surface depends strongly on the velocity of the cooling air; the usual range of the transfer coefficient is

$$\alpha = \ 50 \text{ to } 500 \text{ W}/^\circ\text{C m}^2 \ ;$$

at $\Delta\vartheta = 50^\circ\text{C}$ this results in a power flow by convection of

$$\frac{p_{2c}}{A} = \ 2.5 \text{ to } 25 \text{ kW/m}^2 \ ;$$

the dependence of the heat transfer on the velocity of the cooling air has the consequence that self–cooled motors exhibit at standstill a cooling time constant that is much longer than the thermal time constant of the motor when running. Larger motors operating at variable speed and load are usually provided with separate forced cooling, causing the motor speed to have little effect on the cooling conditions.

To show that the heat transfer by radiation can normally be neglected, a rough estimate of the power flow by radiation is given for comparison.

A "black body" emits to the environment power by radiation according to Boltzmanns law,

$$\frac{p_R}{A} = \sigma T^4 \ ,$$

where

$$\sigma = 5.710^{-8} \text{ W/K}^4 \text{ m}^2$$

is the coefficient of radiation and T the absolute temperature. With $\vartheta = 70^\circ\text{C}$, i.e. $T = 343$ K, the power flow due to radiation is

$$\frac{p_R}{A} = 0.77 \text{ kW/m}^2 \ .$$

This relatively small amount is further reduced by inverse radiation. Also, the radiation coefficient of a motor surface is much lower than that of a "black body".

4.3 Different Modes of Operation

In contrast to internal combustion engines, most electrical drives possess considerable overload-capacity, the limit of which is determined by the power losses rising with the load. Since the temperature follows the power loss with a substantial lag, a temporary overload is admissible without exceeding the temperature limits, provided that the motor is initially below the nominal temperature. This effect may be utilised when selecting a drive for a particular application. Let us look at some modes of operation frequently occurring in practice.

4.3.1 Continuous Duty

This mode of operation exists if the load torque is constant during an extended period of time, corresponding to a multiple of the thermal time constant, so that the temperature reaches its steady–state value, Fig. 4.2 a. This is the normal continuous duty, e.g. with paper mill drives or boiler feed pumps. The motor must be so chosen that its nominal output power equals or exceeds the continuous load. With traction drives, other definitions are also in use, such as a one–hour– or five–minutes–rating of the drive motors.

4.3.2 Short Time Intermittent Duty

In this case the time of operation is considerably less than the thermal time constant and the motor is allowed to cool off before a new load cycle begins. This duty applies for example to some crane drives or small motors in household appliances. Under these conditions the motor can be overloaded until the temperature at the hottest point reaches the permissible limit; however, the internal temperature rise is particularly large with high overload. The simplified thermal model provides only a rough estimate under these conditions.

In Fig. 4.3 the temperature transient is drawn, following a short time power loss p_{1s} which may be considerably in excess of the nominal power loss p_{1n}. The temperature rises rapidly towards an extrapolated steady-state value $\Delta\vartheta_\infty$; when $\Delta\vartheta_1$, is reached, the motor is disconnected, giving it the opportunity to cool. In the example it is assumed that the motor is self–cooled causing the thermal time constant at standstill to have an increased value.

The maximum temperature $\Delta\vartheta_1$, at the time of disconnection is

$$\Delta\vartheta_1 = \Delta\vartheta_\infty[1 - e^{-t_1/T_\vartheta}] \leq \Delta\vartheta_{max} , \tag{4.7}$$

where

$$\Delta\vartheta_\infty = \Delta\vartheta_{max} \frac{p_{1s}}{p_{1n}} \tag{4.8}$$

is the extrapolated end temperature during overload. Hence

$$\frac{p_{1s}}{p_{1n}} \leq \frac{1}{1 - e^{-t_1/T_\vartheta}} \approx \frac{T_\vartheta}{t_1} , \quad t_1 \ll T_\vartheta , \tag{4.9}$$

is the maximum relative power loss during the time t_1.

Increasing the output power while simultaneously reducing the time t_1 of operation is of course only feasible within the available torque range; this is illustrated in Fig. 4.4. At the left, a typical curve relating torque to power loss is drawn, while the other curve depicts Eq. (4.9). This indicates that the power loss rises much faster than the torque (assuming constant speed). The shaded contour corresponds to the admissible short time overload range.

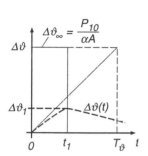
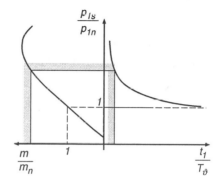

Fig. 4.3. Thermal transient during short time duty

Fig. 4.4. Short time overload limits

With motors exhibiting a pull out effect there is an absolute torque limit, whereas the losses continue to rise up to the stalled condition; on commutator machines, the commutator is usually the weakest link.

To realise high torque pulses which exceed even the short time overload capacity described by Fig. 4.4, a mechanical energy storage device, for instance a flywheel, may be used; of course, the inertia must be accelerated to sufficient speed before the load can be applied. This is the usual solution with highly pulsating loads such as punching presses.

4.3.3 Periodic intermittent duty

With some types of load, for example elevators, automatic machine tools, mine hoists or rolling mill drives typical load cycles exist which are repeated periodically. A very simple case is seen in Fig. 4.5, where the power loss p_{1i} follows a rectangular periodic pattern.

This results in temperature fluctuations, the mean of which rises until a steady state condition is reached. With the thermal time constants T_1, T_2 valid during On- and Off-times t_1, t_2, the following relations hold:

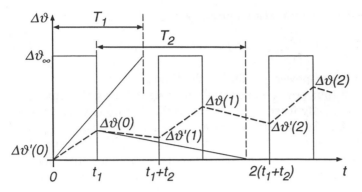

Fig. 4.5. Periodic intermittent duty cycle

$$\Delta\vartheta'(\nu + 1) = \Delta\vartheta(\nu)\, e^{-t_2/T_2}\,, \tag{4.10}$$
$$\Delta\vartheta(\nu + 1) = \Delta\vartheta'(\nu + 1)\, e^{-t_1/T_1} + \Delta\vartheta_\infty[1 - e^{-t_1/T_1}]\,.$$

Eliminating $\Delta\vartheta'(\nu + 1)$ yields

$$\Delta\vartheta(\nu + 1) = \Delta\vartheta(0) + a\,\Delta\vartheta(\nu)\,, \tag{4.11}$$

where

$$\Delta\vartheta(0) = [1 - e^{-t_1/T_1}]\,\Delta\vartheta_\infty\,. \tag{4.12}$$
$$a = e^{-(t_1/T_1 + t_2/T_2)} < 1\,.$$

Equation (4.11) represents a linear recursion between two successive peak values of the temperature. The solution is a geometrical progression

$$\Delta\vartheta(\nu) = \frac{1 - a^{\nu+1}}{1 - a}\,\Delta\vartheta(0) =$$
$$= \frac{1 - e^{-t_1/T_1}}{1 - e^{-(t_1/T_1 + t_2/T_2)}}\,[1 - e^{-(\nu+1)(t_1/T_1 + t_2/T_2)}]\,\Delta\vartheta_\infty\,. \tag{4.13}$$

Because of $t = t_1 + \nu\,(t_1 + t_2)$ the peak values lie on an exponential function having a time constant

$$T_\vartheta \approx \frac{t_1 + t_2}{t_1/T_1 + t_2/T_2} \quad \text{with} \quad T_1 < T_\vartheta < T_2\,. \tag{4.14}$$

The steady state peak temperature follows from Eq. (4.13)

$$\Delta\vartheta(\infty) = \frac{1 - e^{-t_1/T_1}}{1 - e^{-(t_1/T_1 + t_2/T_2)}}\,\Delta\vartheta_\infty\,. \tag{4.15}$$

With forced ventilation, a special case exists with

$$T_1 = T_2 = T_\vartheta\,.$$

5. Separately Excited DC Machine

5.1 Introduction

Direct current (DC) motors have been dominating the field of adjustable speed drives for over a century; they are still the most common choice if a controlled electrical drive operating over a wide speed range is specified. This is due to their excellent operational properties and control characteristics; the only essential disadvantage is the mechanical commutator which restricts the power and speed of the motor, increases the inertia and the axial length and requires periodic maintenance. With alternating current (AC) motors, fed by variable frequency static power converters, the commutator is eliminated, however at the cost of increased complexity of control. This is one of the reasons why controlled AC drives could not immediately supplant DC drives, once the semiconductor–technology had sufficiently advanced.

The principle of a DC machine operating in steady state is assumed to be known [4, 64], but let us recall some basic facts.

In Fig. 5.1a a schematic cross-section of a two-pole DC machine is shown, containing the fixed stator S and the cylindrical rotor, called armature A. While rotor and pole shoes are always laminated to reduce the iron losses caused by the varying magnetic field, the rest of the stator is laminated only in larger machines, when the motor is required to operate with rapidly varying torque and speed or when a static power converter with highly distorted voltages and currents is employed as the power supply. The main poles MP are fitted with the field windings, carrying the field current i_e which drives the main flux Φ_e through stator and rotor. A closed armature winding is placed in the axial slots of the rotor and connected to the commutator bars; it is supplied through the brushes and the commutator with the armature current i_a. This creates a distributed ampereturns (m.m.f.) wave, fixed in space and orientated in the direction of the quadrature axis, orthogonal to the main field axis, so that maximum torque for a given armature current is produced.

In view of the large airgap in the quadrature direction the resulting armature flux Φ_a is much smaller than the main flux Φ_e. It can be reduced even further by placing compensating windings CW in axial slots on the pole shoes and connecting them in series with the armature. Their opposing ampereturns all but cancel the quadrature field excited by the armature and

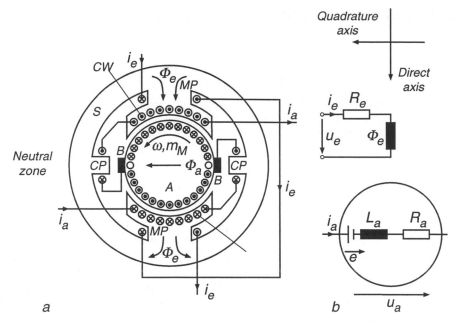

Fig. 5.1. (a) Cross–section and **(b)** schematic circuit of a DC machine

remove the undesired armature reaction, which otherwise tends to distort the
even distribution of the main field under the poles along the circumference
of the rotor. Compensating windings are common only on larger machines
or converter-fed motors for heavy duty applications such as traction or steel-
mill drives. Compensated DC machines can withstand higher overloads than
uncompensated machines; also the armature current may rise much faster
and larger current harmonics are acceptable without detrimental effects on
commutation, i.e. sparking of brushes. This is of particular importance if the
machine is supplied by a static converter.

The commutating poles CP, placed between the main poles and also
carrying the armature current, have the task of locally modifying the field in
the neutral zone, to achieve rapid and sparkfree commutation. This is done
by inducing a suitable voltage in the armature coil temporarily shorted by
the brushes.

The principle of commutation is illustrated in Fig. 5.2, where the closed
armature winding of a two–pole DC motor is schematically drawn, showing
the brush positions at two consecutive instants of time. Clearly, when the
feeding points of the winding are shifted by the motion of the commutator
relative to the brushes, the current in the pertinent armature conductors is
inverted; this results in a rectangular current wave, the frequency of which
is proportional to the speed of the motor; at the same time, a stationary
ampereturns distribution, fixed in space, is created, Fig. 5.1 a. Hence the

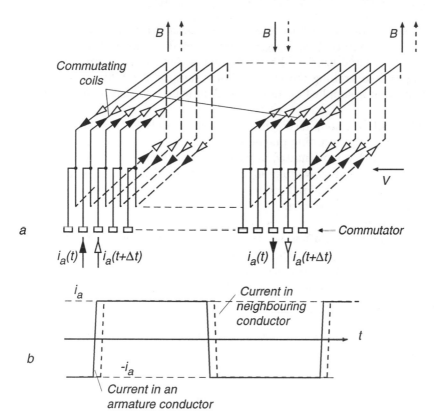

Fig. 5.2. Commutation of current in an armature coil, (a) Winding and (b) Currents in two adjacent armature conductors

commutator acts like a multi-phase electromechanical switching converter which is synchronised by the rotor angle.

While the current is being inverted, the commutating coil is temporarily short-circuited by the brushes overlaping adjacent commutator bars. Since each coil, embedded in slots and surrounded by iron, exhibits some inductance, the commutation is a continuous process which requires a finite time; this limits the speed at which the motor can operate without excessively sparking brushes. For a two pole motor with 100 commutator bars operating at 3000 rev/min, the overlap takes about 200 μs. If, due to excessive speed, the short circuit of the commutating coil is opened before the commutation is completed, a voltage is induced which causes sparking at the trailing edges of the brushes.

The electrical torque m_M is exerted at the rotor surface; more precisely, by tangential magnetic forces acting on the sides of the slots. It is proportional to the product of main flux and armature current. The voltages in the armature circuit consist of an induced voltage (e.m.f.) proportional to main flux and speed, in addition to resistive and inductive voltage drops in the armature-, commutating- and compensating windings as well as the armature connections and brushes. The main flux is thus of prime importance for operating and controlling the DC machine; greatest flexibility exists, when the field current is supplied by an independent source so that it can be varied at will.

In Abb. 5.1 b a simplified scheme of a separately excited DC machine is shown, where the resulting magnetic fluxes are concentrated into lumped inductances which are orientated in parallel and orthogonal to the main field axis. The total armature voltage u_a thus consists of the voltage e, induced by rotation, the voltage of self-induction $L_a \, di_a/dt$ and the ohmic drop $R_a i_a$, where L_a, R_a are the resulting parameters in the armature circuit. With larger machines, $R_a i_a \ll u_a$ is valid at normal speed and load. Also, the power required for excitation, i.e. the losses in the field winding, amount to only a few percent of the power converted in the armature.

With motors having permanent magnet excitation the power losses in the field winding and the need for a separate field power supply are avoided at the cost of operational flexibility. Servomotors for machine tool feed drives are usually of the permanent magnet type.

The DC machine as a dynamic system, including the interactions of the electromagnetic and the mechanical effects, is dealt with in the next section.

5.2 Mathematical Model of the DC Machine

The equivalent circuit of the separately excited DC machine can be represented in schematic form by Fig. 5.3 containing concentrated components and showing no longer a geometrical resemblance to the actual system.

The pertinent differential equations are

$R_a \, i_a + L_a \dfrac{di_a}{dt} + e = u_a$ armature circuit

$e = c_1 \, \Phi_e \, \omega$ induced voltage (e.m.f.)

$J \dfrac{d\omega}{dt} = m_M - m_L$ Newtons law, concentrated mass

$m_M = c_2 \, \Phi_e \, i_a$ electrical torque

$R_e \, i_e + N_e \dfrac{d\Phi_e}{dt} = u_e$ field circuit

$\Phi_e = f(i_e)$ static magnetising curve, omitting hysteresis

$\dfrac{d\epsilon}{dt} = \omega$ angular velocity.

Eliminating Φ_e from the second and fourth equation results in

$e \, i_a = \dfrac{c_1}{c_2} \, m_M \, \omega \, ,$

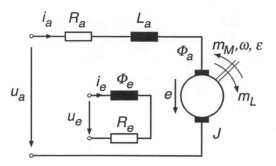

Fig. 5.3. Equivalent circuit of a separately excited DC machine

where $e\,i_a$ is the instantaneous electrical power converted in the armature to the mechanical power $m_M\,\omega$. Both must be identical, hence $c_1 = c_2 = c$.

Normalisation with the nominal values

ω_0 nominal speed at no-load (base speed),
Φ_{e0} nominal flux
$u_{a0} = e_0 = c\,\Phi_{e0}\,\omega_0$ nominal armature voltage
$i_{a0} = \dfrac{u_{a0}}{R_a}$ extr. current at standstill with nominal voltage
     (8–10 times nominal current)
$m_0 = c\,\Phi_{e0}\,i_{a0}$ extrapolated torque at standstill
$u_{e0} = R_e\,i_{e0}$ nominal field voltage

results in the following non-dimensional state equations

$$T_a \frac{d}{dt}\left(\frac{i_1}{i_{a0}}\right) = \frac{u_a}{u_{a0}} - \frac{i_a}{i_{a0}} - \frac{\omega}{\omega_0}\frac{\Phi_e}{\Phi_{e0}} \;, \quad T_a = \frac{L_a}{R_a} \;; \tag{5.1}$$

$$T_{e0} \frac{d}{dt}\left(\frac{\Phi_e}{\Phi_{e0}}\right) = \frac{u_e}{u_{e0}} - f_e\left(\frac{\Phi_e}{\Phi_{e0}}\right) \;, \quad T_{e0} = \frac{N_e\,\Phi_{e0}}{u_{e0}} \;; \tag{5.2}$$

$$T_m \frac{d}{dt}\left(\frac{\omega}{\omega_0}\right) = \frac{i_a}{i_{a0}}\frac{\Phi_e}{\Phi_{e0}} - \frac{m_L}{m_0} \;, \quad T_m = \frac{J\,\omega_0}{m_0} \;; \tag{5.3}$$

$$T_\varepsilon \frac{d}{dt}\left(\frac{\varepsilon}{\varepsilon_0}\right) = \frac{\omega}{\omega_0} \;, \quad T_\varepsilon = \frac{\varepsilon_0}{\omega_0} \;. \tag{5.4}$$

$i_e/i_{e0} = f_e(\Phi_e/\Phi_{e0})$ is the inverse normalised magnetisation curve, ε_0 is an arbitrary reference angle.

The voltages u_a, u_e are assumed to be independently controllable, m_L is the applied load torque. The output variables of the four integrators correspond to the four storage quantities, subject to continuous change only. Due

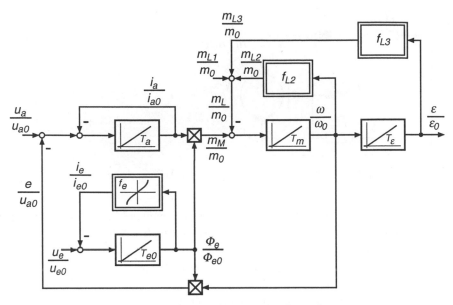

Fig. 5.4. Block diagram of a separately excited DC machine

to the larger airgap in the quadrature axis and the possible presence of a compensating winding $T_a \ll T_{e0}$ holds. Magnetic saturation in the quadrature direction is usually negligible.

The dynamic system, described by Eqs. (5.1)–(5.4), is represented as a block diagram in Fig. 5.4, containing linear integrators with their time constants and nonlinear algebraic functional blocks; it is assumed that the load torque consists of an independent component m_{L1} acting as a disturbance and two components as nonlinear functions of speed and position respectively; when the load torque shows no such dependence, the corresponding functions are zero.

$$\frac{m_L}{m_0} = \frac{m_{L1}}{m_0} + f_{L2}\left(\frac{\omega}{\omega_0}\right) + f_{L3}\left(\frac{\varepsilon}{\varepsilon_0}\right) . \qquad (5.5)$$

5.3 Steady State Characteristics with Armature and Field Control

The steady state condition, valid with constant input quantities u_a, u_e, m_L is obtained by setting the derivatives in Eqs. (5.1)–(5.3) equal to zero,

$$\frac{u_a}{u_{a0}} - \frac{i_a}{i_{a0}} - \frac{\omega}{\omega_0}\frac{\Phi_e}{\Phi_{e0}} = 0 , \qquad (5.6)$$

$$\frac{u_e}{u_{e0}} - f_e\left(\frac{\Phi_e}{\Phi_{e0}}\right) = 0 ,\tag{5.7}$$

$$\frac{i_a}{i_{a0}}\frac{\Phi_e}{\Phi_{e0}} - \frac{m_L}{m_0} = 0 .\tag{5.8}$$

The angle of rotation ε changes linearly with time; it is of no consequence for the time being. For the following discussion the normalised flux

$$b = \frac{\Phi_e}{\Phi_{e0}} \le 1 ,$$

which is limited by saturation of the iron core, is assumed as an independent control input, instead of u_e/u_{e0}. This results in

$$\frac{\omega}{\omega_0} = \frac{1}{b}\frac{u_a}{u_{a0}} - \frac{1}{b^2}\frac{m_L}{m_0} ,\tag{5.9}$$

$$\frac{i_a}{i_{a0}} = \frac{1}{b}\frac{m_L}{m_0} .\tag{5.10}$$

Hence the steady state behaviour of the controlled motor,

$$\frac{\omega}{\omega_0} = f_1\left(\frac{u_a}{u_{a0}}, \frac{m_L}{m_0}\right) ,$$

$$\frac{i_a}{i_{a0}} - f_2\left(\frac{m_L}{m_0}\right)$$

is characterised by linear functions, connecting speed, torque and armature current.

Depending on whether u_a or Φ_e, i.e. u_e, serves as control input, the term armature- or field-control applies.

5.3.1 Armature Control

Assuming $b = \Phi_e/\Phi_{e0} = 1$, the nonlinear effect of the two multiplications in Fig. 5.4 vanishes and a particularly simple set of linear control curves results,

$$\frac{\omega}{\omega_0} = \frac{u_a}{u_{a0}} - \frac{m_L}{m_0} ,\tag{5.11}$$

$$\frac{i_a}{i_{a0}} = \frac{m_L}{m_0} ,\tag{5.12}$$

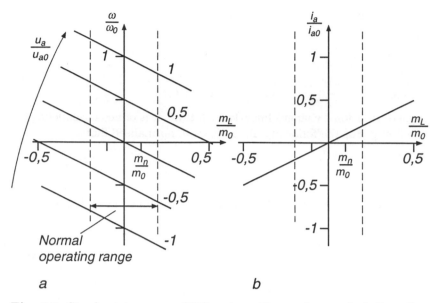

Fig. 5.5. Steady state curves of DC motor with armature control, $\Phi_e = \Phi_{e0}$

they are drawn in Fig. 5.5 as straight lines. The torque/speed curves are valid in all four quadrants permitting continuous control and reversal of torque and speed.

Since the armature voltage u_a is referred to its nominal value u_{a0}, only the range $-1 \leq u_a/u_{a0} \leq 1$ is of interest. Excessive voltage would cause sparking of the brushes and possibly commutation failure by formation of an arc short-circuiting the brushes (flashover).

The armature current is proportional to the torque, independent of voltage and speed. Armature current and torque are normalised with their extrapolated values at standstill. Therefore the normal operating region in Abb. 5.5 b is contained in a narrow band, e.g.

$$-0.2 \leq \frac{m_L}{m_0} = \frac{i_a}{i_{a0}} \leq 0.2 \,,$$

even when allowing a temporary overload of about twice nominal torque. Outside this range, the curves are likely to be distorted by armature reaction and there may be commutation problems, particularly on machines without a compensating winding.

5.3.2 Field Control

The second option for controlling the motor is by changing the main flux Φ_e. Due to saturation Φ_e cannot be increased much beyond Φ_{e0} - otherwise the motor designer would not have fully utilised the potential of the motor - hence it is only feasible to weaken the field,

$$-1 \le b = \Phi_e/\Phi_{e0} \le 1 .$$

If the armature power supply permits operation in all four quadrants of the u_a, i_a-plane, it is sufficient to restrict field control to positive values, $b_{min} < b \le 1$.

The purpose of field weakening is to raise the speed at light load, as seen from Eq. (5.9), accepting the disadvantage of increasing the armature current for a given torque, Eq. (5.10). Therefore it is normally not appropriate to employ field weakening unless the possibilities of armature control have already been fully exploited.

Inserting $u_a/u_{a0} = \pm 1$ into Eq. (5.9) yields a new set of steady state curves with the normalised field factor b as control parameter; the characteristics are again straight lines, however with increased slope (Fig. 5.6). The speed/torque curves intersect with the axes at

no load: $m_L = 0$, $\dfrac{\omega_{NL}}{\omega_0} = \dfrac{1}{b}$,

standstill: $\omega = 0$, $\dfrac{m_{st}}{m_0} = b$.

With reduced flux, the no-load speed rises, while the (extrapolated) stalled torque is reduced. Hence the slope of the curve is varying with $1/b^2$; this inherently "soft" load behaviour is undesirable with drives to be controlled for constant speed.

As is seen in Fig. 5.6 b, the armature current and the armature losses increase when the motor is operated at a given torque with reduced field; therefore, as already mentioned, field weakening makes only sense if the desired operating point in the torque-speed plane cannot be reached by armature voltage control at full field or if the armature voltage is fixed. At less than maximum armature voltage, operation at reduced field is uncommon, except in special circumstances, for instance when several motors are supplied from a common armature bus voltage and only minor speed variations are needed as on some paper mill drive systems. Neither is field weakening a solution for the start-up of motors since at standstill the armature current is not affected by the main flux. Fig. 5.6 a indicates also that a reduction of the field does not necessarily lead to higher speed; because of the steeper torque/speed curves the result of field-weakening may even be reduced speed at higher armature current which is of course most undesirable. The maximum speed at a given torque is obtained by differentiating Eq. (5.9) with the condition $u_a = u_{a0}$,

$$\left. \frac{d\frac{\omega}{\omega_0}}{db} \right|_{m_L=\text{const}} = -\frac{1}{b^2} + \frac{2}{b^3}\frac{m_L}{m_0} = 0 ,$$

which leads to a practical lower limit for the field

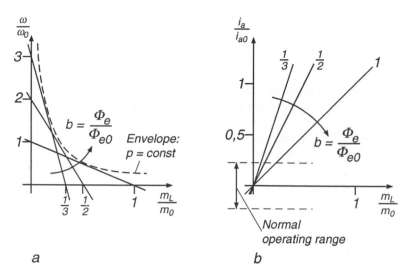

a   b

Fig. 5.6. Steady–state curves of DC motor in field control range, $u_a = u_{a0}$

$$b_{min} = 2\,\frac{m_L}{m_0}\,, \tag{5.13}$$

where the undesirable speed reduction begins. Note that m_0 is the extrapolated stalled torque for nominal armature voltage and field. Inserting Eq.(5.13) into Eq. (5.9) and assuming $u_a = u_{a0}$ yields a hyperbolic function

$$\frac{\omega}{\omega_0} = \frac{1}{4}\frac{m_0}{m_L} \tag{5.14}$$

which represents an envelope for the family of torque/speed-curves with parameter b. Maximum power is obtained at the point, where the straight line is tangent to the hyperbola. To avoid the situation mentioned before, the motor should never be operated at the right hand side of the tangent point.

DC machines without compensating windings show a pronounced armature reaction in the field weakening range because the field distribution in the airgap is no longer stabilised by saturation. Therefore the validity of the torque/speed-curves is restricted to reduced values of the normalised armature current; higher currents have detrimental effects on commutation and may result in sparking of the brushes. This calls for a reduction of the maximum armature current at reduced field and hence a power derating.

Despite these drawbacks, field weakening is a valuable means for increasing the speed in the low torque region; typical applications are spindle drives on machine tools to allow cutting at full power, even with a small diameter workpiece or tool. Other examples are reversing rolling mills for rods or rails, where the first few passes call for high torque to squeeze the short and wide slabs whereas in the later passes, when the material is thin and long, low torque but high rolling speeds are desired for saving time. A similar situation exists with coilers where high speed and low torque are needed when the diameter of the coil is small, while the reverse is true when a large coil diameter has built up.

After eliminating the field factor b from Eqs. (5.9), (5.10) with the condition $u_a/u_{a0} = 1$, an expression for the mechanical power is derived

$$\frac{\omega}{\omega_0} \frac{m_L}{m_0} = \frac{i_a}{i_{a0}} \left(1 - \frac{i_a}{i_{a0}}\right), \tag{5.15}$$

which only depends on the normalised armature current. Hence, with limited armature current, the motor can be operated at variable speed with a given maximum power, indicating that field weakening has a similar effect as a variable gear for increasing the speed of the load. For the reasons already mentioned and because of the mechanical stresses, the maximum speed achieved by field weakening is seldom higher than two or three times base speed depending on the design and the size of the machine. However, there are exceptions, such as on vehicle or servo drives, where a much higher speed ratio may be necessary.

The similarity of a speed increase by field control and by a variable gear is corroborated by their effects on the mechanical time constant. At nominal field, $\Phi_e = \Phi_{e0}$, the nominal time constant is $T_m = J\omega_0/m_0$, where ω_0 is the base (no load) speed and m_0 the extrapolated stalled torque at nominal armature voltage and field. With reduced field, the no load speed is raised to $b^{-1}\omega_0$, while the stalled torque is lowered to $b\,m_0$. Hence the mechanical time constant is a quadratic function of the field factor,

$$T_m\,(b \leq 1) = \frac{J\omega_0}{m_0}\,b^{-2} = T_m\,b^{-2} .$$

A similar effect is observed if the load is coupled to the motor through a step-up gear having the speed ratio b^{-1}; as explained in Sect. 1.3, this would increase the effective inertia of the load, as seen from the motor side, by b^{-2}.

5.3.3 Combined Armature and Field Control

From the preceding paragraph it is seen that armature and field-control of DC motors each have their merits but that their normal applications exclude each other. Below base speed ω_0 the main flux is best kept at nominal value Φ_{e0} while the speed is varied through the armature voltage u_a. This is called the base speed- or armature control range. When the nominal armature voltage $\pm u_{a0}$ is reached, a further speed increase is only possible by lowering the main flux, thus creating separate field-control regions which extend the base speed range in each direction of rotation. This is depicted in Fig. 5.7.

It is seen that a reversal of the no-load speed requires reversed armature voltage. In principle it would also be possible to reverse the motor speed

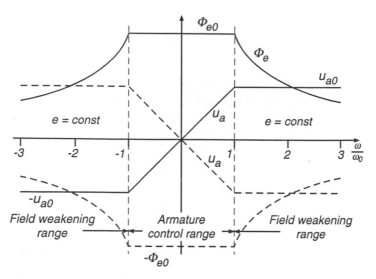

Fig. 5.7. Control ranges of separately excited DC motor in steady state

with the same polarity of the armature voltage by reversing the field (dashed curves); at the same time the torque would be reversed so that the motor operates in the third quadrant. However, this method of "field reversal" which was frequently used in the past for cost reasons is rarely employed today as it calls for the de-excitation and subsequent re-excitation of the motor; this requires time in view of the large magnetic energy stored in the field circuit. If high forcing voltage is applied to the field winding to accelerate the flux reversal, considerable voltages are also induced in the armature winding endangering the commutator. Also, the armature current must be blocked during the time of flux reversal causing an interruption of the torque. With larger motors, the transition time with zero torque may be up to 1s which is not acceptable for a fast reversing drive. It is for these reasons that drives which are expected to operate with both directions of torque and speed ("four quadrant drive"), are today usually equipped with an armature power supply providing both polarities of voltage and current ("four-quadrant DC power supply") and capable of operation in all quadrants of the u_a/i_a - plane, Sect. 9.1; this has the advantage that the polarity of the main flux can remain unchanged while the motor can operate continuously in all quadrants of the m_M/ω plane.

At low power, the combined armature-field control shown in Fig. 5.7 can be realised by a circuit containing resistors. With larger machines this is not feasible in view of the high losses, so that separate power supplies are needed for feeding the armature and the field windings. Field weakening may be performed automatically by an auxiliary control loop which, by adjusting the field supply voltage, limits the induced armature voltage according to

$$\frac{e}{u_{a0}} = \frac{\omega}{\omega_0}\frac{\Phi_e}{\Phi_{e0}} \leq 1 \; ; \tag{5.16}$$

this causes the desired variation of the flux in the field weakening range (see also Sect. 7.3).

The operating regions with armature and field control in the torque/speed plane are displayed in Fig. 5.8. The curves i_a =const. are represented by parallel straight lines in the armature control regime; in the field weakening region, the torque at a given armature current is reduced, because of $m_M \approx \Phi_e i_a$. For the same reason, the curves for u_a =const. become steeper when the flux is reduced. In the upper right hand corner a typical commutation limit is indicated belonging to a reduced armature current limit when the field is weakened.

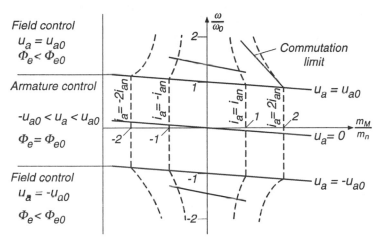

Fig. 5.8. Operating regions of separately excited DC motor in torque/speed plane

The figure clearly shows that the DC motor represents a linear control plant only in the armature control region; beyond base speed, there are considerable non-linearities even in steady state condition.

Simultaneous field– and armature–control is sometimes proposed as a means of reducing motor losses through diminished copper losses in the field windings and iron losses in the armature; of course, this is only applicable at very light load when the increased conduction losses in the armature are not masking this effect. Also, this may cause an objectionable delay in building up motor torque, should an unexpected load surge occur, because the flux would first have to be raised to full value. Hence, this practice can only be a marginal option in few applications, such as battery supplied electric car drives, where energy conservation is of overriding importance and where the dynamic drawbacks are not objectionable.

5.4 Dynamic Behaviour of DC Motor with Constant Flux

In order to have a clearer view of the transient behaviour of a DC machine the block diagram in Fig. 5.4 is simplified by assuming the armature voltage to be an impressed and independently variable input quantity, the load torque m_L as an independent disturbance and the flux Φ_e a suitably chosen parameter. This results in Fig. 5.9; the angle of rotation is again omitted for the time being, as it does not affect the operation of the motor.

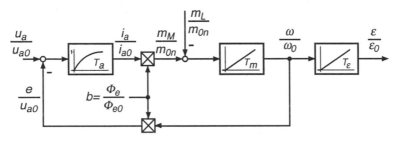

Fig. 5.9. Linear block diagram of DC motor

Assuming the main flux Φ_e to be a constant parameter, a linear system results which may be described by transfer functions; with the Laplace-transforms

$$L(x(t)) = \int_0^\infty x(t)\, e^{-st}\, dt = X(s), \qquad s = \sigma + j\omega \,,$$

of the nondimensional variables

$$L\left(\frac{u_a}{u_{a0}}\right) = U_a(s), \quad L\left(\frac{\omega}{\omega_0}\right) = \Omega(s) \,,$$

$$L\left(\frac{i_a}{i_{a0}}\right) = I_a(s), \quad L\left(\frac{m_L}{m_0}\right) = M_L(s)$$

and the normalised field parameter

$$\frac{\Phi_e}{\Phi_{e0}} = b \leq 1$$

the following equations are obtained

$$\Omega(s) = F_1(s)\, U_a(s) + F_2(s)\, M_L(s) = \frac{b\, U_a(s) - (T_a\, s + 1)\, M_L(s)}{T_m\, s\, (T_a\, s + 1) + b^2} \,,$$

$$(5.17)$$

$$I_a(s) = F_3(s)\, U_a(s) + F_1(s)\, M_L(s) = \frac{T_m\, s\, U_a(s) + b\, M_L(s)}{T_m\, s\, (T_a\, s + 1) + b^2}\ . \tag{5.18}$$

$U_a(s)$ is the independent input or actuating variable, $M_L(s)$ is a disturbance, also assumed to be impressed. $\Omega(s)$ and $I_a(s)$ stand for output variables, which appear in the frequency domain as linear combinations of the input variables $U_a(s)$ and $M_L(s)$.

All partial transfer functions of the multivariable plant have the same denominator,

$$N(s) = T_m\, s\, (T_a\, s + 1) + b^2 = T_m\, T_a\, (s - s_1)\, (s - s_2)\,, \tag{5.19}$$

which is identical with the characteristic polynomial of the pertinent linear differential equation. The zeros of $N(s)$ are the eigenvalues of the system,

$$s_{1,2} = \frac{1}{2\, T_a}\left[-1 \pm \sqrt{1 - (2b)^2\, T_a/T_m}\right]\ ; \tag{5.20}$$

the natural frequency is

$$\omega_e = \sqrt{s_1\, s_2} = \frac{b}{\sqrt{T_a\, T_m}}\,, \tag{5.21}$$

and the damping ratio

$$D = \frac{-\,\mathrm{Re}(s_1)}{|s_1|} = \frac{1}{2b}\sqrt{\frac{T_m}{T_a}}\ . \tag{5.22}$$

Field weakening ($b < 1$) causes the motor to respond more sluggishly to control or disturbance inputs as is manifested by a lower natural frequency

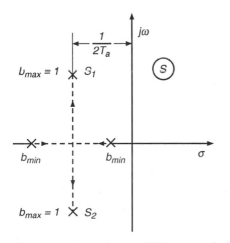

Fig. 5.10. Root locus of DC motor for different field factors $b \le 1$

and increased damping ratio. Fig. 5.10 shows the root locus, i.e. the locus of the eigenvalues in the complex plane, for different values of the parameter b. Real eigenvalues, corresponding to asymptotic transients $(D > 1)$ exist for

$$T_m > (2b)^2 T_a .$$

In Fig. 5.10 it is assumed that the motor exhibits below base speed $(b_{max} = 1)$ a damped oscillatory response. This may in practice be the case if the inertia of the drive is deliberately kept small as with servo drives or rolling-mill reversing drives but in most cases the eigenvalues of a drive are real because of the added inertia of the load.

In Fig. 5.11 the step responses of the motor speed and the armature current are depicted when the motor, initially at standstill, is started without load $(m_L = 0)$ at $t = 0$ by switching on the armature voltage $u_a = 0.1\,u_{a0}$. After attaining its steady state speed, load is applied at t_1 in the form of constant torque, $m_L = 0.02\,m_0$, corresponding approximately to 0.2 of nominal torque.

The transients in Fig. 5.11 have been computed for three different values of b, which result in the damping ratios $D = 1/\sqrt{2}, 1, \sqrt{2}$. It is seen that field weakening makes the motor much more sensitive to load; at higher torque the speed under load would even drop below the value at full flux $(b = 1)$, as indicated in Fig. 5.6 a. However, the initial slope of the speed at t_1 is the same in all cases being solely dependent on the nominal mechanical time constant T_m,

$$\frac{d}{dt}\left(\frac{\omega}{\omega_0}\right)\bigg|_{t_1} = -\frac{m_L}{m_0}\frac{1}{T_m} \neq f(b) .$$

This is immediately understood from inspection of Fig. 5.9, as the current i_a, i.e. the motor torque, is delayed by the armature inductance L_a.

The steady-state value of the armature current for no-load condition (neglecting friction and windage) is zero; this is apparent from the factor s in the numerator of the transfer function F_3, Eq. (5.18), which indicates a differentiating effect due to the cancellation of the applied armature voltage u_a by the induced voltage e at no load.

If the load torque contains a component rising with speed, Fig. 5.12 a,

$$\frac{m_L}{m_0} = \frac{m_{L1}}{m_0} + k_L\frac{\omega}{\omega_0} , \tag{5.23}$$

the block diagram in Fig. 5.12 b results; the integrator now possesses proportional negative feedback, assuming the properties of a lag element. The differentiating effect in Eq. (5.18) disappears in the presence of load torque because armature current is needed to generate the motor torque. The partial transfer function F_3 connecting armature voltage and current is then

$$\frac{I_a}{U_a}(s) = \frac{T_m\,s + k_L}{T_m T_a\,s^2 + (T_m + k_L\,T_a)\,s + b^2 + k_L} , \tag{5.24}$$

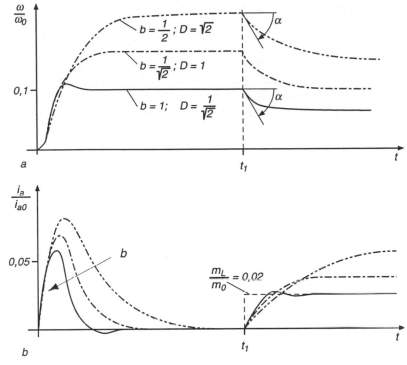

a

b

Fig. 5.11. Starting and load transients of DC motor for different field factor $b \leq 1$

with $k_L > -b^2$ and $k_L > T_m/T_a$ required for stability.

The damping ratio D_1 may be referred to the damping ratio D at no load

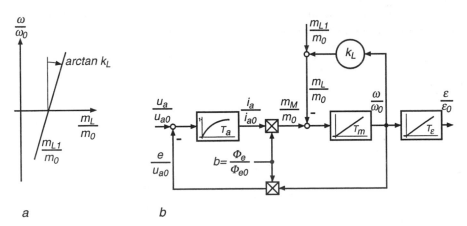

a   b

Fig. 5.12. DC motor with speed- dependent load

$$D_1 = D \frac{1 + \frac{1}{(2D)^2} \frac{k_L}{b^2}}{\sqrt{1 + \frac{k_L}{b^2}}} \ . \tag{5.25}$$

Some drives are subject to disturbances by periodic components in the supply voltage or the load torque, causing continuous oscillations of armature current and speed around their steady state values; the transfer function permits a simple computation. When a torque consists of a constant and a sinusoidally varying component

$$\frac{m_L}{m_0} = \frac{m_{L1}}{m_0} + \frac{\widehat{\Delta m_L}}{m_0} \cos \omega_m t \ ,$$

each variable of the linear drive will in steady state contain a periodic component of the same frequency that may be computed by linear superposition. With the complex expression

$$\frac{\Delta m_L}{m_0} \cos \omega_m t = \frac{\widehat{\Delta m_L}}{2 m_0} [e^{j \omega_m t} + e^{-j \omega_m t}]$$

we find for instance from Eq. (5.18) the peak value of the alternating current component $(k_L = 0)$

$$\frac{\widehat{\Delta i_a}}{i_{a0}} = \left| \frac{b}{T_m T_a (j \omega_m)^2 + T_m j \omega_m + b^2} \right| \frac{\widehat{\Delta m_L}}{m_0} \ . \tag{5.26}$$

Similarly the phase of the oscillatory current may be determined. If the frequency of a periodic disturbance lies near an eigenvalue of a poorly damped drive, a resonant condition could exist causing large amplitudes of the oscillatory current and speed deviations.

The dynamic behaviour of separately excited DC motors will be further discussed in later sections in connection with control problems.

6. DC Motor with Series Field Winding

This motor differs from the one previously discussed only by the design and the connection of the field winding which, according to Fig. 6.1, carries now part or all of the armature current. R_{a1} is the armature resistance, possibly increased by an external resistor, R_p is an adjustable shunt resistor for field weakening.

Because of their operating characteristics, series-wound motors of larger rating are restricted to traction drives. In urban transportation, such as trolley buses or subways, motors up to about 200 kW are used but for main line locomotives motors of 1000 kW are common. In principle, series wound motors can be fed with either direct or alternating current but the constructional design would be quite different; for example, a laminated stator and special provisions with regard to commutation are required for the AC motor. When AC traction was developed at the begin of the century, the current induced by transformer action in the armature coil temporarily shorted by the brushes during commutation, Fig. 5.2, caused problems, which were then solved by choosing a lower supply frequency; this is the reason for the $16^2/_3$ Hz power supply used on central European railways until today. Since compact and rugged power electronic components have become available, main line locomotives with AC supply were often equipped with rectifier- fed DC motors. In future, all these schemes are likely to be supplanted by AC motors fed from variable frequency inverters.

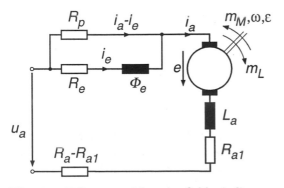

Fig. 6.1. DC motor with series field winding

Small high-speed AC series motors, so called universal motors, have a wide field of application in household appliances and power tools.

The typical operating characteristics of a series-wound motor could also be realised with a suitably controlled separately excited motor; this can even be preferable in view of the added flexibility, for example when a traction motor is supposed to serve as a regenerative brake.

6.1 Block Diagram of a Series-wound Motor

The equations describing the dynamics of the series-wound motor are similar to those derived in Chap. 5. The field and the armature windings are orthogonal to each other and hence are magnetically decoupled. Due to the large airgap in the armature axis, saturation is neglected and the magnetic circuit in quadrature direction is assumed to be linear.

$$L_a \frac{di_a}{dt} = u_a - R_a\, i_a - R_p\, (i_a - i_e) - e\,, \tag{6.1}$$

$$N_e \frac{d\Phi_e}{dt} = -R_e\, i_e + R_p\, (i_a - i_e)\,, \tag{6.2}$$

$$e = c\,\Phi_e\,\omega\,, \tag{6.3}$$

$$m_M = c\,\Phi_e\, i_a\,, \tag{6.4}$$

$$J \frac{d\omega}{dt} = m_M - m_L\,. \tag{6.5}$$

A nominal operating point (index 1) is defined at nominal armature voltage u_{a1}, without field weakening ($R_p \to \infty$) and with no external armature resistor ($R_a = R_{a1}$). At this reference point, the following steady state equations hold:

$$e_1 = c\,\Phi_{e1}\,\omega_1 = \eta_1\, u_{a1}$$

$$i_{a1} = i_{e1} = (1 - \eta_1)\, \frac{u_{a1}}{R_{a1} + R_e}\,,$$

$$m_1 = c\,\Phi_{e1}\, i_{a1}\,.$$

$\eta_1 = e_1/u_{a1}$ is the electrical efficiency at the nominal operating point (not accounting for iron- and friction losses). These quantities are used for normalising the Eqs. (6.1– 6.5).

$$\frac{i_e}{i_{a1}} = f_e\left(\frac{\Phi_e}{\Phi_{e1}}\right) \tag{6.6}$$

is the magnetisation curve in the longitudinal axis of the machine.

As a result, three dimensionless (except time) first order state equations for the three energy storage variables i_a, Φ_e, ω, are obtained:

$$T_{a1} \frac{d}{dt} \left(\frac{i_a}{i_{a1}} \right) = \frac{1}{1 - \eta_1} \left(\frac{u_a}{u_{a1}} - \eta_1 \frac{\Phi_e}{\Phi_{e1}} \frac{\omega}{\omega_1} \right) - \frac{R_a}{R_{a1} + R_e} \frac{i_a}{i_{a1}}$$

$$- \frac{R_p}{R_{a1} + R_e} \frac{i_a - i_e}{i_{a1}} , \qquad T_{a1} = \frac{L_a}{R_{a1} + R_e} , \qquad (6.7)$$

$$T_{e1} \frac{d}{dt} \left(\frac{\Phi_e}{\Phi_{e1}} \right) = - \frac{R_e}{R_{a1} + R_e} \frac{i_e}{i_{a1}} + \frac{R_p}{R_{a1} + R_e} \frac{i_a - i_e}{i_{a1}} ,$$

$$T_{e1} = \frac{N_e \Phi_{e1}}{(R_{a1} + R_e) i_{a1}} , \qquad (6.8)$$

$$T_{m1} \frac{d}{dt} \left(\frac{\omega}{\omega_1} \right) = \frac{\Phi_e}{\Phi_{e1}} \frac{i_a}{i_{a1}} - \frac{m_L}{m_1} , \qquad T_{m1} = \frac{J \omega_1}{m_1} . \qquad (6.9)$$

A graphical representation of these normalised equations is seen in Fig. 6.2; the continuous state variables appear again at the outputs of integrators. The effect of the parallel resistor R_p in coupling the two magnetic circuits is clearly recognisable.

Without field weakening, i.e. $R_p \to \infty$, $i_e = i_a$ follows; armature and field winding are then forming part of a common electrical circuit, hence

$$\frac{d}{dt} (N_e \Phi_e + L_a i_a) = u_a - e - (R_a + R_e) i_a . \qquad (6.10)$$

Normalisation, as before, yields

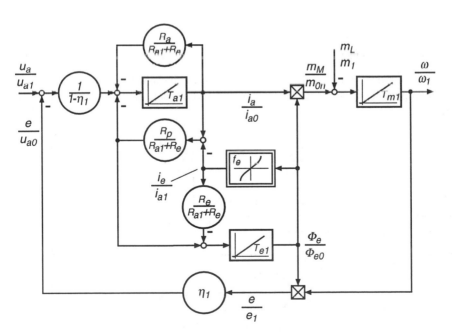

Fig. 6.2. Block diagram of a series-wound DC motor

$$T_{e1} \frac{d}{dt}\left(\frac{\Phi_e}{\Phi_{e1}} + \frac{L_a\, i_{a1}}{N_e\, \Phi_{e1}}\frac{i_a}{i_{a1}} \right) = \frac{1}{1-\eta_1}\left(\frac{u_a}{u_{a1}} - \eta_1 \frac{\Phi_e}{\Phi_{e1}}\frac{\omega}{\omega_1} \right)$$
$$- \frac{R_a + R_e}{R_{a1} + R_e}\frac{i_a}{i_{a1}}.$$

Because of the larger airgap in the quadrature axis,

$$q = \frac{L_a\, i_{a1}}{N_e\, \Phi_{e1}} = \frac{\Psi_{a1}}{\Psi_{e1}} \ll 1.$$

With Eq. (6.6), noting $i_e = i_a$, Eq. (6.10) is transformed into

$$T_{e1} \frac{d}{dt}\left[\frac{\Phi_e}{\Phi_{e1}} + q\, f_e\left(\frac{\Phi_e}{\Phi_{e1}} \right) \right] = \frac{1}{1-\eta_1}\left(\frac{u_a}{u_{a1}} - \eta_1 \frac{\Phi_e}{\Phi_{e1}}\frac{\omega}{\omega_1} \right)$$
$$- \frac{R_a + R_e}{R_{a1} + R_e}\frac{i_a}{i_{a1}}. \tag{6.11}$$

The associated block diagram is drawn in Fig. 6.3 for $q \ll 1$. A characteristic feature of the series-wound motor is the dependence of the flux on the armature current, which causes it to vary with the load. Since most of the armature voltage is balanced by the induced voltage, $u_a \approx e = c\,\Phi_e\,\omega$, high speed is to be expected at light load; this is reflected by the steady state curves discussed next.

To gain a better insight, saturation of the iron may be ignored which in Figs. 6.2 and 6.3 is achieved by omitting the nonlinear functional block, Eq. (6.6),

$$\frac{i_a}{i_{a1}} = f\left(\frac{\Phi_e}{\Phi_{e1}} \right) = \frac{\Phi_e}{\Phi_{e1}}. \tag{6.12}$$

At high current, when saturation cannot be neglected, the motor characteristics approach those of a separately excited motor, having constant main flux.

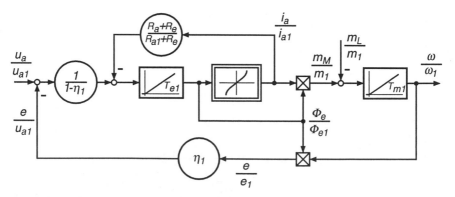

Fig. 6.3. Block diagram of series-wound DC motor without field weakening

6.2 Steady State Characteristics

The steady state behaviour of the motor is again found by putting the right hand sides of Eqs. (6.8 - 6.9) equal zero; with Eq. (6.12), a linear magnetic circuit is assumed in the longitudinal direction of the machine. Elimination of i_e/i_{a1}, results in

$$\frac{u_a}{u_{a1}} = \left[\eta_1 \, b \, \frac{\omega}{\omega_1} + (1 - \eta_1) \, r \right] \frac{i_a}{i_{a1}} \,, \tag{6.13}$$

where

$$b = \frac{i_e}{i_a} = \frac{R_p}{R_e + R_p} \tag{6.14}$$

is a factor characterising field weakening and

$$r = \frac{R_a + b \, R_e}{R_{a1} + R_e} = \frac{R_a + \frac{R_e \, R_p}{R_e + R_p}}{R_{a1} + R_e} \tag{6.15}$$

is the ratio of the effective armature circuit resistance to the nominal resistance without parallel and series resistors. Eq. (6.13) indicates that the armature circuit of an unsaturated series-wound motor appears like a resistive circuit, whose effective resistance depends on speed; this is of course due to the induced armature voltage e which is proportional to armature current and speed. Likewise we find from Eqs. (6.5, 6.9) with $m_M = m_L$

$$\frac{i_a}{i_{a1}} = \sqrt{\frac{1}{b} \frac{m_L}{m_1}} \,. \tag{6.16}$$

Eliminating i_a/i_{a1} from Eqs. (6.13), (6.16) finally yields the torque/speed curves of the unsaturated series-wound motor

$$\frac{\omega}{\omega_1} = \frac{\frac{u_a}{u_{a1}}}{\eta_1 \sqrt{b \frac{m_L}{m_1}}} - \frac{1 - \eta_1}{\eta_1} \frac{r}{b} \,, \tag{6.17}$$

where the normalised armature voltage u_a/u_{a1} may be continuously variable whereas field weakening (b) and armature resistance (r) are subject to discontinuous control actions; the load torque m_L/m_1 represents again a disturbance.

A set of these curves is shown in Fig. 6.4, calculated for $\eta_1 = 0.9$. As expected, the torque-speed curves are very steep at light load. This is desirable for vehicle drives, because the motor can be operated over a wide speed range and sharing the load on multi-motor drives is no problem. The similarity of the curves to the dashed hyperbola pertaining to constant power – which is the ideal characteristic of a vehicle drive without gear change – is striking.

The speed rise at no-load could of course be a source of danger in other than traction applications.

The armature current/torque curves, also depicted in Fig. 6.4, show a similar increase at field weakening as was observed with the separately excited motor. In view of the higher copper losses in the armature and the more critical commutation at reduced field, the amount of field weakening must be restricted here too.

On street-cars, trolley buses and subways, where normally a constant direct voltage is used for supply, as well as with battery operated vehicles, switched armature circuit resistors may be employed for control. In recent years electronic choppers have been introduced which permit continuous variation of the mean armature voltage. At higher speed the field may be weakened by parallel resistors. In addition, if several drive motors are present, they can be connected in series or parallel, thus providing a simple way of changing the voltage in a few steps. With city trains or main line locomotives having an AC catenary the armature voltage can be varied in fine steps by transformer tap changer or continuously by power electronic devices; rectification may or may not be included.

To reverse the torque, which is proportional to $\Phi_e\, i_a$, a reversal of the field winding is required; the machine can so be operated in the motoring, i.e. the first and third quadrants of the torque/speed plane. When operating in the second and fourth, i.e. the generating quadrants, an unstable condition exists which is manifested by a horizontal asymptote of the torque/speed curve at a critical speed, Eq. (6.17),

$$\frac{\omega_c}{\omega_1} = -\frac{1-\eta_1}{\eta_1}\frac{r}{b} < 0 . \tag{6.18}$$

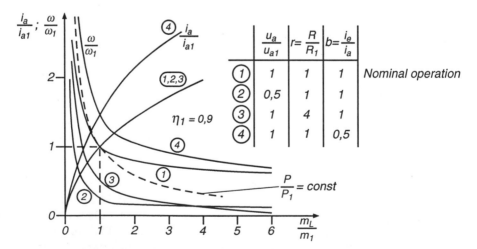

Fig. 6.4. Steady state characteristics of a series-wound DC motor

This constitutes a state of self-excitation, where the effective armature resistance, contained in the brackets in Eq. (6.13) becomes zero which means that the armature, being driven in the second or fourth quadrant by an active load torque, is supplying the losses of the armature circuit. The result is a steep increase of the braking torque produced by the machine. The critical speed ω_c is a function of the total armature circuit resistance and field weakening but is independent of the applied armature voltage.

The effect of self-excitation is seen in Fig. 6.5 showing a calculated transient with an unsaturated machine. At first, the motor is accelerated in the first quadrant at no load with a small value of the armature voltage u_a until time t_1, when a constant load torque m_L is applied which decelerates and eventually reverses the motor into the fourth quadrant. When the critical speed $\omega_c < 0$ is reached, self-excitation in the form of a nonlinear transient sets in with speed oscillations and high peak torques which in practice could endanger the motor shaft. Since this effect does not depend on the supply voltage u_a, it can be utilised as a supply-independent electric brake in vehicles; this is described in Fig. 6.6, [B2].

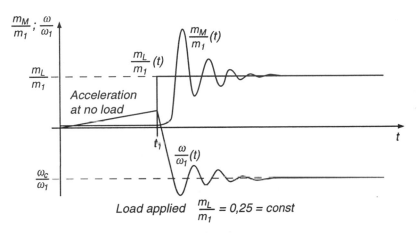

Fig. 6.5. Transient of self–excitation of a series-wound motor

It is assumed that the field winding has been reversed and the circuit is closed through a braking resistor. The motor is running with $\omega > 0$, so that braking calls for operation in the second quadrant of the torque/speed plane; it is controlled by the effective armature circuit resistance, Eq. 6.15. A detailed analysis of the braking transient requires the inclusion of iron saturation, but the principle may be deduced qualitatively from the curves in Fig. 6.6 b, where $e(i_a)$ is the induced armature voltage at a given overcritical speed and $R\,i_a$ is the voltage across the total braking resistor including the resistance R_{a1} of the armature itself. In the initial phase, the current is driven by the remanence voltage e_R, without external voltage being applied. The

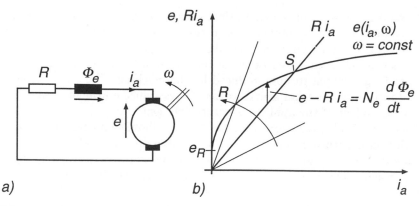

Fig. 6.6. Short circuit electric brake with series-wound DC motor.
(a) Circuit; (b) Characteristics

rate of rise and the eventual stable operating point S are determined by the differential equation

$$N_e \frac{d\Phi_e}{dt} = e\,(i_a, \omega) - R\,i_a \,,$$

again neglecting the self-inductance of the armature compared with that of the field winding. In the steady-state condition the machine then operates as a generator, converting mechanical power to heat in the predominantly external braking resistor R.

On main line locomotives, the drive motors are often reconnected as separately excited generators during braking operation, in order to have more direct control. The auxiliary field power may be taken from a battery, if supply-independent braking is specified.

An example of controlling a DC series motor is discussed in Sect. 9.2.

7. Control of a Separately Excited DC Machine

7.1 Introduction

In Chap. 5 the steady state and dynamic behaviour of a separately excited DC machine with adjustable armature and field voltage has been explained; this discussion is now extended by considering the machine as part of a feedback control system. The reason for this is that in practice the choice of a DC drive is normally motivated by the possibility of operating over a wide range of the torque/speed plane with low losses and matching the behaviour of the motor to the needs of the mechanical load. To achieve the desired operating characteristics in the presence of supply- and load-disturbances, feedback control is usually necessary. Another reason why DC drives are normally contained in feedback loops is that the armature of a larger motor represents a very small impedance which – when supplied with nominal voltage – would result in an excessive current of up to 10 times nominal value. Under normal operating conditions this is prevented by the induced armature voltage e, which cancels most of the applied voltage u_a so that only the difference is driving the armature current i_a; these two quantities are performing the actual electromechanical energy conversion. In transient operation, for example while accelerating or braking, there is always the danger of excessive current due to the rapidly changing armature voltage or speed; the same is true with a steady state overload on the motor. It is therefore important to provide a fast current or torque limit to protect the motor, the power supply and the load; this is best realised by feedback control establishing an effective safeguard against electrical or mechanical stresses. At the same time an unequivocal criterion is gained for distinguishing tolerable temporary overcurrents from currents caused by a functional fault which call for immediate clearance by a breaker or by fuses.

In most cases the user of a controlled drive wants to be able to select a reference speed ω_{Ref} which the motor should maintain as long as it is not overloaded. When overload occurs, the motor should produce maximum torque in the prescribed direction; twice nominal torque is often specified as a short term limit. This results in steady state characteristics of the type shown in Fig. 7.1.

For many applications, such as elevators, these specifications are insufficient because there are operating conditions when the acceleration or the

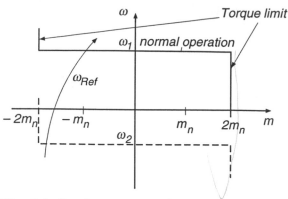

Fig. 7.1. Steady-state torque/speed curves of a controlled drive with torque limit

angular position must be controlled according to prescribed references. Also on machine tools or robots, the feed drives in the different axes must accurately follow prescribed angular commands as functions of time, in order to move the tool or workpiece along a desired spatial trajectory.

All these requirements can be fulfilled with the general scheme shown in Fig. 7.2 where the armature and field of the motor are fed from separate controllable power supplies. Rotating generators in the form of the well known Ward Leonard scheme have been the common choice in the past but, after brief interludes by magnetic amplifiers and mercury-arc converters, power electronic converters employing solid state semiconductor switching elements (various types of thyristors or power transistors) are now the standard solution.

With a "four quadrant" armature supply allowing both polarities of voltage and current, the power flow to the motor can be reversed, with the machine operating in all four quadrants of the torque/speed plane. This is the case if a rotating DC generator driven by a line-fed AC motor serves as a power supply; the same can be achieved with switched converters. With magnetic amplifiers, containing only diodes instead of switches, regeneration was not possible. Of course, there is always the option of electrical braking by placing resistors in the armature circuit.

The steady state power required for the field winding is only a few percent of that for the armature; a "single quadrant" rectifier, controlled or uncontrolled is adequate for feeding the field winding.

Fig. 7.2 contains a block "control equipment" which is controlled by reference signals ε_{Ref}, ω_{Ref}, α_{Ref}, $|i_{a\ max}|$ as well as feedback signals ε, ω, α, i_a, as needed for specific control tasks. The speed signal is normally obtained from a small permanent magnet tachometer generator, delivering a speed-proportional voltage from which a signal for acceleration can also be derived. When higher accuracy is specified, a digital solution is preferable where the speed signal is taken from a magnetic or optical sensor generating pulses for

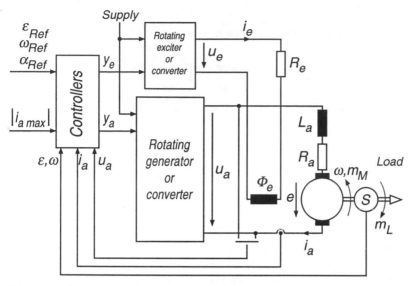

Fig. 7.2. General schematic of DC drive control

each angular increment. The frequency of the pulse train is then proportional to speed and counting the individual pulses provides a measure for the angular position change; the sign of the rotation is determined from the sequence of pulses generated by a dual track sensor with the two signals being 90° out of phase, Chap. 15.2.

The controllers transform these input signals into the actuating variables y_a, y_e which serve as control inputs for the two adjustable power supplies.

In the following paragraphs some proven control procedures for DC drives are discussed, making no specific reference to the type of power supply used, since this topic is dealt with in later sections.

7.2 Cascade Control of DC Motor in the Armature Control Region

Let us first look at the case when the motor operates below base speed, $-\omega_0 \le \omega \le \omega_0$. As explained in Chap. 5, the main flux is kept at the nominal value Φ_{e0} to have the lowest possible armature current for a given torque and to guard against unexpected load surges.

In Fig. 7.3 the schematic diagram of the armature circuit is drawn; e_a is the internal voltage source of the power supply. This synthetic voltage which is controlled by y_a differs from the armature voltage u_a due to the internal impedance R_i, L_i of the supply circuit; this must be taken into account when normalising the motor parameters,

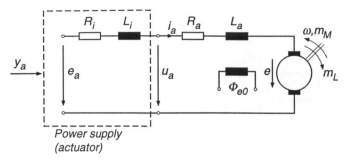

Fig. 7.3. DC motor with armature power supply

$$i_{a0} = \frac{e_{a0}}{R_a + R_i}, \qquad T_a = \frac{L_a + L_i}{R_a + R_i}.$$

With the results of Sect. 5.3 and $b = 1$, the block diagram of the motor assumes the form shown in Fig. 7.4 a, where the controllable voltage source e_a is represented by a first order lag element having the voltage gain G_{as} and the time constant T_{as}, which strongly depends on the type of power supply. When a rotating generator is used, T_{as} corresponds to its field time constant and may be in the range of 0.1 to 1.0 s, whereas the corresponding value would be 1 to 5 ms, if the residual delay of a line-commutated converter is to be modelled. Naturally this difference greatly affects the dynamics of the control system; it must be taken into account, when designing the controllers. Fig. 7.4 b is an equivalent structure, redrawn for noninteracting access to the armature current.

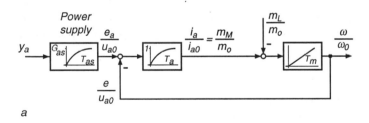

a

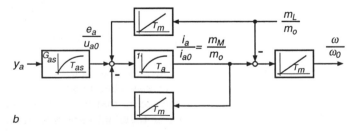

b

Fig. 7.4. Block diagrams of DC motor with armature power supply, $\Phi_e = \Phi_{e0}$

The armature time constant T_a usually has values between 10 ms and 100 ms; it is determined by the impedance of the complete armature circuit including possibly a smoothing choke that may be needed for reducing the current ripple caused by switching converters; with small servo motors having an iron-free rotor the armature time constant could be less than 1 ms. The mechanical time constant T_{mn} comprises the total inertia of the drive (assuming rigid coupling); therefore it covers a wide range from a few milliseconds (servo motors and reversing rolling mill drives) to several seconds (paper mill drives and mine hoists).

There is general agreement that the most effective control scheme for drives is a cascaded or nested structure with a fast inner control loop, limiting torque or, on a DC motor, armature current, to which an outer speed control loop is superimposed [K24, L79, S85]. This multi-loop system proves very flexible when the control plant possesses a chain like structure; for example, in order to control position, it can be extended by a position loop superimposed on the existing speed control loop. Also, if acceleration is of importance, a corresponding inner control loop may be added.

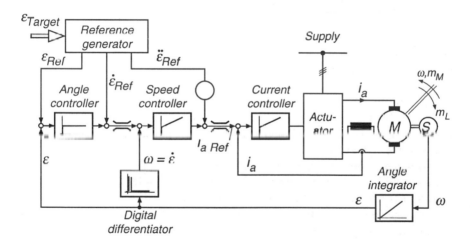

Fig. 7.5. Cascade control structure of drive including position-, speed-, and current control with feed-forward from a reference model

The idea of cascade control is exemplified by Fig. 7.5, where a lumped inertia drive is equipped with most of the control loops mentioned; it is seen that the sequence: torque-acceleration-speed-position is a natural one as it conforms to the structure of the plant. The torque (or armature current) loop, forming the innermost control function may be regarded as approximately creating an impressed current source for the armature. The current controller has to deal primarily with the dynamics of the power supply and the armature, cancelling the effect of the induced armature voltage; thus it provides the fastest available control action. By limiting the reference $m_{M\,Ref}$ (or $i_{a\,Ref}$), the inner control loop also assumes a protective function. An acceleration control loop which is not shown in Fig. 7.5 could be desirable with high dynamic performance drives, such as servo or elevator drives; its main purpose would be to counteract the effects of load torque m_L as well as changes of inertia, by generating – within the admissible range – a suitable current reference $i_{a\,Ref}$. The hierarchical structure is then continued with a speed controller which controls the closed inner loop in conjunction with the subsequent integrator; the same idea is once more extended when going to a position control loop with the position controller generating the speed reference.

Of course, this multi-loop control structure can only function under the assumption that the bandwidth of the control increases towards the inner loops with the current loop being the fastest and the position loop the slowest; clearly, the position controller can only perform well if the speed loop is quickly executing the commands it is given, etc. Following this reasoning, the design of the controllers proceeds in the direction from the lower to the higher control levels, at every stage approximating by a simple model whatever has been achieved before. Obviously, the design of the control system is greatly simplified by such an iterative procedure, where only a part of the complete plant is dealt with at a time. The fact that each feedback variable can be limited by clamping the pertinent reference is a major advantage of cascade control; also, field work is considerably simplified, since the control loops can be put into operation and tested one after another.

It may be argued that a control system containing several inner loops is likely to respond slower to changes of the reference than an equivalent single loop control (provided that it can be stabilised). Therefore, when dynamic performance is important, e.g. with position controlled servo drives, the inner loops should be activated in parallel by feed-forward reference signals derived from an external reference generator as indicated in Fig. 7.5; as long as the reference generator is outside the feedback loops, the stability of the control system remains unaffected. At the same time all disadvantages with regard to dynamic response are avoided. For load disturbances entering the controlled plant, cascade control is superior to any single loop control because already the next controller will detect the disturbance and counteract it. Cascade control involving torque, speed, position etc. may be considered as a special form of state variable control, making use of all the available, physically important quantities.

It should be mentioned that cascade control represents a very powerful control technique that is not restricted to drives; in fact there is hardly any control system put into operation today that does not contain inner loops. Of course, this principle, known for some time, could only be widely applied after the cost of control amplifiers was dramatically reduced by transistor operational amplifiers and electronic integrated circuitry.

When comparing the structure of the plant in Fig.7.5 with the block diagram of a DC machine in Fig. 7.4 a it is seen that the machine-internal feedback of the induced voltage is not immediately compatible with cascade control because it represents a deviation from the nested loop structure of the plant. However, by redrawing Fig. 7.4 a in the form of Fig. 7.4 b, a non-interacting structure is obtained; the fact that the load torque is now entering the plant in two places is no problem since m_L is assumed to be an independent disturbance. The variables not specified in Fig. 7.4 b have no immediate physical significance.

The part of the plant that lies between e_a/u_{a0} and i_a/i_{a0} may be described by the transfer function

$$\frac{I_a}{E_a}(s) = \frac{T_m\, s}{T_m\, T_a\, s^2 + T_m\, s + 1} \tag{7.1}$$

corresponding to $F_3(s)$ in Eq. (5.18) for $b = 1$. The differentiating effect vanishes with speed dependent load, as was seen from Eq. (5.24).

The design of the current controller is explained with the help of Fig. 7.6 assuming no inherent damping by the load. The feedback signal derived from the current transducer is passed through a low-pass filter to reduce possible ripple; a filter time constant of 2 to 5 ms is usually sufficient.

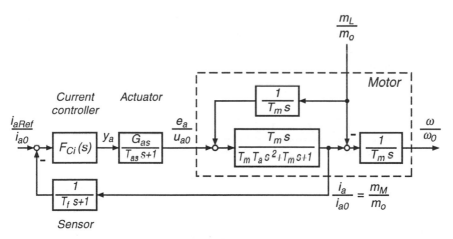

Fig. 7.6. Design of the current controller, $\Phi_e = \Phi_{e0}$

If a power electronic converter is employed as armature supply, a proportional-integral controller (PI) is adequate as will be discussed later. For a rotating generator having a much larger lag, the current controller may be extended by a lead-lag term (PDT) to form a PID-controller; when realising

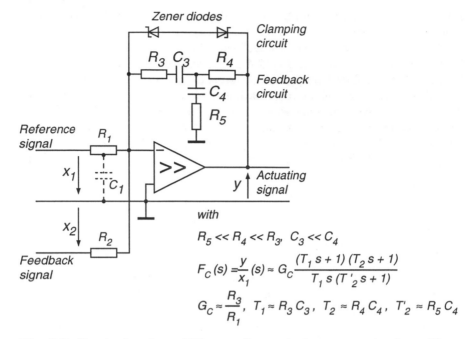

Fig. 7.7. Circuit of analogue PID-controller employing an operational amplifier

the controller function by an analog operational amplifier, this is a minor addition to the RC-feedback network of the amplifier. A typical example of an analogue PID-controller is shown in Fig. 7.7.

With these assumptions the open loop transfer function of the current control becomes

$$F_{0L,i}(s) = G_{Ci} \underbrace{\frac{T_1 s + 1}{T_1 s}}_{\text{PI}} \underbrace{\frac{T_2 s + 1}{T_2' s + 1}}_{\text{PDT}} \underbrace{\frac{G_{as}}{T_{as} s + 1}}_{\text{Supply}}$$

$$\underbrace{\text{PID–controller, } T_2' \ll T_2}$$

$$\cdot \underbrace{\frac{T_m s}{T_m T_a s^2 + T_m s + 1}}_{\text{Armature}} \underbrace{\frac{1}{T_f s + 1}}_{\text{Sensor}} \cdot \quad (7.2)$$

Because of the factor s in the numerator of the armature transfer function, a proportional-acting closed loop results, $\lim_{s\to 0} F_{0L}(s) = F_{0L}(0)$ finite, even though the controller itself comprises an integral term; as a result the closed current control loop will exhibit a steady state error. The reason is of course the e.m.f.-feedback which remains effective also in the transformed block diagram of Fig. 7.4 b. Clearly, when supplying constant current to the armature of a DC motor at no load, the motor speed, the induced voltage and the applied voltage rise linearly with time; with other words, generation of a ramp function with a controller containing a single integration calls for a constant control error.

For the practical application it is normally of no consequence whether the current controller admits a steady state control error during acceleration, because current control represents an inner (auxiliary) function, being part of the superimposed speed loop. The main purpose of the current controller is to linearise the plant and prevent overloading the motor; at constant speed, this is accomplished by a controller with a single integration. It would in principle be possible to use a current controller performing double integration, e.g. a dual PI-controller having the transfer function

$$F'_{C,i}(s) = G_{C,i} \frac{T_1 s + 1}{T_1 s} \frac{T_2 s + 1}{T_2 s} , \tag{7.3}$$

for avoiding a steady state control error during acceleration. However this is not normally done in view of the reduced margin of stability. If maintaining a prescribed acceleration is a necessary condition, as may be the case with machine tool- or elevator-drives, it would be easier to include a separate acceleration control loop as will be shown in Fig. 15.9.

The choice of parameters for the current controller according to Eq. (7.2) follows the usual lines for the design of cascade control systems [K25,38]. In the case of a Ward Leonard scheme containing a rotating generator and with aperiodic damping of the drive ($T_m > 4\,T_a$) the lead time constants T_1, T_2 of the PID-controller could be tuned to T_a and one of the emerging lag time constants of the drive, while the rest can be dealt with by approximation on the basis of the principle of "equivalent lag", e.g. [38].

The closed current control loop is now inserted into the speed loop forming the next higher level of control, as is shown in Fig. 7.8. Provided the current loop is well damped, it is acceptable to replace its closed loop transfer function by a proportional lag element having the gain $G_{CL,i}$ and the time constant $T_{CL,i}$. Since this equivalent lag, being a first order, low frequency approximation of a much more complicated transfer function, cannot be simplified by lead time constants, a PI-controller for speed is the appropriate choice. Hence the transfer function of the open speed loop is approximated by

$$F_{0L,\omega}(s) \approx G_{0\omega} \underbrace{\frac{T_{c,\omega} s + 1}{T_{c,\omega} s}}_{\text{PI- Controller}} \underbrace{\frac{G_{CL,i}}{T_{CL,i} s + 1}}_{\text{Current control}} \frac{1}{T_m s} . \tag{7.4}$$

The parameters $G_{c,\omega}, T_{c,\omega}$ may be chosen in accordance with the "symmetrical optimum" [K23,L33,38], which is a standard design procedure for transfer functions containing a double integration (including the controller).

The main idea is to choose the cross-over frequency at the geometric mean of the two corner frequencies, to obtain maximum phase margin ψ_d which in

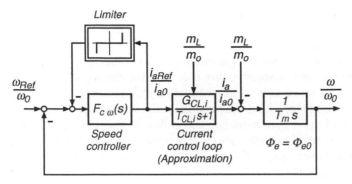

Fig. 7.8. Speed control with approximate representation of current control loop

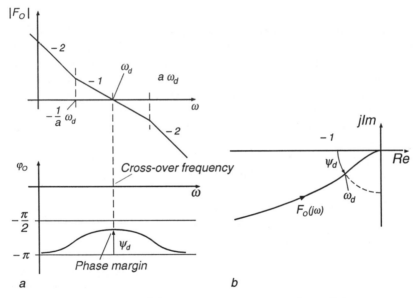

Fig. 7.9. Bode-diagram (**a**) and frequency response locus (**b**) according to "symmetrical optimum"

turn will result in optimum damping of the speed loop. The name stems from the Bode-diagram that shows symmetry with respect to the cross-over frequency, Fig. 7.9.

By normalising with

$$T_{c,\omega} = a^2 \, T_{CL,i}, \qquad a > 1, \tag{7.5}$$

the cross-over frequency is

$$\omega_d = \frac{1}{\sqrt{T_{c,\omega} \, T_{CL,i}}} = \frac{1}{a \, T_{CL,i}}. \tag{7.6}$$

From the cross-over condition

$$|F_{0L,\omega}(j\,\omega_d)| = 1 \tag{7.7}$$

the gain of the speed controller is obtained with Eqs.(7.5),(7.6)

$$G_{c,\omega} = \frac{1}{a\,G_{CL,i}} \frac{T_m}{T_{CL,i}} \,. \tag{7.8}$$

The poles of the resulting transfer function of the closed speed control loop, i.e. the eigenvalues of the speed control loop, can be calculated explicitely. They are, assuming a conjugate complex pair, positioned on a circle in the s-plane with the radius ω_d,

$$s_1 = -\omega_d, \quad s_{2,3} = \omega_d \left[-\frac{a-1}{2} \pm j\,\sqrt{1 - \left(\frac{a-1}{2}\right)^2} \right] \,. \tag{7.9}$$

Hence

$$D = \frac{a-1}{2} \tag{7.10}$$

is the damping ratio of the oscillatory part of the response. Assuming a slightly underdamped response, $D = 1/\sqrt{2}$, we have $a = 1 + \sqrt{2} \approx 2.41$. When aperiodic transients are specified, $D = 1$, leading to $a = 3$, a triple real pole at $s = -\omega_d$ results.

Due to the double integrating term in the open loop transfer function, the closed speed loop exhibits zero control area or, synonymously, zero velocity error [38,88]. This means that the step response is characterised by considerable overshoot eventhough the transients are well damped. To eliminate this effect which is caused by the lead term of the PI-controller, a corresponding lag term

$$\frac{1}{T_{c,\omega}\,s + 1}$$

may be inserted into the reference channel of the speed controller. In practice this can be done by connecting a suitable capacitor c_1 between a center tap of the input resistor and ground, as is indicated in Fig. 7.7.

The speed controller shown in Fig. 7.8 contains a nonlinear feedback block which limits the current reference to the range $-i_{a\,max} \leq i_{Ref} \leq i_{a\,max}$. This serves as an effective protection for the power supply and the drive, producing the desired steady state torque-speed characteristics shown in Fig. 7.10; of course, a short overshoot of the current, for example following a load disturbance, cannot be ruled out by clamping the current reference. If the electronic controller is realised with the help of an operational amplifier, the

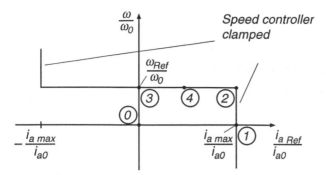

Fig. 7.10. Steady-state torque- speed characteristics of speed control loop with torque limit

clamping can be achieved by placing Zener-diodes in the feedback branch, Fig. 7.7. Limiting the current reference removes in principle all restrictions on the rate of change of the speed reference, which is important if it is produced manually. Whenever the speed controller, due to a rapidly changing command, reaches clamped condition, the speed loop is disconnected while the current control loop remains operative with clamped current reference. As soon as the speed has caught up with the command, the speed controller reverts to linear operation and resumes control.

However, with larger motors under manual control, a rate-of-change limiter is often inserted in the reference channel of the speed controller to ensure that the current limit, which may correspond to twice nominal current, is not reached unnecessarily. A block diagram of this device, together with typical

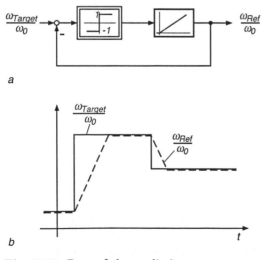

Fig. 7.11. Rate-of-change limiter

transients, is seen in Fig. 7.11; it restricts the occurrence of current limit to load surges, for instance when a reversing rolling mill is mechanically jammed because of a wrong screw- down setting calling for an excessive reduction of the billet thickness. On the other hand, current limit is not reached just because an impatient or careless operator rapidly changes the manual speed command.

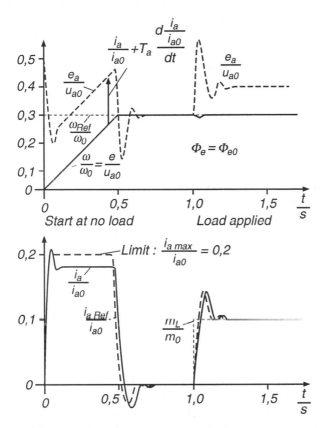

Fig. 7.12. Computed transients of DC motor with current/speed cascade control

Computed transients of a DC drive with current/speed cascade control are seen in Fig. 7.12. The parameters assumed correspond to those of a thyristor-fed drive with a six-pulse-converter, Chap. 8. The controllers have been designed as outlined in the preceding paragraph; if a rotating motor-generator were used instead of the power electronic converter, the response would naturally be much slower, particularly in the current loop.

At first, a starting transient is displayed, caused by a step of the speed reference from zero to $0.3\ \omega_0$. The effect of current limit during the acceleration phase as well as the dynamic control error are clearly seen. This transient corresponds roughly to the transition of the operating point in Fig. 7.10 from 0 to 3 via 1, 2. At time $t = 1$ s the drive, having attained the no-load point (3), is loaded with nominal torque. This causes a transient speed drop but the drive remains in the linear control regime, eventually reaching the operating point 4.

From Fig. 7.12 it is seen that the source voltage e_a rises with the induced voltage, which is proportional to speed; in addition there is the resistive and inductive voltage drop in the armature circuit.

The principle of cascade control has proven its usefulness over the years in innumerable applications; as mentioned before, the two levels of control may be extended in both directions, for example by adding acceleration or position control. Since cascade control is by no means restricted to DC drives, we will come across more examples where this exceedingly flexible method of control can be used to great advantage, see also Chap. 12.

7.3 Cascade Control of DC Motor in the Field-weakening Region

In Sect. 5.3 it was shown that by reducing the main field, a DC motor can be operated above base speed. For good utilisation it is important that the motor should be supplied either with maximum field- and reduced armature voltage or with maximum armature voltage and reduced field voltage; as mentioned before there can be some overlap for convenience of control.

Since the motor may alternately operate in either one of these regions it is appropriate to choose a control strategy which fulfills both requirements and permits a continuous and automatic transition from one operating regime to the other.

A control scheme that has proved useful in practice, is shown in Fig. 7.13 in simplified form [G20, V9]. It contains the current/speed cascade control acting through the armature voltage, as discussed before; in addition, there is an auxiliary control loop with the field voltage as actuating variable, which has the purpose of limiting the induced armature voltage as the speed rises beyond base speed. The adjustable voltage source supplying the field winding is usually realised by a thyristor converter, which is represented in Fig. 7.13 by a lag element. The nonlinear model of the magnetic circuit corresponds to Fig. 5.4.

A feedback signal for the induced voltage can be reconstructed from the measured armature voltage and current according to the voltage equation

$$e = u_a - R_a\, i_a - L_a \frac{di_a}{dt} \; ; \tag{7.11}$$

an analogue computing circuit, not shown in Fig. 7.13, may be used for this purpose. This signal – on reversing drives after rectification – is compared with a constant reference, e.g. $|e_{max}| \approx 0.9\, u_{a0}$; the e-controller then attempts to remove the difference by adjusting the field voltage. Since, however, the reference voltage is out of reach as long as the motor operates below base speed, the controller containing an integral term will be saturated and the field power supply produces maximum field voltage which is the aim in the armature control range. When the speed is rising, the induced voltage $|e|$ approaches the reference value $|e_{max}|$; after a balance has been reached, the e-controller leaves saturation and begins to lower the field voltage to limit the induced voltage, thus automatically weakening the field as desired. Of course, the initiative for a change in field voltage must always come from the armature because this is the direct channel of the speed controller, working through the armature current loop.

The arrangement drawn in Fig. 7.13 represents a complex and highly nonlinear control system, confounding all general design methods; however, considerable simplification results if one succeeds in speeding up the e-control loop, active in the field weakening range only, so that e may be approximated in that operating region by a constant impressed voltage. Such an approximation is not entirely unrealistic if the field power is supplied from a converter having a sufficiently large ceiling voltage (only one polarity of the field current is required). If the stator and the field poles are laminated, the "control plant" may be represented by a simple lag element, having a speed-dependent gain, and the fast dynamics of the actuator; no serious stability problems are

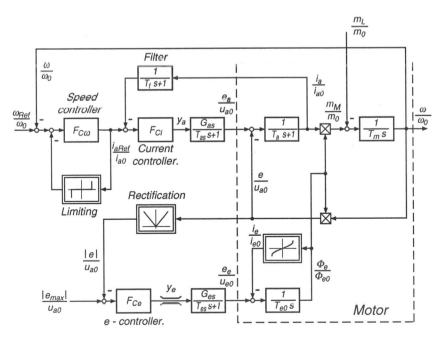

Fig. 7.13. Control of DC motor in the armature and field weakening regions

to be expected even at high controller gain. Neither is the nonlinearity of the magnetic circuit of great concern since the e-controller operates predominantly at reduced flux where saturation is not pronounced. For these reasons a PI-controller is quite adequate; its lead time may be tuned to the field time constant of the motor while the residual delays are taken care of by the remaining I-term of the controller.

Assuming fast e-control in the field weakening range, the simplified equivalent block diagram in Fig.7.14 results. There the effect of the e-control loop is approximated by a constant voltage $|e_{max}|$ acting like an external disturbance and the flux approximately becomes a nonlinear algebraic function of speed,

$$\frac{\Phi_e}{\Phi_{e0}} \approx \left|\frac{\omega}{\omega_0}\right|^{-1} \qquad \text{for} \qquad \left|\frac{\omega}{\omega_0}\right| > 1,$$

affecting the mechanical time constant. In principle, this undesirable change of parameter could be compensated by dividing the speed error by Φ_e/Φ_{e0} at the input of the speed controller, as is indicated in Fig. 7.14; however, in most cases this is omitted and practical results prove that the correction is usually not necessary. Some consideration must of course be given that the speed controller yields acceptable results in the armature as well as the field control regime. As an example, Fig. 7.15 shows a simulated no-load reversing transient between $\pm 1.5\ \omega_0$ that was computed on the basis of the system shown in Fig. 7.13. The effect of the e-control loop is clearly recognisable from the Φ_e- and the e-traces.

It has been mentioned that the use of a fast e-control loop with high forcing voltage across the field winding calls for a laminated motor frame. With large motors it may be necessary to employ a more accurate model of the field circuit for the design of the e-control loop, taking eddy currents in the iron into account [G20]. Also the high voltages induced by transformer action in the armature winding affecting the commutator must be considered. To

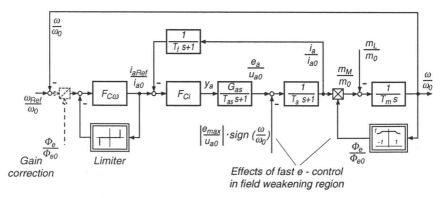

Fig. 7.14. Simplified representation of speed control loop in field weakening range

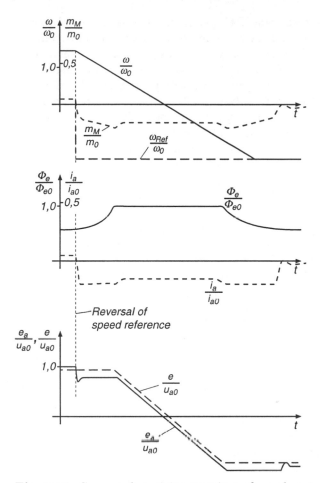

Fig. 7.15. Computed reversing transient of speed control system in Fig. 7.13

eliminate the effects of winding temperature and supply voltage fluctuations on the main flux in the base speed region, an inner field current control loop is often included which maintains a prescribed maximum field current in the armature control range independent of a changing resistance of the field winding and helps to speed up flux control.

7.4 Supplying a DC Motor from a Rotating Generator

In the preceeding sections was mentioned that any adjustable voltage source could serve as power supply for the armature of the DC motor. The classical scheme of DC motor control developed by Ward Leonard [L21, L22, Y4, Y5], employs for this purpose a rotating DC generator, the armature voltage of which is controlled through the field. Combined with up-to-date control, drives of this type are still in use today. The basic circuit is seen in Fig. 7.16; the separately excited DC motor (2) driving the load, for instance a rolling mill or an elevator, is fed from a separately excited DC generator, running at approximately constant speed. While a rolling mill or a gear-less elevator drive would normally operate at very low speed, for example between ± 100 1/min, resulting in high torque and a relatively large frame size, the generator being of similar power rating as the motor can run at any conveniently higher speed ω_1, with the aim of reducing its size and cost. Instead of a DC generator, an AC generator with rectifier could be used for supplying the DC-link; of course, this would rule out regeneration.

The prime mover of the rotating converter is most frequently an AC induction motor (1) with cage or wound rotor, Chap. 10, requiring no continuous control. The power flow of the drive is then reversible with the controlled motor being able to operate in all quadrants of the torque/speed plane. In case of a ship- or vehicle-propulsion scheme a turbine or a Diesel engine could be substituted as prime mover (1) (Turbo- or Diesel-electric drive), while the drive motor (2) would most likely be a series type motor. Resistive braking could be employed instead of regeneration which is of course not possible with a unidirectional prime mover.

With large rolling mill- or mine hoist-drives the total drive is often split up into several machines; for instance, on a large reversing mill, two twin motors may drive the upper and lower rolls without direct mechanical connection apart from the frictional forces through the material being rolled. All machines constituting the rotating converter, such as AC motors, DC generators and, possibly, secondary machines are normally coupled to form a long drive shaft, sometimes 50 m long, which is located in a hall adjacent to the controlled drive and the load.

An advantage of the rotating motor-generator-set is the storage of kinetic energy in the masses rotating with ω_1, which may be accentuated by connecting flywheels to this shaft and increasing the slip of the wound rotor driving motor(1); a converter of this type is then called an Ilgner-set. For large drives (MW) special schemes have been developed to recuperate slip power in the rotor circuit, Sect. 13.3. The kinetic energy of the rotating masses provides a convenient means of dynamically decoupling the AC supply from the load; there may be different motives for this, depending on the application. With reversing rolling mills there has always been the desire to smooth the load fluctuations of 20 MW or more lasting a few seconds; this point of view has been of particular interest with weak supply systems but its importance receded as the supplies have become stronger and more closely interconnected. Conversely, a rotating converter can serve for protection of sensitive loads, such as paper mill drives, against short-time line disturbances. Similar arguments are valid when uninterruptible power for important loads is required, for example computers, elevators or operating rooms in hospitals; with an intermediate motor-generator-set supplying these loads, there is more time for switching to an alternative supply in case of a line disturbance.

The main drawback of the Ward-Leonard-drive is of course that, besides the adjustable speed drive, two additional machines of about the same power rating are needed. Even though they may be smaller in size because of higher speed, large machines will still require heavy foundations and continuous service of bearings and commutators. It is for these reasons that static power converters, particularly of the modern semiconductor type have supplanted rotating motor-generator-sets in new installations.

If the field of a rotating generator is fed from a rotating exciter, both responding relatively slowly, it is with some care quite possible to control the drive motor manually and without any automatic equipment. However, the machine cannot be fully utilised because of the danger of overcurrent; feedback control with its much faster response then offers considerable advantages making it possible to take various side conditions into account so that the

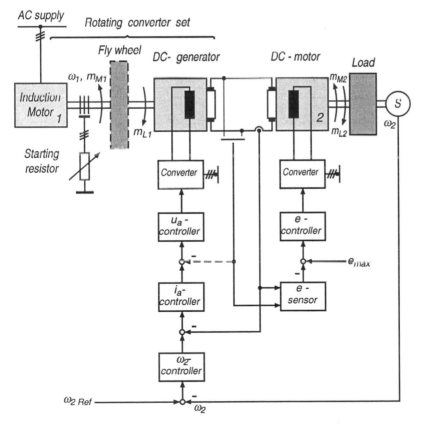

Fig. 7.16. Control scheme of Ward Leonard-drive

drive assumes well defined operating characteristics and may be inserted into automatic production processes. By exciting the generator and the drive motor through static converters with high ceiling voltage (2 to 3 times steady state voltage) the speed of response of the drive may be sufficiently improved for meeting most dynamic specifications.

The control structure shown in Fig. 7.16 is essentially in line with that in Fig. 7.13. The only difference is the possible introduction of an inner control loop for the armature voltage u_a of the generator [S76]. Since the pertinent portion of the control plant comprises only the field time constant of the generator and the residual delay of the electronic field power supply, the response of this control loop can be made quite fast. Its advantages are the linearisation of the inner part of the plant, the reduction of the effective armature impedance of the generator and the possibility of precisely limiting the armature voltage. This is important with large machines where the permissible limits of commutation must be fully used without running the risk of excessively sparking brushes. For the superimposed current controller the closed voltage control loop looks like a linear actuator with constant gain and considerably reduced lag.

8. Static Converter as a Power Actuator for DC Drives

8.1 Electronic Switching Devices

A characteristic feature of the rotating converter discussed in Sect. 7.4 is the consecutive power conversion from electrical (constant AC line voltage) to mechanical (speed of motor-generator) and back to electrical form (variable direct voltage) from which it is eventually transformed by the drive motor into controlled mechanical power. These conversions constitute the advantages of the control scheme (separate electrical circuits, decoupling by rotating masses) as well as its drawbacks (cost of machines, foundations, power losses, servicing, limited speed of response).

This naturally gave rise to early ideas for eliminating the intermediate mechanical stage by supplying the DC drive motor directly from the AC grid through static electronic devices. That such possibilities exist in the form of static switching converters has been known for a long time [A15, W29, 46]. In fact, switching devices are the only available means for high power converters because any other circuit components would result in prohibitive power losses.

The basic element of a power converter is an electronic switch having the properties of a controllable rectifier; the idealised steady state characteristics of different switches are seen in Fig. 8.1. An uncontrolled diode rectifier, Fig. 8.1 a, has two distinct states: With reverse voltage ($u < 0$), there is only negligible leakage current, whereas in forward direction ($i > 0$) a low forward resistance combined with a threshold voltage exists. Controlled electronic switches have at least three different states: A thyristor, Fig. 8.1 b, is blocking reverse current like a diode, but in forward direction ($u, i > 0$) two possible states exist: the device may either be "conducting" or "blocking". The first state is again associated with a low conducting resistance, typically $10^{-3}\,\Omega$, while in blocking condition the device exhibits a very high blocking resistance, for example $10^5\,\Omega$, similar to that in the reverse direction. Hence the thyristor valve may be approximately described by a mechanical switch in series with an uncontrolled rectifier (diode). The main difference between a thyristor and a diode is the existence of the forward blocking state.

The transition from "blocking" to "conducting" may be initiated by a small gate current pulse i_g, while the inverse transition from "conducing" to "blocking" calls for a zero intersection of the main current, "Turn-Off at current zero"; thyristors are widely used for controlling DC drives.

A more recent development are Gate-Turn-Off thyristors (GTO), where gate controlled electronic blocking is also possible, Fig.8.1 c; they are mainly used in force-commutated converter circuits with a DC link, to be discussed in Chaps. 9.2 and 11. Other types of electronic switches which are of interest for special applications are reverse conducting thyristors or voltage controlled transistors with internal diodes, where the two forward states are combined with reverse conduction; an example is the "Insulated gate bipolar transistor" (IGBT), seen in Fig. 8.1 d; there are also applications for symmetrical switches, Fig. 8.1 e, as will be shown in Chap. 11.4.

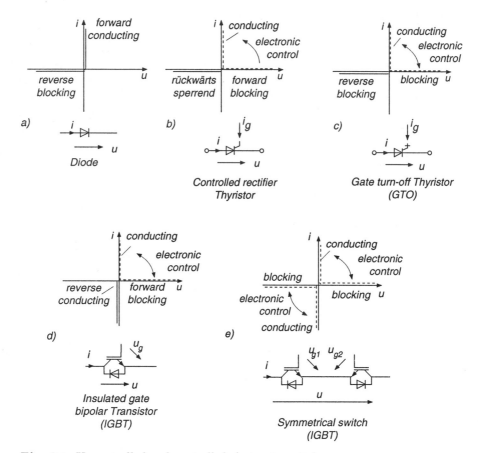

Fig. 8.1. Uncontrolled and controlled electronic switches

In the past the function of controlled electronic switches has been realised with various devices making use of different physical effects; all these early solutions are obsolete today. Since electron tubes with heated cathodes could not supply enough current to drive motors and were plagued with high voltage drop resulting in poor efficiency, they were superseded by gas-filled tubes with heated cathode, called thyratrons; however, power rating, life time and reliability were also rather limited so that they were only useful for small drives. High power in the MW-range, such as needed for rolling mills or mine hoists, could only be supplied by mercury-arc valves having a liquid mercury cathode. These electronic valves were developed to high perfection by the 1950's but the liquid Hg-cathode limited their use in vehicles and remained a disadvantage, as well as the large size and the relatively high voltage drop across the electric arc connecting cathode and anode (20–25 V). Outside a narrow operating temperature range, controlled mercury-arc valves showed a tendency to occasional spontaneous arc-backs, i.e. conduction in reverse direction resulting in short circuits and high over- currents. This required close temperature control by heating or cooling the steel vessels. Also, mercury-arc converters needed relatively long intervals for deionisation which made them unsuitable for operation at higher frequency. After a brief interlude with magnetic amplifiers where magnetic materials with an approximately "square" hysteresis loop were used as "switches", all these difficulties were completely overcome with semiconductor switching devices (Transistor 1948, Thyristor 1957) which have become available at steadily increasing power and with improved performance since about 1960. They have replaced mercury-arc valves even at the highest power ratings in high voltage DC transmissions (HVDC).

The physics of a thyristor are extensively dealt with in the literature, e.g. [17, 24], a simple description may suffice here. In Fig. 8.2 a a small slice of highly purified monocrystalline silicon having four electrons in the valence band is shown, which is doped with materials of valence three or five, for instance Boron or Phosphorus, so that three $p - n$ junctions are created, separating the layers of p - and n - conducting material. The inner layers are relatively weakly doped while the outer ones, forming the anode and cathode terminals, contain more doping material; the inner p_2 layer is also connected to a gating electrode supplied by a low impedance control circuit. The following steady state operating conditions exist:

- With negative voltage between anode and cathode, $u < 0$, the diode junction $p_1 - n_1$ is reversely biased; this corresponds to the reverse blocking state seen in Fig. 8.1 b.
- If positive voltage is applied ($u > 0$) but without gate current ($u_g < 0$), the diode $n_1 - p_2$ has reverse bias which constitutes the state of forward blocking. Again, only negligible current flows in the load circuit.
- With positive voltage, $u > 0$, and with a sufficiently large gate current, $i_g > 0$, the junction $n_1 - p_2$ becomes conducting, with the load current i saturating the region n_1 from both sides with charge carriers. Thus the thyristor assumes very low resistance in forward direction. When i has exceeded a certain latching threshold i_L, the conducting state persists even after the gate current is removed.

Only by reducing the load current i below a lower holding threshold i_H can the thyristor again revert to blocking state; hence it represents a switch that can be closed by a low power control signal, whereas opening it requires the load current to become very small or zero.

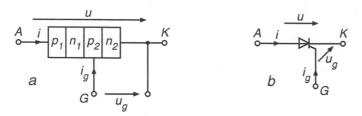

Fig. 8.2. Thyristor model and symbol

Simplified characteristics of a typical power thyristor are shown in Fig. 8.3; the scales of the curves are different to emphasise the details. While the leakage current in both directions could be below 100 mA, the load current in conducting state may be several thousand A. The voltage drop of the conducting thyristor is about 1.5 V and the maximum voltage in blocking states can be more than 8 kV.

When the maximum reverse voltage is exceeded, the leakage current rises rapidly, as with semiconductor diodes, leading to break-down and thermal destruction of the thyristor. In the forward direction ($u > 0$) the effects are more complex. When the voltage reaches a breakover level, which depends on the gate current, the thyristor switches spontaneously to the conducting state. However, turn-on by increased forward blocking voltage is not a suitable mode of control for a power thyristor because the "firing instant" (a relict from plasma discharge valves) is not sufficiently well defined and there may be local overheating of the semiconductor element. Instead, the switching to the On - state should always be effected by a gate current pulse having a short rise time which has the additional advantage that the thyristor fires at a precisely defined instant, independent of the individual thyristor parameters.

Due to the small dimensions of the active semiconductor wafer and its low thermal capacity, the safe use of thyristors calls for observing some restrictions; apart from limits on temperature, voltage and current there are maximum values for the rate of the current rise (di/dt) while switching On, as well as the voltage rate (du/dt) during and after turn-Off. This requires the inclusion of protective circuitry such as a series inductance and a parallel capacitor (snubber). Considering also the switching surges that are unavoidable in power supply systems, it is necessary to maintain a proper safety margin of the operating voltages against the specified peak values. For example, a 1.2 kV thyristor may be specified in a converter connected to the 400 V line where the maximum operating voltage is only 565 V. However, when all the

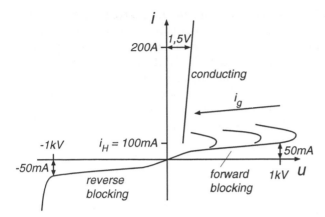

Fig. 8.3. Static and dynamic characteristics of thyristor

recommendations by the supplier are observed, the thyristor has proved to be a reliable and very robust switching element that can stand high short time over-load current, such as a 10 fold surge for one cycle. Persistent overload must be prevented by a fast acting closed loop control and in the limit by fuselinks.

An important design parameter of thyristors, particularly when applied in force commutated converters, is the recovery time t_r. It indicates the time that must elapse from the instant when the load current has reached zero, until the thyristor has attained full blocking capability in forward direction ($u > 0$). The purpose of this interval is to give the residual charge carriers accumulated in region n_1 during the conduction period time to decay through diffusion and recombination; depending on the type of thyristor, reverse voltage and junction temperature, the recovery time ranges from a few μs for a high frequency thyristor to several hundred μs for a high power thyristor designed for low voltage drop.

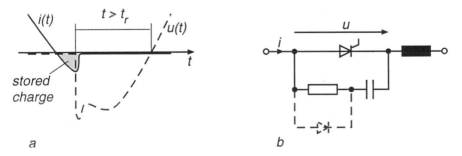

Fig. 8.4. Turn-off transient of thyristor
(a) Current and voltage **(b)** Protective circuit

A typical transient of the voltage $u(t)$ and the current $i(t)$ of a thyristor during turn-Off is depicted in Fig. 8.4 a. As the load current i reaches zero, the thyristor remains conducting for a short period with reverse current, thus clearing the charge-carriers from the junction. After this charge has disappeared, current flow decays rapidly causing high reverse blocking voltage due to the circuit inductance; this voltage peak must be limited by the protective parallel capacitor indicated in Fig. 8.4 b; in the subsequent interval the reverse voltage is determined by the external load circuit. When the recovery time t_r has elapsed, the voltage can again become positive without the danger of breakover and unwanted forward conduction. The recovery time is considerably reduced if reversed voltage $u < 0$ exists after current zero. The resistor in the snubbing circuit, Fig. 8.4 b, is required for limiting the discharge current of the capacitor at turn-On; frequently the resistor has a diode in parallel to make the capacitor more effective at limiting du/dt in forward direction.

8.2 Line-commutated Converter in Single-phase Bridge Connection

Thyristors, which are available in a variety of voltage and current ratings, are combined to converter circuits to serve as highly efficient electronic power supplies for a wide power range. Of particular importance for the supply of DC motors are the so called line-commutated converter circuits fed from a single- or three-phase AC supply of constant voltage and frequency. They are suitable for supplying adjustable direct voltage and current to the armature and field winding of the motor, either directly or through a transformer.

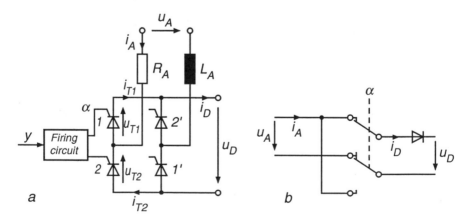

Fig. 8.5. Single phase line-commutated converter (a) circuit (b) mechanical equivalent

The single-phase circuits discussed first are of interest, apart from railway traction drives, for low power applications up to a few kW. The circuit shown in Fig. 8.5 a contains four thyristors, which are made conducting or "fired" in diagonal pairs during every alternate half cycle of the line voltage u_A. Provided there is a continuous direct current i_D, the function of the circuit may be described by a mechanical commutating switch, connecting the DC circuit with alternate polarity to the line voltage

$$u_A = \hat{u}_A \sin \tau, \qquad \omega t = \tau. \tag{8.1}$$

When assuming at first that the direct current is smooth, $i_D = I_D$, being maintained by a fictitious current source, and that the impedance of the AC circuit is negligible, with u_A coming from an ideal voltage source, the voltage $u_D(\tau)$ between the DC terminals consists of periodic sections of the AC voltage or its inverse; this is seen in Fig. 8.6 for different firing angles α, which determine the instant at which the other diagonal pair of thyristors is made conducting. The firing angle represents the switching delay with respect to a "natural" firing instant that would be valid for a circuit consisting of diodes; for the thyristors 1, 1' the "natural" firing instants are $\tau = 0, 2\pi, \ldots$ when the voltage $u_A(\tau)$ and — with thyristors 2, 2' conducting — the voltages $u_{T1}(\tau)$, $u_{T1'}(\tau)$ become positive. The maximum firing delay is $\alpha_{max} = \pi$; for $\alpha > \pi$, the thyristors 1, 1' cannot be made conducting since they are reversely biased. This is also seen from the curve of the voltage across the thyristor, $u_{T1}(\tau)$, drawn in Fig. 8.6 c for different firing angles α. Later the range of the control angle will have to be further narrowed.

The trace of the thyristor current $i_{T1}(\tau)$, also shown in Fig. 8.6 c, clearly indicates the switching action of the thyristors; at any time at least one of the quantities $u_{T1}(\tau)$, $i_{T1}(\tau)$ is zero. In steady state condition with $\alpha =$ const. the voltages and currents in thyristors 2, 2' pertaining to the opposite diagonal pair are identical, being shifted by a half period.

Finally, in Fig. 8.6 d the input current $i_A(\tau)$ for the three different values of the firing angle α is drawn. As a consequence of the periodic commutation between the two pairs of switches the impressed continuous current I_D appears at the line side of the converter as a square wave alternating current $i_A(\tau)$, whose phase lag φ with respect to the line voltage $u_A(\tau)$ is prescribed by the firing delay, $\varphi = \alpha$. The active power (mean of the instantaneous power) at the line side is — assuming sinusoidal line voltage — determined by the fundamental component $i_{A1}(\tau)$ of the line current

$$P_{AC} = U_A I_{A1} \cos \varphi = \frac{2\sqrt{2}}{\pi} U_A I_D \cos \alpha, \tag{8.2}$$

where U_A, I_{A1} are RMS-values of the line voltage and the fundamental current component. For $0 < \alpha < \frac{\pi}{2}$, the power flows from the AC - to the DC - side, while for $\frac{\pi}{2} < \alpha < \alpha_{max}$ the power flow is reversed.

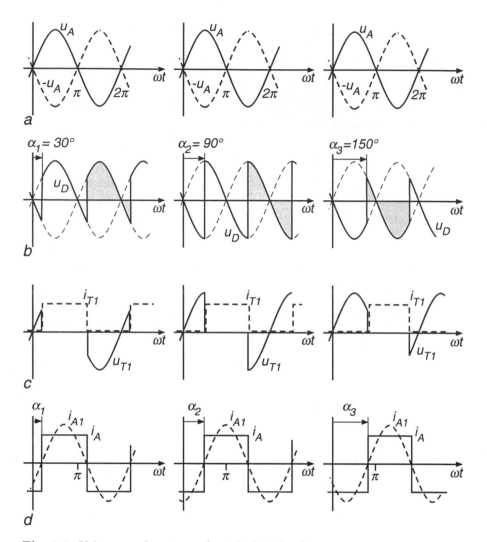

Fig. 8.6. Voltages and currents of an ideal single phase converter

The first mode corresponds to controlled rectifier-, the other to inverter operation. In Fig. 8.7 a phasor diagram is showing the phasors of the line voltage $\underline{U}_A$ and of the fundamental component of the line current, $\underline{I}_{A1}$. Clearly, a controlled converter appears as an inductive load at the line side, drawing lagging reactive current; this is due to the control principle allowing only delayed firing. Firing advance is not possible with line-commutated converters because the firing pulses would be applied to thyristors having reverse bias voltage which prevents them from conducting; advance firing is only possible with force commutated converters, where an interval with forward

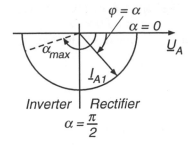

Fig. 8.7. Phasor diagram of a single phase converter

blocking voltage $u_T > 0$ is created for $\alpha < 0$, during which the thyristor can be fired. Forced commutation calls for additional components and thyristors with short recovery time which involves increased cost and complexity; this will be discussed later.

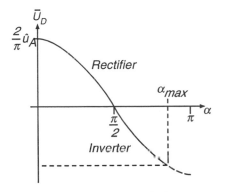

Fig. 8.8. Control curve of phase controlled converter

The possibility of reversing the power flow by electronic control is of importance if the converter is to supply a DC motor since it permits regeneration. It is pointed out, however, that firing control affects not only the active but also the reactive power flow at the line side. At $\alpha = \frac{\pi}{2}$ the reactive power assumes a maximum value (for $I_D = $ const.) while the active power is zero; this is the case when the motor produces torque at standstill.

Identical results are obtained when examining the DC side of the converter, where the mean voltage is

$$\overline{u_D} = \frac{1}{\pi} \int_{\alpha}^{\alpha+\pi} \hat{u}_A \sin\tau \; d\tau = \frac{2}{\pi} \hat{u}_A \cos\alpha = \frac{2\sqrt{2}}{\pi} U_A \cos\alpha = U_{D0} \cos\alpha$$

$$(8.3)$$

This is described by the control characteristic $\overline{u_D}(\alpha)$ seen in Fig. 8.8; for reasons discussed later the curve is only valid in the interval $0 < \alpha < \alpha_{max}$, with $\alpha_{max} \approx 150°$. Because of $i_D = I_D = \text{const.}$, the mean of the output power is

$$P_{DC} = \overline{u_D}\,I_D = \frac{2\sqrt{2}}{\pi}\,U_A\,I_D\,\cos\alpha\,, \tag{8.4}$$

which is in agreement with Eq. (8.2). This is so because the idealised switching converter consumes no power; it functions by shifting the 180°-blocks of the direct current with respect to the line voltage. The transferred mean power changes sign in accordance with the direct voltage as the firing angle α exceeds $\frac{\pi}{2}$. Since the current cannot change sign because of the unidirectional thyristors, the line-commutated converter circuit in Fig. 8.5 represents a controlled power supply operable in the two quadrants of the I_D, $\overline{u_D}$-plane, as seen in Fig. 8.9. The control is performed by electronically delaying the firing pulses to the thyristors by the firing angle α.

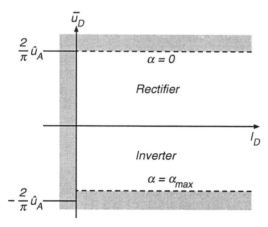

Fig. 8.9. Two quadrant operation of line-commutated converter

Clearly, a switched converter causes interactions with the line, involving lagging reactive power as well as current harmonics. Because of the supply- and the transformer leakage impedances this causes voltage fluctuations and harmonic distortion which must be taken into consideration as the rating of the converter increases. While the harmonics problem may be relieved by going to multi-phase circuits, the reactive power is a characteristic feature of all line-commutated converters with delayed firing control; it is independent of the number of phases and can only be counteracted by reactive compensation. At high power both effects may require correction by harmonic filters having a leading input power factor at line frequency.

Controlling a converter by electronic phase shift of firing pulses requires little control power and involves only very short delay. This is seen in Fig. 8.10 showing the output voltage $u_D(\tau)$ with the firing angle being switched between $\alpha = 30°$ and $\alpha = 150°$. Since only one diagonal pair of thyristors can be fired in each half period, there is a maximum delay of 10 ms with a 50 Hz supply until the converter output responds to a command calling for a change in firing angle. Hence a static converter represents an actuator of high dynamic performance combining
– high power gain
– large output power and low losses
– very fast response.
These features may yet be enhanced by employing multi-phase circuits, as will be shown in the next section.

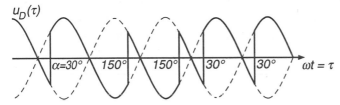

Fig. 8.10. Output voltage of single phase converter with changing firing angle

One drawback, though of little practical importance, is the fact that the steady-state and the dynamic behaviour of the converter is highly nonlinear, corresponding to a nonlinear sampled data system with one firing transient in each half period. From Fig. 8.10 it is apparent that, following a changing firing command, the result may not become effective immediately because there are variable waiting intervals. This is particularly pronounced when the firing angle is increased for inverter operation because it is necessary to allow for the relatively "slow" reversal of the line voltage. When going to rectifier operation, i.e. reducing the firing angle, the converter may respond instantaneously. An accurate analysis of dynamic converter operation is quite complex; for small perturbations linearised difference equations can be derived but large signal dynamics are governed by non-linear difference equations for which there is no analytical solution [F12,L34,P46]. Fortunately the converter is usually followed by an inductive load having low-pass transfer characteristics, such as the armature circuit of a DC machine which smoothes the current; therefore the converter dynamics may be approximated by very simple models when designing the inner current control loop.

Another point where high power converters may occasionally cause problems is a direct consequence of its advantages mentioned before; the power supplied to the load must instantaneously come from the AC line. There is no decoupling effect by internally stored energy because an ideal converter contains only loss-free switches. As was mentioned before, the importance of these effects has receded as the capacity of the interconnected supply systems has increased and is more closely meshed.

The control of the converter is normally performed by the output signal of an electronic controller influencing the phase of the firing pulses through a firing control device. Its basic function is explained with the help of Fig. 8.11. The output signal $y_1(t)$ of the innermost controller, usually the current controller, is compared with a saw-tooth-shaped periodic ramp function y_T which is synchronised by the zero intersections of the line voltage u_A. Whenever the difference becomes positive for the first time in each half period, the diagonal pair of thyristors, Fig. 8.5, having forward bias voltage is fired. The selection is done by alternately activating one of two monostable trigger circuits MT_1, MT_2. One of the trigger circuits produces a short pulse which, after amplification, is passed through a small isolating transformer to the gates of the thyristors to be fired (only one of the output windings is drawn). In high voltage applications, such as high voltage DC-transmission or static reactive power compensators, optical fibres may be used for transmitting the firing pulses, thus preventing electromagnetic interference. It is seen that the comparison of the control signal y_1 with the saw-tooth function represents a sampling method, at the same time providing a voltage-to-phase conversion.

The basic principle of control described can be modified in various ways. If, instead of a saw-tooth function, sections of cos-functions are employed, as is indicated by dashed lines in Fig. 8.11 b, the control curve $\overline{u_D}(y_1)$ seen in Fig. 8.8 becomes a straight line, i.e. the converter assumes constant voltage gain $d\overline{u_D}/dy_1$. The same effect is achieved, at least in steady state, by inserting an arcsin-function generator $y_1 = \arcsin y$ into the input channel of the firing circuit; this is indicated in Fig. 8.11 a. Control devices for converters are today available in a variety of integrated circuits.

In order to gain a better insight into the principle of converter operation, some simplifications that are not fully realised in practice have been introduced; one is the assumption of a smooth current I_D, the other of an impressed voltage u_A, neglecting the line impedance. These points will now be discussed in a little more detail. When a resistor R_D is connected to the output of the converter in Fig. 8.5, $u_D = R_D i_D$ holds in the DC branch, with the current having the same waveform as the voltage. Hence the conducting thyristors will be blocked when the driving voltage u_A becomes negative; this is so because i_D cannot change sign and the passive DC circuit cannot – even instantaneously – feed power back to the line. Thus the current will exhibit the pulsating shape seen in Fig. 8.12; during the zero current intervals the voltage across the series connected non-conducting valves is divided according to their blocking resistances.

The discontinuity of the load current is accentuated if a back voltage $e(\tau) > 0$ is connected in series with the resistor R_D. Now a thyristor pair can

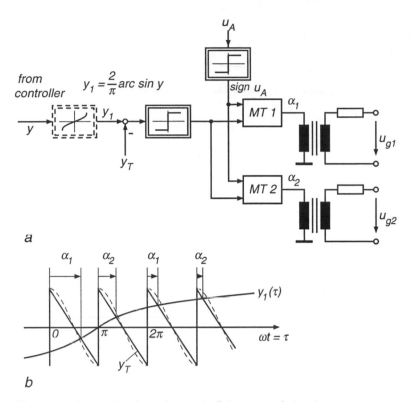

Fig. 8.11. Principle of an electronic firing control circuit.
(a) Block diagram; **(b)** Mode of operation

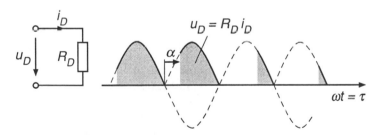

Fig. 8.12. Single phase converter with resistive load

only begin to conduct if the line voltage exceeds the back voltage, $|u_A| > e$, because otherwise the thyristors would be reversely biased; the current ceases to flow as soon as this condition is violated; Fig. 8.13 shows the effect.

A further complication arises, if the DC circuit contains an inductance L_D in addition to the resistor R_D and the back voltage e, which is the case of a converter feeding the armature of a DC motor. Now the current i_D may become continuous even for $\alpha > 0$ because the armature inductance with its

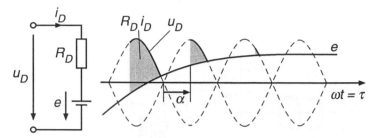

Fig. 8.13. Single phase converter with back voltage and resistive load

stored energy can maintain the current i_D during part of the interval where $|u_A| < e$. This is drawn in Fig. 8.14 for different values of the firing angle α and assuming $e = E = $ const.. The internal impedance of the line voltage is still neglected.

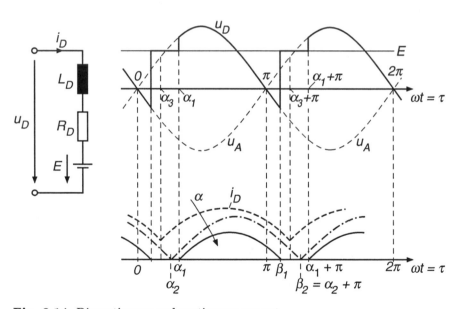

Fig. 8.14. Discontinuous and continuous current

Let us assume an initial zero current state, $i_D = di_D/dt = 0$ and consequently $u_D = E$. At time $\omega t = \tau = \alpha_1$, with the condition $\hat{u}_A \sin \alpha_1 > E$, the thyristors 1, 1' are fired so that the current i_D begins to rise with finite slope and quickly exceeds the latching value i_L. With the thyristors 1, 1' conducting and $u_D(\tau) = u_A(\tau)$, the following differential equation holds,

$$\omega L_D \frac{di_D}{d\tau} + R_D i_D = \hat{u}_A \sin \tau - E, \qquad \alpha_1 \leq \tau \leq \beta_1. \qquad (8.5)$$

The solution, satisfying the initial condition $i_D(\alpha_1) = 0$, and with the abbreviation $L_D/R_D = T_D$ is

$$i_D(\tau) = \frac{\hat{u}_A}{R_D} \frac{1}{\sqrt{1 + (\omega T_D)^2}}$$
$$\cdot \left[\sin(\tau - \arctan \omega T_D) - e^{-(\tau - \alpha_1)/\omega T_D} \sin(\alpha_1 - \arctan \omega T_D) \right]$$
$$- \frac{E}{R_D} [1 - e^{-(\tau - \alpha_1)/\omega T_D}] . \quad (8.6)$$

At $\beta_1 < \alpha_1 + \pi$ the current i_D becomes zero, $i_D(\beta_1) = 0$, see Fig. 8.14, at which time the thyristors return to blocked condition; the current remains zero until $\tau = \alpha_1 + \pi$, when the thyristors 2, 2' are fired and the same transient is repeated. Thus in steady-state the direct current is discontinuous, consisting of separate pulses. The mean of the current is a function of all parameters,

$$\overline{i_D} = \frac{1}{\pi} \int_{\alpha_1}^{\beta_1} i_D(\tau) \, d\tau = F[\alpha, \hat{u}_A, \omega T_D, R_D, E] . \quad (8.7)$$

When evaluating Eq. (8.5) the impedance (R_D, L_D) must be augmented by a small line-side impedance as is shown later.

A similar expression holds for the mean voltage

$$\overline{u_D} = \frac{1}{\pi} \int_{\alpha_1}^{\alpha_1 + \pi} u_D(\tau) \, d\tau . \quad (8.8)$$

This integral may be simplified with the help of Eq. (8.5) considering the periodic boundary condition

$$i_D(\alpha_1) = i_D(\alpha_1 + \pi) = 0 ; \quad (8.9)$$

thus it assumes the form

$$\overline{u_D} = E + R_D \overline{i_D} \quad (8.10)$$

where $\overline{i_D}$ is given by Eq. (8.7). This load characteristic $\overline{u_D}(\overline{i_D})$ for $\alpha = \text{const.}$ and different values of the back voltage E is highly nonlinear as long as the current is discontinuous; this is due to the variable period of conduction, $\beta_1 - \alpha_1 < \pi$, as is seen in Fig. 8.14.

When gradually reducing the firing angle, $\alpha < \alpha_1$, with the same initial conditions as before, the amplitude and the duration of the current pulses increase until a limiting case occurs, $\alpha = \alpha_2$, where one pair of thyristors is fired at the moment the other pair ceases conducting, $\beta_2 = \alpha_2 + \pi$; the current then only touches zero once every half period.

If the firing angle is still further reduced, $\alpha < \alpha_2$, the current will not have reached zero by the time the other thyristors are fired. This then constitutes the region of continuous conduction, where the current is transferred in a commutation transient directly from the outgoing thyristors 2, 2' to the incoming thyristors 1, 1' and the current is no longer discontinuous.

The bordering case to continuous conduction follows with the periodic boundary condition

$$i_D(\alpha_2) = i_D(\alpha_2 + \pi) = 0 \tag{8.11}$$

from Eq. (8.6),

$$\frac{\hat{u}_A}{\sqrt{1 + (\omega T_D)^2}} (1 + e^{-\pi/\omega T_D}) \sin(\alpha_2 - \arctan \omega T_D)$$
$$+ E(1 - e^{-\pi/\omega T_D}) = 0 . \tag{8.12}$$

Solving for α_2, the condition for continuous conduction can be written as

$$\alpha < \alpha_2 = \arctan \omega T_D - \arcsin\left[\frac{E\sqrt{1 + (\omega T_D)^2}}{\hat{u}_A} \tanh\left(\frac{\pi}{2\omega T_D}\right)\right] . \tag{8.13}$$

For each value of the back voltage E a firing angle α_2 exists, at which the range of continuous conduction begins. With $\omega T_D \gg 1$, i.e. assuming a perfectly smooth current, Eq. (8.13) yields the obvious result

$$\overline{u_D} = \frac{2}{\pi} \hat{u}_A \cos \alpha \geq E , \tag{8.14}$$

which means that the mean output voltage $\overline{u_D}$ must exceed the back voltage for continuous load current to flow.

All the preceding equations are valid for $E \gtrless 0$, that is, for rectifier as well as inverter operation.

As was mentioned before, continuous current flow occurs if the firing angle is reduced beyond the bordering case, $\alpha_3 < \alpha_2$; lowering the back voltage E or increasing the lag term ωT_D would have a similar effect. With the current flowing continuously, Eq. (8.5) is valid during the full half period following the firing of the thyristors 1, 1', $\alpha_3 \leq \tau \leq \alpha_3 + \pi$; in steady state the periodic boundary condition is

$$i_D(\alpha_3) = i_D(\alpha_3 + \pi) > 0 ; \tag{8.15}$$

hence the solution may be written as

$$i_D(\tau) = \frac{\hat{u}_A}{R_D} \frac{1}{\sqrt{1+(\omega T_D)^2}}$$
$$\cdot \left[\sin(\tau - \arctan \omega T_D) - e^{-(\tau-\alpha_3)/\omega T_D} \sin(\alpha_3 - \arctan \omega T_D)\right]$$
$$-\frac{E}{R_D}[1 - e^{-(\tau-\alpha_3)/\omega T_D}] + i_D(\alpha_3) e^{-(\tau-\alpha_3)/\omega T_D} , \quad (8.16)$$

where the initial value $i_D(\alpha_3)$ in steady-state is obtained with the help of Eq. (8.15).

If the mean value of the current is mainly of interest as is the case with drives, a much simpler solution exists for continuous current in steady-state. By integrating Eq. (8.5) over a half period and inserting the condition (8.15) for periodicity, we find

$$\overline{i_D}(\alpha_3) = \frac{1}{\pi} \int\limits_{\alpha_3}^{\alpha_3+\pi} \frac{1}{R_D} (\hat{u}_A \sin\tau - E) \, d\tau = \frac{1}{R_D} (\overline{u_D}(\alpha_3) - E) , \quad (8.17)$$

where the average voltage $\overline{u_D}$ is given by Eq. (8.3). Hence the converter acts as a linear controllable voltage source only if the current in the load branch is continuous.

The mean current $\overline{i_D}(\alpha_2)$, at which discontinuous operation sets in, Eq. (8.13), is an important design parameter for the possible selection of a smoothing choke, resulting in an adequate value of the normalised time constant ωT_D. As is shown in Fig. 8.16 this current also depends on the back voltage E, its peak occurs around $E \approx 0$, where the AC component contained in $u_D(\tau)$ assumes a maximum. With large drives the specification may call for the maximum discontinuous current to be less than 10% of nominal current, $\overline{i_D}(\alpha_2) < 0.1 I_{Dr}$, to limit the ripple current which produces additional losses in the armature and may impair commutation. However, with medium and smaller drives the smoothing choke is frequently omitted for reasons of cost, so that only the armature inductance is left for filtering the current. This may result in discontinuous current of half nominal value, affecting not only the losses but also the design of the current controller, as will be explained in Sect. 8.5.

Another important effect neglected so far is the finite commutation interval caused by the impedance in the AC circuit, which consists mainly of the leakage inductance and winding resistance of the transformer; also, line inductances may have been inserted for protective reasons to limit the rate of change of currents (including fault currents) through the thyristors. At medium and larger power rating, the line impedance is predominantly inductive, $R_A \approx 0.1\omega L_A$, which for approximate calculations justifies neglecting the resistive component.

The line-side inductance L_A, even though it is small when compared with the inductance L_D in the load circuit, causes the line current to become a state variable subject to continuous change only; hence $i_A(\tau)$ and the currents in the thyristor branches can no longer exhibit the discontinuities seen in Fig. 8.6. Instead, the rise and decay of the thyristor currents and, correspondingly, the reversal of the alternating current $i_A(\tau)$ now assume the form of continuous commutating transients requiring a finite time, called the commutation interval τ_c or overlap; naturally, this effect occurs only with continuous load current because in discontinuous current mode the currents begin in each period with zero initial condition.

For example it may be assumed that the thyristors 2, 2' of the circuit in Fig. 8.15 a are initially conducting and that at $\tau = \alpha$, i.e. with positive line voltage u_A, the thyristors 1, 1' are fired. Since the currents i_D and i_A are continuous on account of L_D and L_A, the thyristors 2, 2' remain also conducting at first, so that the four thyristors represent temporarily a short circuit connection, $u_D = u_B = 0$, decoupling the line current from the load circuit. Hence, from Kirchhoffs laws, it follows with $\omega t = \tau$,

$$i_{T1} + i_{T2} = i_D \,, \tag{8.18}$$

$$i_{T1} - i_{T2} = i_A \,, \tag{8.19}$$

$$\omega L_A \frac{di_A}{d\tau} + R_A i_A = u_A \,, \tag{8.20}$$

$$\omega L_D \frac{di_D}{d\tau} + R_D i_D = -E \,. \tag{8.21}$$

The initial conditions are

$$i_{T1}(\alpha) = 0 \,,$$

$$i_{T2}(\alpha) = -i_A(\alpha) = i_D(\alpha) \,.$$

Because of the much smaller inductance L_A, the line current changes rapidly whereas the load current may be assumed to be approximately constant during the short commutation interval, $i_D(\tau) \approx i_D(\alpha)$. Hence, with the abbreviation $T_A = L_A/R_A$ we find for $\alpha \leq \tau \leq \alpha + \tau_c$

$$i_A(\tau) \approx \hat{i}_A \sin(\tau - \arctan \omega T_A)$$
$$-[i_D(\alpha) + \hat{i}_A \sin(\alpha - \arctan \omega T_A)] e^{-(\tau - \alpha)/\omega T_A} \tag{8.22}$$

where

$$\hat{i}_A = \frac{\hat{u}_A}{\sqrt{R_A^2 + (\omega L_A)^2}}$$

is the peak value of the (steady state) short circuit current at the line terminals. From Eqs. (8.18), (8.19) the thyristor currents are obtained,

$$i_{T1}(\tau) \approx \frac{1}{2} [i_D(\alpha) + i_A(\tau)] \,,$$

$$i_{T2}(\tau) \approx \frac{1}{2} [i_D(\alpha) - i_A(\tau)] \,.$$

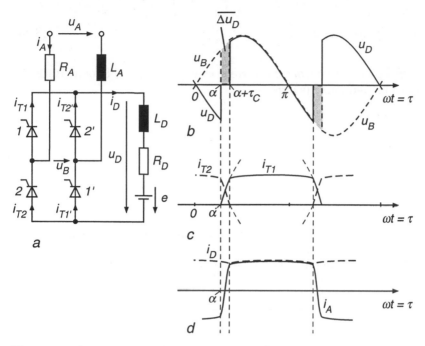

Fig. 8.15. Commutation of a single phase converter

These results apply only as long as the four thyristors are conducting. The commutation terminates at time $\alpha + \tau_c$, when

$$i_{T2}(\alpha + \tau_c) = 0 \tag{8.23}$$

after this instant, Eq. (8.5) is again valid.

The various voltages and currents, when including the line impedance, are seen in Fig. 8.15 a,b, assuming continuous current i_D. All the currents are now continuous functions of time; in contrast, the voltages $u_D(\tau)$ and $u_B(\tau)$ exhibit characteristic discontinuities during commutation, when all thyristors are conducting.

As a consequence of the finite commutation interval, the line current experiences a further shift in time resulting in an additional component of reactive power due to commutation, increasing the reactive power caused by the delayed firing control as discussed before, Fig. 8.7

Another effect of the line impedance is a loss in output voltage due to the temporary short circuit,

$$u_D(\tau) = 0, \qquad \alpha \le \tau \le \alpha + \tau_c ; \tag{8.24}$$

it causes a drop of the mean voltage by

$$\overline{\Delta u_D} \approx \frac{1}{\pi} \int_{\alpha}^{\alpha+\tau_c} u_A \, d\tau \, , \qquad\qquad (8.25)$$

which may be estimated from the current change in the AC circuit. With the simplification mentioned before,

$$R_A \ll \omega L_A \, ,$$

we find from Eq. (8.20)

$$\overline{\Delta u_D} \approx \frac{1}{\pi} \int_{\alpha}^{\alpha+\tau_c} u_A \, d\tau \approx \frac{1}{\pi} \omega L_A \, i_A(\tau) \, \Big|_{\alpha}^{\alpha+\tau_c}$$

$$= \frac{1}{\pi} \omega L_A \left[i_A(\alpha + \tau_c) - i_A(\alpha) \right] . \quad (8.26)$$

Hence the voltage-time-area lost during commutation is related to the flux change in the line inductance.

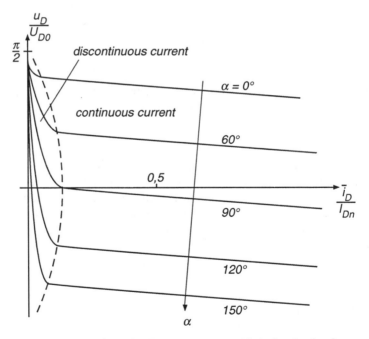

Fig. 8.16. Characteristics of single phase converter with inductive load

If a sufficiently large smoothing inductance L_D is employed, such that the ripple current contained in $i_D(\tau)$ may be neglected, $i_A(\alpha + \tau_c) \approx -i_A(\alpha) \approx \overline{i_D}$; Eq. (8.26) then yields a voltage drop proportional to the load current

$$\overline{\Delta u_D} \approx \frac{2}{\pi} \omega L_A \overline{\overline{i_D}} \, , \tag{8.27}$$

which indicates that the line-commutated converter acts like a controlled voltage source having an apparent internal resistance proportional to the reactance of the AC line; of course, this drop of the terminal voltage does not involve real power losses, as it is caused by commutation at the AC side of the converter. As a consequence, the load curves of the converter $\overline{u_D}(\overline{i_D})$ for $\alpha = $ const. are drooping in the continuous current region (Fig. 8.16), even though all ohmic resistances apart from that in the load branch have been neglected. In contrast to the intermittent current range where the curves are strongly nonlinear, the characteristics in the continuous current region are linear, exhibiting only a slight droop. By normalising Eq. (8.27) with the no-load voltage U_{D0} and nominal current I_{Dn}, we find

$$\frac{\overline{\Delta u_D}}{U_{D0}} \approx \frac{\omega L_A I_{Dn}}{\sqrt{2} U_A} \frac{\overline{i_D}}{I_{Dn}} = k \frac{\overline{i_D}}{I_{Dn}} \, . \tag{8.28}$$

The normalised impedance factor k is mainly determined by the leakage reactance of the line transformer; a common value is $k \approx 0.05$ to 0.10.

The commutation interval τ_c, which usually lasts only a few degrees, depends also on the firing angle α; this is seen from a rough estimate of the integral in Eq. (8.25)

$$\overline{\Delta u_D} \approx \frac{1}{\pi} \int_{\alpha}^{\alpha+\tau_c} \hat{u}_A \sin \tau \, d\tau \approx \frac{\sqrt{2}}{\pi} \tau_c U_A \sin \alpha \, , \tag{8.29}$$

which shows that for constant load current the overlap is minimal at $\alpha = \frac{\pi}{2}$, when the AC voltage has its peak amplitude. A more accurate result is obtained by evaluating the integral, Eq. (8.25),

$$\tau_c \approx \arccos \left[\cos \alpha - \frac{\sqrt{2} \omega L_A}{U_A} \overline{i_D} \right] - \alpha \, . \tag{8.30}$$

The knowledge of the overlap is of particular importance when the converter operates as inverter, i.e. with large firing delay α, because then it must be assured that the commutation is completed and in addition the outgoing thyristors have recovered and are ready to block forward voltage when the line voltage changes sign; otherwise an unintentional refiring may occur. Since the back voltage is negative during inverter operation, $E < 0$, this would connect in series two voltages (E, u_A) having (for a half period) the same sign; the result would be a short circuit condition with a fast current rise which is only limited by the smoothing inductance. This malfunction is called a commutation failure.

In Fig. 8.17 this disturbance is described in more detail. The inverter is assumed to operate initially at the safe firing angle $\alpha_1 = 150°$, when an

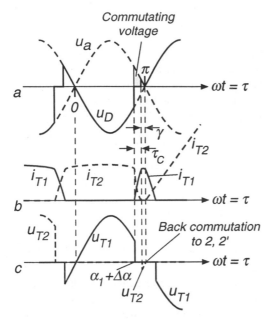

Fig. 8.17. Commutation failure due to late firing of thyristors 1, 1'

additional firing delay $\Delta\alpha$ for the thyristors 1, 1' is introduced. Due to the low voltage $u_A(\tau)$ in this interval the commutation proceeds quite slowly as it takes time to accumulate the necessary voltage-time-area. The commutating transient is eventually completed at $\alpha + \tau_c$, when the current through thyristors 2, 2' has become zero. There is still some negative blocking voltage $u_{T2} < 0$ across the outgoing thyristors but it is small and the time until the voltage u_A changes sign may be too short for the thyristors 2, 2' to recover their forward blocking capability. In other words, the time equivalent of the extinction angle $\gamma = \pi - (\alpha + \tau_c)$ no longer exceeds the recovery time t_r which is increased by the low reversed bias voltage across the outgoing thyristors.

The consequence could be a spontaneous refiring of thyristors 2, 2' and back-commutation, leading to the series connection of $u_A < 0$ and $E < 0$ through thyristors 2, 2' and a subsequent steep rise of the current. The next regular commutation of the inverter is possible one period after the attempted firing of thyristors 1, 1', where the controller must now assure a sufficiently advanced timing of the firing pulses to successfully complete the commutation despite the increased current. Otherwise a circuit breaker or fuse would have to clear the fault.

It is seen that the danger of a commutation failure increases with the magnitude of the current and the line-side reactance, since both effects prolong the commutation interval τ_c. In order to fully utilise the voltage of the inverter, a sufficient magnitude of the extinction angle γ may be assured by a separate control loop. However, in view of the additional complexity and the remaining uncertainty in case of a sudden voltage dip of the AC line voltage (which would unexpectedly increase the overlap), extinction angle control is seldom used except with very high power installations, such as HVDC, where operation of the inverter at the maximum firing angle is important for power factor considerations and the cost of the control system is less significant [F21, 77]. The usual practice with drives is to limit the firing angle to "safe" values, for instance $\alpha < \alpha_{max} = 150°$, and to prevent excessive currents by a fast acting control loop.

8.3 Line-commutated Converter in Three-phase Bridge Connection

With the exception of traction drives having a single phase supply the single phase circuit discussed in the preceding section is normally used only for low power applications, such as the supply of field windings. Beyond a few kW there are strong incentives for three-phase converter circuits. The bridge connection shown in Fig. 8.18 is the circuit most commonly used with thyristors or power transistors; it requires two more thyristors but has a number of important advantages:

- The three-phase line is symmetrically loaded in steady state.
- The line currents have a lower harmonic content; as a consequence there is less distortion of the line voltages than with a single phase circuit.
- The same is true for the direct voltage $u_D(\tau)$ which contains ripple components of higher frequencies and lower amplitudes, thereby permitting a reduction of the filter components and causing lower losses in the load.
- The dynamic performance of three-phase converters is superior because thyristors are fired at shorter intervals; hence the reduced delay for control signals permits more rapid control.

In contrast to the single-phase bridge circuit (Fig. 8.5) where the thyristors had to be fired in pairs, the commutations in the three-phase circuit (Fig. 8.18) alternate between the upper and lower row of thyristors, so that in steady state six regularly spaced firing transients occur during each line period. Hence the three-phase bridge is a "six-pulse converter" , whereas the single phase bridge with two firing instants per period belongs to the class of "two-pulse converter" circuits.

With Hg-arc valves, having a much higher voltage drop than thyristors, center tap circuits without series connection of valves had been in widespread use; they were the natural choice with multi-anode valves with a common Hg-cathode. However, since the introduction of thyristors as individual switching devices with small physical dimensions and low voltage drop these aspects have become obsolete.

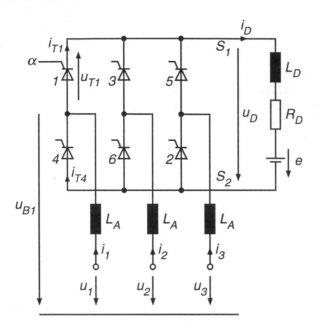

Fig. 8.18. Three-phase line-commutated converter bridge

The AC voltages in Fig. 8.18 are assumed to represent the line-to-neutral voltages of a balanced three-phase system. The thyristors 1, 3, 5 connect the AC line-terminals in cyclic sequence to the upper DC bus S_1, the valves 2, 4, 6 to the lower bus S_2. In normal steady-state operation and with continuous direct current, each thyristor in the upper and lower row carries the current for 120°; also, the two thyristors connected to one AC line-terminal do not normally conduct at the same time, as this constitutes a short circuit of the DC-terminals.

Figures 8.19 and 8.20 show various voltages and currents in steady-state conditions with the firing angles $\alpha_1 = 45°$ and $\alpha_2 = 135°$, belonging to rectifier- and inverter-operation respectively. The line inductances L_A are neglected for the time being so that no overlap between subsequent thyristors occurs; the constant direct current $i_D = I_D$ is again assumed to be generated by a current source. The conducting thyristors are indicated in each 120° interval, always comprising one of the upper and one of the lower row; the firing alternates between these two groups; hence the output voltage $u_D(\tau)$ consists of 60° sections of sequential line-to-line voltages, for example $u_{12} = u_1 - u_2$. As long as thyristor 1 is conducting, thyristor 3 sees a negative bias voltage for $u_2 - u_1 < 0$ and can only be fired after $u_2 - u_1$ has become positive.

This "natural" firing instant ($\tau = \pi/6$ for thyristor 1) serves as a reference for the firing angle α, which can again vary from $\alpha = 0$ (maximum output voltage, rectifier) to $\alpha = \alpha_{max}$ (minimum output voltage, inverter).

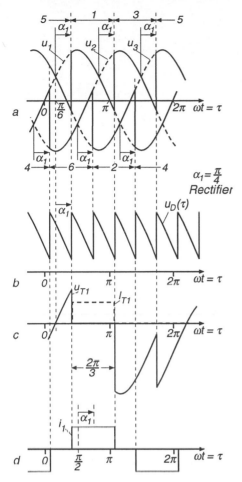

Fig. 8.19. Voltages and currents of a 6- pulse converter at $\alpha_1 = 45°$

An appropriate safety margin $\pi - \alpha_{max}$ is again required which must exceed the maximum overlap plus recovery time t_r to prevent commutation failures.

With the definitions for the line-to-neutral voltages

$$u_1(\tau) = \sqrt{2}\,U_A \sin\tau\,, \qquad u_2(\tau) = \sqrt{2}\,U_A \sin\left(\tau - \frac{2\,\pi}{3}\right),$$

$$u_3(\tau) = \sqrt{2}\,U_A \sin\left(\tau - \frac{4\,\pi}{3}\right)\,;$$

and the assumed simplifications the mean of the output voltage is

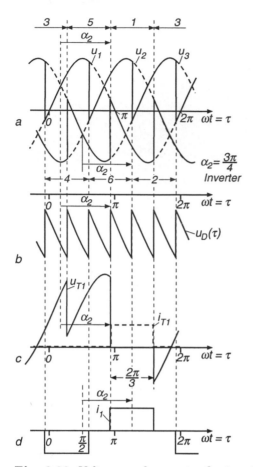

Fig. 8.20. Voltages and currents of a 6- pulse converter at $\alpha_1 = 135°$

$$\overline{u_D} = \frac{3}{\pi} \int\limits_{\frac{\pi}{6}+\alpha}^{\frac{\pi}{2}+\alpha} (u_1 - u_2)\, d\tau = \frac{3\sqrt{6}}{\pi} U_A \int\limits_{\frac{\pi}{6}+\alpha}^{\frac{\pi}{2}+\alpha} \sin\left(\tau + \frac{\pi}{6}\right) d\tau$$

$$= \frac{3\sqrt{6}}{\pi} U_A \cos\left(\tau + \frac{\pi}{6}\right)\Big|_{\frac{\pi}{2}+\alpha}^{\frac{\pi}{6}+\alpha} = \frac{3\sqrt{6}}{\pi} U_A \cos\alpha = U_{D0} \cos\alpha \ .$$

$$(8.31)$$

Here too, the control curve follows a cos-function. When the converter is connected to the 230/400 V three-phase line, the maximum mean output voltage is $U_{D0} = 540$ V.

The output voltage $u_D(\tau)$ in Figs. 8.19 and 8.20 contains harmoncis, the frequencies of which are multiples of $6f$, i.e. 300 Hz for a 50 Hz line.

Voltage and current of a thyristor are plotted in Fig. 8.19 c for rectifier- and in Fig. 8.20 c for inverter-operation. Reverse bias voltage ($u_{T1} < 0$)

dominates in the first, forward blocking voltage ($u_{T1} > 0$) in the second case. This indicates that rectifier operation is insensitive to overload, while at the inverter limit the danger of commutation failure always lurks in the background, because negative voltage is required across the outgoing thyristor for a sufficient time to let it recover its forward blocking capability.

In Figs. 8.19 d and 8.20 d the current i_1 in one of the AC terminals is drawn, the fundamental component of which is again lagging the pertinent line-to-neutral voltage u_1 by the firing angle α (neglecting overlap). A comparison with Fig. 8.6 d shows that the harmonics content of the line currents is reduced; whereas all odd harmonics of line frequency f were present in the single-phase circuit, there are additionally no odd harmonics of frequency $3f$ in the three-phase circuit. Hence the lowest order harmonics of the line currents, assuming perfect symmetry of line voltages and firing sequence, are the 5th and the 7th in the three-phase circuit.

So far, it had again been assumed that the commutation between the conducting thyristors occurs instantaneously; in reality, it requires finite time because of the presence of leakage inductances in the supply lines and the thyristor branches. As mentioned before, line inductances are often introduced deliberately to limit the rise of the current during commutation; if no transformer for supplying the converter is used, this is mandatory. Again the line impedance is assumed to be purely inductive, $R_A \ll \omega L_A$.

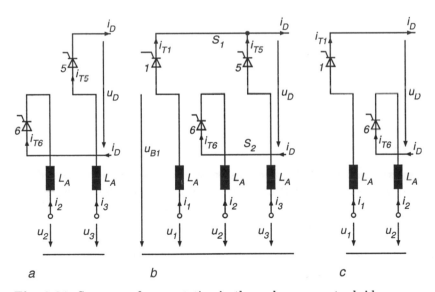

Fig. 8.21. Sequence of commutation in three-phase converter bridge

The commutation described in Fig. 8.21 proceeds in three stages: Initially the thyristors 5 and 6 may carry the constant current $i_D = I_D$ (Fig. 8.21 a); at time $\tau = \frac{\pi}{6} + \alpha$ thyristor 1 is fired leading to the situation in Fig. 8.21 b, where two AC terminals are connected through thyristors 1 and 5; due to the small inductances L_A a short circuit current rapidly builds up, transferring the current I_D from thyristor 5 to thyristor 1. As soon as i_{T5} has become zero, thyristor 5 blocks and the circuit finds itself in the position shown in Fig. 8.21 c. The overlap during which three thyristors are simultaneously conducting extends normally over a few degrees only; however there may be exceptions if the line inductances are unusually large.

The commutation according to Fig.8.21 b is described by the following equations, valid for $\frac{\pi}{6} + \alpha \leq \tau \leq \frac{\pi}{6} + \alpha + \tau_c$,

$$i_{T1} + i_{T5} = i_{T6} \approx I_D , \tag{8.32}$$

$$\omega L_A \frac{d}{d\tau}(i_{T1} - i_{T5}) \approx u_1 - u_3 . \tag{8.33}$$

Elimination of i_{T5} results in

$$2 \omega L_A \frac{di_{T1}}{d\tau} \approx u_1 - u_3 = \sqrt{6} U_A \sin(\tau - \pi/6) , \tag{8.34}$$

leading with the initial condition $i_{T1}(\frac{\pi}{6} + \alpha) = 0$ to

$$i_{T1}(\tau) \approx \frac{\sqrt{6} U_A}{2 \omega L_A} \left[\cos \alpha - \cos \left(\tau - \frac{\pi}{6} \right) \right] , \tag{8.35}$$

and

$$i_{T5}(\tau) \approx I_D - i_{T1}(\tau) . \tag{8.36}$$

As soon as i_{T5} is zero,

$$i_{T5} \left(\frac{\pi}{6} + \alpha + \tau_c \right) = 0 , \tag{8.37}$$

the commutation is concluded and Eqs.(8.32), (8.33) are no longer valid.

Figure 8.22 shows a sequence of commutations as obtained by digital simulation of the converter circuit [E6, M17] where for the sake of clarity the commutation intervals have been enlarged. Because of the equal inductances in each line terminal, the potential of the upper bus S_1 corresponds to the mean of the two line-to-neutral voltages affected by the short circuit condition. Hence, the output voltage during overlap is

$$u_D(\tau) \approx \frac{1}{2} (u_1 + u_3) - u_2 , \tag{8.38}$$

which causes a loss in output voltage according to the shaded area in Fig. 8.22; its mean value is

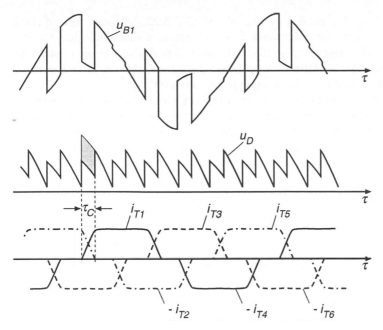

Fig. 8.22. Commutation transients of three-phase converter

$$\overline{\Delta u_D} = \frac{3}{\pi} \int_{\frac{\pi}{6}+\alpha}^{\frac{\pi}{6}+\alpha+\tau_c} \frac{1}{2}\left(u_1 - u_3\right) d\tau . \tag{8.39}$$

Integrating Eq. (8.34) with the limits $0 \leq i_{T1} \leq I_D$ results in

$$\int_{\frac{\pi}{6}+\alpha}^{\frac{\pi}{6}+\alpha+\tau_c} \frac{1}{2}\left(u_1 - u_3\right) d\tau = \omega\, L_A\, I_D ; \tag{8.40}$$

the voltage drop caused by the commutation is again a linear function of the load current,

$$\overline{\Delta u_D} = \frac{3}{\pi}\, \omega\, L_A\, I_D ; \tag{8.41}$$

a fictitious ohmic resistor in series with the controllable voltage source $\overline{u_D}(\alpha)$ would have a similar effect. The load characteristics of the three-phase converter in Fig. 8.23 show a corresponding droop in the continuous current range.

As was explained in the case of the single-phase converter, a load circuit consisting of a finite inductive impedance in series with a back voltage E can lead to discontinuous current flow, causing the curves $\overline{u_D}(\overline{i_D}, \alpha)$ to become highly nonlinear; this is seen at the left of Fig. 8.23. With an appropriate choice of the smoothing inductance L_D the region of discontinuous current can be restricted to light load so that it is not noticeable in normal operation; the requirement for the filter time constant T_D to achieve continuous current is considerably relieved with a three-phase converter.

However, if the smoothing reactor is further reduced or completely omitted for cost reasons, discontinuous current flow can again occur even under load. In view of the different motor dynamics this may require special provisions when designing the control, as will be explained in Sect. 8.5.

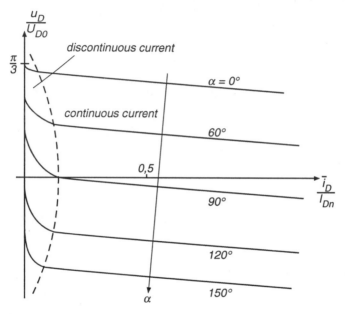

Fig. 8.23. Load curves of a 6- pulse converter

The temporary short circuits of the AC terminals during commutation have of course an immediate effect on the waveforms of the AC voltages, possibly influencing neighbouring equipment. One of the terminal voltages u_{B1} with respect to neutral, Fig. 8.18, is plotted in Fig. 8.22. In a weak supply system having a large internal impedance, the danger of inadvertent feedback to the converter through the synchronising signals exists which may even cause instability [A11]. Apart from the obvious, but not always practical, solution of filtering the synchronising signals, there are various "supply-independant" firing schemes with internal oscillators that permit stable operation even with highly distorted line voltages [H61, M16]. Three-phase firing circuits are also available in the form of integrated circuits (chips).

The dynamic behaviour of the six-pulse converter is similar to that of the two-pulse converter with the exception that now the steady-state firing intervals are $\pi/3$, instead of π, permitting faster access to the output voltage $u_D(\tau)$. The waiting interval is now between 0 and $1/6\ f$ corresponding to 3.3 ms for a 50 Hz-supply.

The firing control of a three-phase converter is simply an extension of the firing scheme for single phase converters (Fig. 8.11), employing several time-displaced saw-tooth functions to sample the output signal of the controller. For example, Fig. 8.24 displays voltage transients $u_D(\tau)$, neglecting commutation, which follow a step change of the firing angle from rectifier- to inverter operation and back. When the firing is delayed, the current carrying thyristors simply remain conducting until the line voltage has changed its sign, while in opposite direction an immediate response is possible. It may even occur, as seen in Fig. 8.24, that some thyristors are skipped because of a multiple commutation with several – in principle all – thyristors temporarily conducting, with the ones connected to the highest and lowest line voltage potential eventually prevailing.

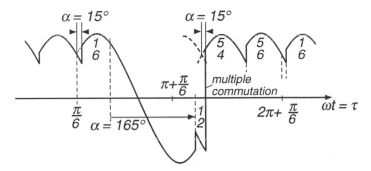

Fig. 8.24. Control transients of a 6- pulse converter

A feature of the three-phase bridge connection is that two thyristors have to be fired simultaneously for start-up and in discontinuous current mode, one in each row; otherwise the current could not begin to flow. This is achieved either by employing wide pulses ($\Delta\tau > \pi/3$) so that they overlap or by generating for each thyristor two short firing pulses, $\pi/3$ apart. Normally the second pulse would be applied to a thyristor that is already conducting and be of no consequence. If the first solution is preferred, the firing pulse may be chopped into narrow pulses to reduce the voltage-time-area and the size of the pulse transformers.

The input current of the converter, drawn in Figs. 8.19 d and 8.19 d without taking commutation into account, is the same as the current that would flow in the primary winding of a Y/Y-connected transformer supplying the converter. However, the waveform of the primary current changes if one of the

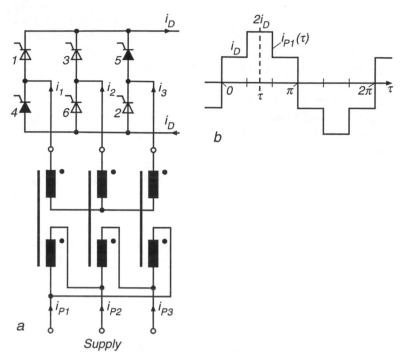

Fig. 8.25. Three-phase converter supplied by $\mathsf{Y}/\triangle$- connected transformer. **(a)** Circuit; **(b)** Waveform of primary current

transformer windings are $\triangle$-connected as is explained in Fig. 8.25. Assume that the constant load current I_D is initially flowing through the thyristors 4 and 5, closing its path through the Y-connected secondary windings of the transformer. When neglecting the magnetising current and assuming – for sake of simplicity – equal numbers of turns, the following equations then hold for the primary currents

$$
\begin{aligned}
i_{p1} &= i_3 - i_1 = 2\,i_D \;, \\
i_{p2} &= i_1 - i_2 = -i_D \;, \\
i_{p3} &= i_2 - i_3 = -i_D \;.
\end{aligned}
\qquad (8.42)
$$

Permutation of the six possible states yields the primary current waveform $i_{p1}(\tau)$ shown in Fig. 8.25 b which also contains the 5th and the 7th as the lowest order harmonics. With the appropriate choice of the number of turns the Y- and the $\triangle$-connected windings could be interchanged without altering the current waveforms.

Frequently in the past, though less commonly now, the output voltage obtainable with a converter having one thyristor per branch was insufficient so that thyristors had to be connected in series. Some advantages may then be gained if — instead of series-connecting individual thyristors — complete

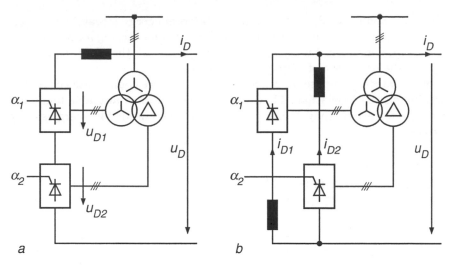

Fig. 8.26. 12-pulse-converter consisting of two 6-pulse converters in series- and parallel connection

three-phase circuits are connected in series, as seen in Fig. 8.26 a. Rather than taking precautions for equal voltage sharing between series connected blocking thyristors one now relies on the output voltages of the transformer to produce the correct voltages. Also, if a $Y/\triangle$-connection with its inherent 30° phase shift is chosen for the secondaries there is an added benefit in the reduction of harmonics on the load-side, where only multiples of 12 f occur, as well as at the line-side, where the harmonics are those of a 12-pulse converter, with the 11th and the 13th being the lowest harmonics; there is a further improvement in dynamic response because the maximum waiting interval is now 30°. Another similar circuit is seen in Fig. 8.26 b where two six-pulse converters supplied from a $Y/\triangle$-connected transformer are connected in parallel, employing two separate chokes or an interphase transformer for summation of the currents. The principle of achieving 12-pulse operation by series connection of complete converters is applied also on HVDC transmissions with hundreds of MW, where many thyristors are still series connected in each converter branch.

This brief introduction to the fundamentals of line-commutated converters seems adequate for dealing with most drive problems. A more detailed analysis such as harmonics, circuitry or protective measures is given in the specialised literature [50, 58]. It should have become clear that multiphase static power converters are electrical actuators of superb dynamic performance providing an excellent tool for controlling large output power (MW) with low power control signals (mW) in very short time (ms). This combination is unmatched by any other type of power actuator. On the other hand it should be noted that due to the switching action static power converters are inherently nonlinear. Since they have no internal energy storage the output power is instantaneously transferred into the AC line causing harmonics and reactive power in the supply system; strong grids are therefore required to support the operation of large power converters.

Even shorter response times are feasible with force commutated converter circuits, discussed in Chap. 11; while the power ratings available today are adequate for drives, they do not yet reach the levels needed in energy supply systems.

8.4 Line-commutated Converters with Reduced Reactive Power

In the preceding sections it was shown that line-commutated converters draw reactive power from the line; this is inherent in the principle of control by delayed firing which results in lagging current at the AC side of the converter. Since the phase shift of the fundamental component of the line current is directly related to the firing angle α, the situation is particularly severe at $\alpha \approx \frac{\pi}{2}$, where the line current is, apart from the losses, purely reactive. This condition is unavoidable, if the converter is supplying the armature of a DC motor because when the motor starts from standstill the induced voltage is initially zero and the converter must be controlled close to $\alpha = \frac{\pi}{2}$; hence a large motor, for example a mine hoist drive starting with twice nominal current causes a huge surge of reactive power on the supply side. There is no easy remedy because capacitors cannot be switched in and out as the speed changes and the cost of rotating or static reactive power compensators would be too high.

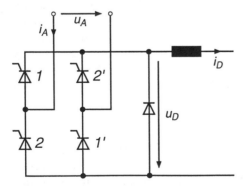

Fig. 8.27. Single-phase converter with a shunting diode

The situation may be tolerable with strong supply systems but is unacceptable on traction drives, where the reactive current in combination with the large inductance of the power supply comprising the catenary would cause heavy voltage fluctuations, in addition to the harmonic distortion. Mainly for the benefit of traction drives special circuits and control techniques have been developed with the aim of limiting the reactive power at reduced output voltage. Some of these circuits contain diodes as well as thyristors [D25, S57].

A simple though uneconomical solution is the use of a diode parallel to the load, as shown in Fig. 8.27 for a single-phase bridge circuit. The diode has the effect of cutting off the negative edges of the output voltage, essentially acting as bypass during the intervals when u_D would otherwise be negative. Thus the trailing end of the line current (during interval $\pi < \tau < \pi + \alpha$ in Fig. 8.6 d does not flow through the converter and the line as it is short circuited through the diode. The result is an offset of the fundamental component of the line current in leading direction and a correspondent improvement of the power factor. This has, of course, the effect that the circuit can no longer act as inverter; the conditions

$$i_D(\tau) \geq 0 \,, \qquad u_D(\tau) \geq 0$$

restrict the operation to a single quadrant in the $\overline{u_D}$, $\overline{i_D}$-plane; with traction drives employing series wound motors this would not be objectionable.

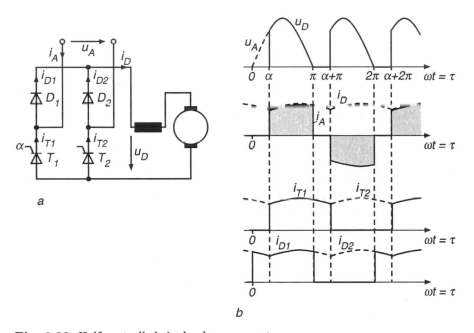

Fig. 8.28. Half-controlled single-phase converter.
(a) Circuit diagram; **(b)** Waveforms of voltages and currents

Another circuit with similar properties is shown in Fig. 8.28 a, which belongs to a class of half-controlled converter circuits containing diodes in place of some of the thyristors. Here too, the line current cannot flow against the line voltage and thus cause a temporary power flow from the DC to the AC side, since the opposite diode would conduct and act as bypass. Figure 8.28 b shows a simplified diagram of the voltages and currents, neglecting the commutation intervals. Again the trailing end of the line current is cut-off, thereby advancing the fundamental component and improving the power factor. Clearly, this circuit could be easily extended to form a half-controlled three-phase bridge connection.

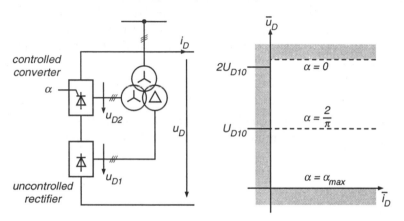

Fig. 8.29. Boost- and -buck connection of controlled and uncontrolled converter

A technique for improving the power factor at the input of the converter is outlined in Fig. 8.29 a, showing a "boost and buck"-scheme of a three-phase rectifier containing only diodes, and a three phase thyristor converter; by employing a transformer with $Y/\triangle$ secondary windings the harmonics situation may be improved as well. This connection makes use of the fact that a converter operating at $\alpha = 0$ or $\alpha = \alpha_{max}$ has a tolerable power factor while operation at $\alpha \approx \frac{\pi}{2}$ should be avoided, if ever possible. The total mean output voltage, assuming continuous current, covers the range

$$U_{D\,10} + U_{D\,20}\ \cos \alpha_{max} \leq \overline{u_D} \leq U_{D\,10} + U_{D\,20}\ . \tag{8.43}$$

By selecting the voltages according to the condition

$$U_{D\,10} + U_{D\,20}\ \cos \alpha_{max} = 0\ ,$$

the complete first quadrant of the $\overline{u_D}$, $\overline{i_D}$-plane is within reach. That this circuit has a better reactive power balance than a single converter circuit follows from the fact that for minimum and maximum voltage, i.e. $\overline{u_D} \approx 0$ and $\overline{u_D} = \overline{u_D}_{max}$, both sections operate with good power factor. It is only in the centre, $\overline{u_D} \approx U_{D10}$, where the controlled converter, accounting for about half the total power, is fired at $\alpha \approx \frac{\pi}{2}$. At full output voltage, $\alpha = 0$, the circuit corresponds to Fig. 8.26 a, causing approximately 12-pulse line interactions.

The principle of boost-and-buck operation may be generalised by splitting the full voltage range $\pm U_0$ and assigning the bands of $\pm U_0/n$ to n series-connected thyristor converters which are supplied from separate transformer secondaries. The control is then arranged on a sequential basis such that only one section is phase-controlled with $0 < \alpha < \alpha_{max}$, while all $n-1$ other sections are operating either with $\alpha = 0$ or with $\alpha = \alpha_{max}$. Sequential phase control is of particular interest if the converter needs to be subdivided anyway for reasons of the high voltage required. This technique has also found application for rail traction drives [K70].

8.5 Control Loop
Containing an Electronic Power Converter

Line-commutated converters are ideal actuators for electrical drives; a main feature, apart from their simplicity, proven reliability and practically unlimited output power is their excellent dynamic performance. Also, electronic controllers can be directly connected to the firing circuit which consists of integrated circuitry and logic devices operating at signal level.

When a control loop containing a converter is to be designed, the question arises, how the dynamic behaviour of the converter should be described by a mathematical model. The difficulties stem from the fact that the firing of a thyristor is a discrete process - the firing angle is not a continuous function of time - and that the steady-state and dynamic behaviour of the converter is highly nonlinear. Pulse- and phase-modulated systems of this or similar kinds have been frequently dealt with in the literature but no coherent theory exists which is simple enough for general application [G31, L3, P46, S24].

A detailed analysis of dynamic processes in line-commutated converters is rather involved; the transients are described by nonlinear difference equations which defy an analytical solution. Linearisation of the equations is of course possible but the validity is then restricted to the vicinity of the chosen operating point. However, as a first order approximation it is found that the converter dynamics can be entirely neglected as long as the control plant, fed by the converter, has a low-pass characteristic, the controller contains an integrating term and the control loop is to be well damped [F12, L35]; these conditions are usually met on drive applications. Near the stability limit this simplification is of course no longer valid; there it is necessary to revert to the accurate model equations.

Another simple model that has served well in practice is based on the observation that a variable waiting period elapses until a small change of the input signal y_1 to the firing circuits (Fig. 8.11) shows an effect by shifting the first firing pulse. As the waiting period depends on the instant at which the input signal is changed, varying between zero and $20/6$ ms $= 3.3$ ms for a six-pulse converter operating on the 50 Hz supply, a mean equivalent delay time $T_T = 1.67$ ms is postulated [52]. This heuristic model produces useful results, as long as well damped transients of the closed loop are the objective but again, as the stability limit is approached, considerable deviations from the predictions must be expected.

The most effective method of neutralising the nonlinear and discontinuous behaviour of the converter is to enclose it in a tight feedback control loop which then imposes characteristics of a nearly linear actuator, such as unity gain. By electronically limiting the current reference $i_{D\ Ref}$ a fast and very effective overload protection of the converter as well as the motor and the mechanical load is achieved.

This is outlined in Fig. 8.30, showing a converter with inductive impedance and a back voltage e in the load circuit. With the help of a current controller the output current is tracking the current reference i_{DRef}, independent of the back voltage which acts as a load disturbance. On a DC drive, where the load circuit is represented by the armature, possibly in series with a smoothing inductor, the parameters could be such that the voltage drop $\overline{\Delta u_D}$ at nominal current is only a few percent of nominal voltage and that for $\alpha = 0$, $e = 0$ about eight or ten times nominal current would flow. A function-generator, as indicated in dashed lines, could be included to linearise the steady state control characteristic of the converter (Fig. 8.8); frequently the flat ends of the control curve only serve as a temporary voltage reserve for transients. For the subsequent simulation, Fig. 8.32, a firing circuit with an internally synchronised oscillator was modelled.

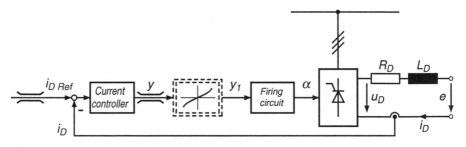

Fig. 8.30. Converter with current control

The circuit in Fig. 8.30 is now transformed into the linearised block diagram in Fig. 8.31, where the converter with its control circuit is approximated by an equivalent mean delay, which for a 6-pulse converter and 50 Hz-supply is $T_T \approx 1.7$ ms. For smoothing the current feedback signal, $T_F = 2$ ms may be assumed with a six-pulse converter . This value depends also on whether

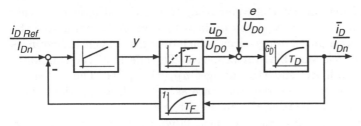

Fig. 8.31. Simplified block diagram of current control loop

the current sensor is installed on the AC or DC side of the converter. Sensing the line current is simpler because current transformers with an output rectifier offer inexpensive isolation; a DC current sensor in the load circuit, which may be faster, more accurate and responding to the sign of $\overline{i_D}$, is likely to be more costly. The use of low level control signals, typically 10 V, which are electrically isolated from line potential is a necessary condition for safety reasons. When designing the controller, the delay of the converter and the filter time constant, which are both small in comparison with the time constant T_D, are lumped together to form an equivalent lag $T_e \approx 4$ ms.

A proven design method for the current control loop was discussed in Sect. 7.2. Instead of the PID-controller that would be desirable with a slowly responding rotating generator, a PI-controller is quite adequate here. In fact, a PID-controller, whose step response begins with an impulse, may cause inconsistent results in conjunction with the discrete firing circuit so that the additional lead term is usually omitted.

Hence the open loop transfer function of the current control scheme is

$$F_{OL}(s) \approx G_{c,i} \frac{T_i\, s + 1}{T_i\, s} \frac{1}{T_e\, s + 1} \frac{G_D}{T_D\, s + 1}, \tag{8.44}$$

where $G_{c,i}$ is the gain and T_i the integrating time constant of the current controller. A typical time constant of the load circuit is $T_D = 30$ ms. Because of $T_e \ll T_D$ the transfer function may be simplified in the region of the cross-over frequency, which determines the band width of the closed loop,

$$F_{OL}(s) \approx G_{c,i}\, G_D \frac{1}{T_i T_D\, s^2} \frac{T_i\, s + 1}{T_e\, s + 1}. \tag{8.45}$$

This is in line with the "symmetrical optimum" mentioned in Sect. 7.2. The best choice of parameters is [K18, 37]

$$T_i = a^2\, T_e, \qquad G_{OL} = G_{c,i}\, G_D = \frac{1}{a} \frac{T_D}{T_e}, \tag{8.46}$$

where $a = 1 + 2D$ follows from the desired damping ratio D of the complex pair of eigenvalues; $D = \frac{1}{2}$ or $D = 1/\sqrt{2}$ is frequently chosen.

The design of the current controller leads to the closed loop transients shown in Fig. 8.32. The curves were obtained by accurately simulating the operation of the three-phase bridge circuit, including the leakage inductances of the transformer and the thyristor branches. The effects of the commutation are seen on both AC and DC voltages $u_{B1}(\tau)$, $u_D(\tau)$. The transients include step changes of the current reference as well as large synthetic load disturbances caused by inverting the back voltage e.

Despite the severe simplifications made in designing the current controller, the transients decay rapidly and are well damped. The equivalent lag of the closed loop for small changes of the reference is $T_{eCL} \approx 10$ ms, which is in line with the model computations for the symmetrical optimum. As the current loop represents the innermost control loop of a nested structure, it provides a good starting point for the higher level control system.

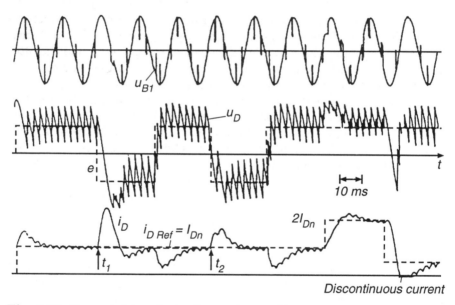

Fig. 8.32. Computed transients of current control loop

When studying the details of Fig. 8.32 more closely, differences become apparent at time t_1, t_2, following the inversion of the back voltage e. The reason is that at t_1 the back voltage was reversed immediately after a thyristor had been fired, while at t_2 the reversal took place slightly before a firing pulse was due. Hence a waiting interval was unavoidable at t_1, whereas at t_2 the converter could respond immediately. The difference in overshoot of the current is a consequence of the discontinuous and nonlinear nature of the converter. The control is only partially successful in compensating this effect, because the disturbance is acting directly on the plant.

The transients in Fig. 8.32 are better damped than would be expected on the basis of the assumed parameters ($a = 2$, $D = 1/2$). The most likely reason is the fact that the assumed mean delay represents a somewhat pessimistic approximation of the dynamic response of the converter; another cause is the simplification

$$\frac{1}{T_D\,s + 1} \approx \frac{1}{T_D\,s} \, ,$$

when applying the rules of the symmetrical optimum.

The operating conditions assumed for Fig. 8.32 were so selected that continuous current prevails except for one brief interval, where the load current becomes discontinuous. As mentioned before, the current is in general not always continuous; for example, if the smoothing reactor is omitted, the region of discontinuous current may cover a large portion of the operating range.

If a current control loop, the controller of which has been designed on the basis of continuous current, is operated in the region of discontinuous current, the transients become very sluggish and the quality of control deteriorates. The reason for this is that the gain of the converter is greatly reduced when the current becomes discontinuous; in addition, the load circuit is no longer characterised by the relatively large time lag T_D, as the current begins at zero in every firing interval, i.e. every 3.3 ms with a 6-pulse converter. Therefore the PI-controller, Eq. (8.44), is poorly tuned to the controlled plant.

If consistency of the dynamic response is important, which is the case for instance with servo drives on machine tools, the current controller should be readjusted during discontinuous current operation, i.e. use of an adaptive current controller becomes desirable [B90, S35, S48].

The fact that the voltage gain of the converter is reduced with discontinuous current is qualitatively explained with the help of Fig. 8.33 for a 2-pulse converter.

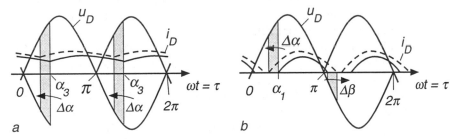

Fig. 8.33. Voltage gain of converter for continuous and discontinuous current

If with continuous current ($\alpha = \alpha_3$) the firing angle is reduced by a small amount $\Delta\alpha$, the positive voltage-time area, $\alpha < \tau < \pi$, is increased, while the negative voltage-time area, $\pi < \tau < \pi + \alpha$, is reduced. With discontinuous

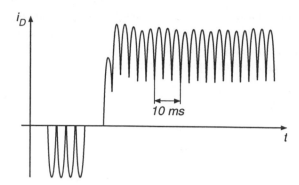

Fig. 8.34. Step response of current control loop with adaptive current controller

current ($\alpha = \alpha_1$) the situation is different; the interval, where $u_D(\tau) > 0$, is still increasing when α is reduced by $\Delta\alpha$, but so is the time, where $u_D(\tau) < 0$. Hence, the voltage gain $d\overline{u}_D/d\alpha$ is much smaller than in the case of continuous current.

The block diagram of the current control loop for discontinuous current corresponds to that shown in Fig. 8.31 but the normalised voltage gain of the converter is now considerably lower; also, the lag element with time constant T_D is now replaced by a lag having a time constant approximately equal to the firing interval $T_D \approx 3.3$ ms, which is about $1/10$ of the former value. Clearly this calls for a different design of the controller; for example, an integral controller with the transfer function

$$F'_{c,i}(s) = \frac{1}{T'_i s} \qquad \text{with} \qquad T'_i = G'_D \left(T_e + T'_D \right), \tag{8.47}$$

characterised by a small time constant T'_i, i.e. high gain, could now be expected to produce fast and well damped closed loop transients.

In order to adapt the controller to the new situation, the condition of intermittent current must be detected which requires a few additional electronic circuits but presents no major difficulty.

The realisation of an adaptive current controller is particularly attractive if the control is performed by a microprocessor because then the controller parameters can be adapted by changes in the program, i.e. by software. In Fig. 8.34 a measured oscillogram of a current transient with a 6-pulse converter is shown, where the entire control, including the firing, was performed by a microprocessor [M4, M5]. In both regions, with continuous as well as discontinuous current, satisfactory transients are now obtained. Besides the adaptive current control, the initial condition of the integral controller channel was also preset on the basis of the current reference and the reconstructed back voltage to prevent a large first current overshoot which would be objectionable on servo drives. This flexibility is an important feature of digital control by microprocessor.

9. Control of Converter-supplied DC Drives

Static converters are ideal electronic actuators for DC drives because of their practically unlimited output power and excellent controllability. The speed of response is usually adequate to handle the electromechanical transients occurring in drives. Line-commutated, phase controlled converters or, as they are also called, converters with natural commutation, are the most frequent choice for industrial applications, where a three-phase supply is available; this is due to the simplicity of the circuits requiring a minimum number of active and passive components.

For vehicle drives, where no AC catenary or independent AC supply is available or where the reactive current and the harmonics caused by a line-commutated converter would be unacceptable, it may be necessary to employ forced commutated converters having a more complex circuitry and involving higher losses; a special situation exists also with small DC servo drives, where the response of a line commutated converter may be insufficient to cope with the stringent dynamic demands and where a chopper converter supplied by a DC link and operated with a higher switching frequency is necessary.

9.1 DC Drive with Line-commutated Converter

When connecting a line-commutated converter, for example the three-phase bridge circuit in Fig. 8.18, to the armature of a DC machine (Fig. 7.2), it should be kept in mind that the converter can only operate in two quadrants of the $\overline{u_D}/\overline{i_D}$-plane as seen in Fig. 8.23. Because of $\overline{u_D} = \overline{u_a}$, $\overline{i_D} = \overline{i_a}$ it follows from Eqs. (5.6, 5.8) that, assuming constant sign of the flux Φ_e, the machine can run in both directions, but the torque is unidirectional. The current/speed cascade control scheme shown in Fig. 9.1 is therefore capable of operating in two quadrants of the torque/speed plane, hence it is called a two-quadrant drive, Fig. 9.2.

Since the converter cannot produce negative output current, there is no point in admitting a command signal calling for negative armature current; with an integrating current controller this would only cause unnecessary waiting intervals when the current reference becomes positive again ("controller wind-up"). Thus, the output signal of the speed controller should be limited at the upper as well as the lower end, $0 \leq i_{a\,Ref} \leq i_{a\,max}$.

Braking of the drive in the 2nd quadrant by regeneration is not immediately feasible; of course, there is always the possibility of non-regenerative braking with the help of external resistors.

Two-quadrant drives of this type are suitable for unidirectional loads, the torque of which contains a large component of friction, such as paper- or printing machines, calenders, also for pumps or fans; in principle, it would also be applicable to hoists without self-locking gears, but reverse torque is usually needed there too because of counter weights or for helping to accelerate in the lowering direction.

With the driving torque being proportional to the product of armature current i_a and flux Φ_e there are two options for achieving reverse torque, i.e. four-quadrant operation: reversing the armature current or the main flux. Both methods are in use, though with markedly different preference.

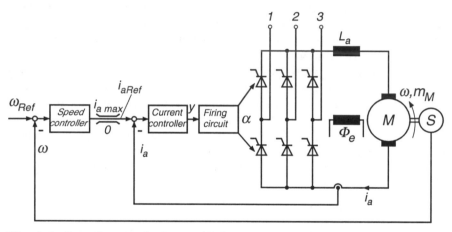

Fig. 9.1. Cascade control scheme of DC motor with two-quadrant converter

The armature can be reversed with the help of a reversing contactor or, as shown in Fig. 9.3, four thyristors, of which one diagonal pair is conducting. Since thyristors cannot be blocked electronically, armature reversal can take place only at zero armature current; hence, a switching command is only effective when the conditions

$$i_{D\,Ref} = 0, \qquad i_D = 0 \tag{9.1}$$

are simultaneously fulfilled. The first signal indicates that the superimposed speed controller calls for a reversal of torque while the second confirms that the current controller has reduced the armature current to zero, possibly after the converter has been operating temporarily as an inverter. Since there may be some uncertainty with the second condition in the face of intermittent current, it is important to wait a short interval with the converter blocked, before the switching command is actually issued; the firing pulses must be inhibited until the commutation is completed and the outgoing thyristors are blocked.

If the current transducer senses the load current at the converter side of the reversing switch (as is shown in Fig. 9.3) or with current transformers at the AC side, the current control loop is unaffected by the switching state. However, the sign in the speed control loop is reversed due to the fact that a given output current of the converter now produces reverse torque; to offset this effect, one more inversion is needed at the input of the speed controller, as is also indicated in Fig. 9.3.

Another possibility for reversing the torque of a DC motor is to reverse the flux Φ_e while leaving the direction of the armature current unchanged. This can be done by reversing the field through contactors or a bidirectional converter (of considerably reduced power than for the armature). However, since considerable energy is stored in the magnetic field, flux reversal requires much more time than inverting the armature; even when applying 3 times nominal field voltage, the reversing transient may take up to a second on large drives, during which time the armature current must be kept at zero; hence, a noticeable zero-torque interval occurs, which is detrimental for the response of the control. A motor subjected to this duty must be fully laminated; also, the voltage induced by transformer action in the armature must be considered. For these reasons field reversing is rarely applied today.

There is now a general preference for fully electronic reversal of the armature current with the help of bidirectional or dual converters. The cost of these schemes has been reduced as a result of advanced semiconductors and manufacturing techniques. The most common circuit of this type is shown in Fig. 9.4; it consists of two six-pulse bridge-converters C_1 and C_2, connected in opposition. Clearly, to exclude short circuits between line terminals, only

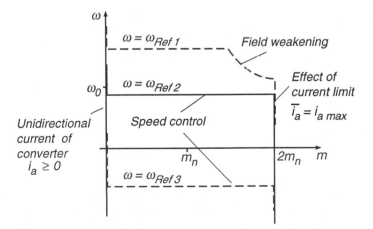

Fig. 9.2. Operating range of a two-quadrant drive

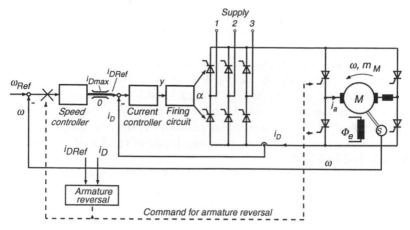

Fig. 9.3. Four quadrant operation of a DC drive by armature reversal

one of the converters can be allowed to conduct at any one time; hence only one thyristor of each pair produces conduction- and switching- losses so that the pair can be mounted on the same heat sink. However, having opposite polarity, they must be electrically insulated. At lower power ratings, complete thyristor modules are available, having the necessary interconnections built-in.

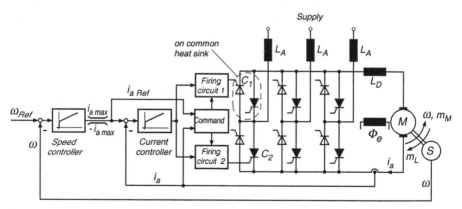

Fig. 9.4. DC drive employing a three-phase dual converter without circulating current

The current controller supplies an input signal to both firing circuits but only one is allowed to produce pulses for firing the thyristors; the selection of the active converter is performed by a command module on the basis of the polarity of the current reference $i_{a\,Ref}$ and the feedback signal i_a. As in the case of the armature reversal, the speed controller determines the need for a torque reversal, for instance $i_{a\,\text{Ref}} < 0$ and the current controller responds by lowering the armature voltage $\overline{u_D} = \overline{u_a}$ with the consequence that the armature current eventually reaches zero, $i_a = 0$. As soon as this condition holds, the firing pulses for the conducting converter C_1 are blocked by the command module, while those of the incoming converter C_2 are activated. Hence the switching conditions are

$$[i_{a\,\text{Ref}} > 0 \wedge i_a = 0] \rightarrow C_1 \text{ On}, \qquad C_2 \text{ Off},$$
$$[i_{a\,\text{Ref}} < 0 \wedge i_a = 0] \rightarrow C_1 \text{ Off}, \qquad C_2 \text{ On}.$$

Additional features must be incorporated to enhance the safety of operation and to improve the control transients. For example, a waiting interval of 2 to 5 ms may be included before firing pulses are issued to the incoming converter, to be certain that the current in the outgoing converter is really zero. Activating the firing pulses of the incoming converter before all opposing thyristors are safely blocked would cause a line-side short circuit current that cannot be suppressed by control but must be cleared by fuse links or by a breaker. On the other hand, the firing pulses for the outgoing converter must not be blocked as long as current is flowing because, should the converter operate as inverter, this could lead to a commutation failure, as explained in Fig. 8.17.

Therefore, the decision to switch from one converter to the other is of considerable consequence; in particular, the state $i_a \equiv 0$ must be ascertained safely and with a high resolution. For instance, a converter having a nominal current of 1000 A may contain two arms where at the critical time a residual current of 100 mA is flowing; if now the opposite converter is turned on, a short circuit is still likely. Sometimes two current sensors are employed on large drives, one having a linear characteristic over the full current range, to be used for feedback control, while the other serves as a zero-current detector for the command module; its characteristic would have to be linear in the low current region only. Also note that the currents through the snubber circuits may affect the current measurements, increasing the noise of the signal. Clearly, great emphasis must be given to the reliability of the sensors and control components if this circuit is to operate safely under practical conditions. Indeed, this has initially been a problem that has delayed the general use of this exceedingly simple converter circuit. In the meanwhile the problems have been overcome and reversible converters with antiparallel thyristors are produced in large numbers in the form of very compact units covering a wide power range from a few kW to about 10 MW. For the higher ratings modular designs with air- or water-cooled heat sinks are available, Sect. 11.4.

Besides the control scheme shown in Fig. 9.4 there are a number of variations; for example, only one firing circuit might be used with the output pulses being switched from one converter to the other. Also, the drive is usually extended to include field weakening as discussed in Sect. 7.3. The field winding would then be supplied from a separate unidirectional converter of smaller size.

There are other ways to further improve the performance of the control. The adaptive current controller has already been mentioned in Sect. 8.5; also, the current controller may be given a suitable initial condition, to optimise the current transient after the zero current interval, when the incoming converter is activated. The best initial condition depends on the existing current reference and the magnitude of the back voltage which is known from the firing angle existing before the switching operation. Sophisticated techniques of this type are particularly attractive if the control is realised with microprocessors, because they are then implemented by software, without increasing the complexity of the control circuitry [M4, M5, S20]. The reversing transient, shown in Fig. 8.34 was executed with a microprocessor being programmed as an adaptive current controller for continuous and discontinuous current; the variable initial condition was precalculated before firing pulses were sent to the incoming converter. The brief current pause, seen in Fig. 8.34, is typical for this type of reversing scheme; when comparing this time to the long zero-current interval required for the armature reversal by contactor, the advantages of the fully electronic solution are obvious.

Another feature, which is easily implemented with microprocessor control is the compensation of the variable gain in the speed loop which occurs in the field weakening range as was mentioned in Sect. 7.3. Further sophistication may be added by predictive digital control, all but eliminating any control transients [L28].

The zero current interval, required for safely switching the armature current from one converter in Fig. 9.4 to the other, is very short but there are applications, such as some rolling mill drives, where even 5 ms may not be tolerable and where a completely continuous transition of armature current and torque is specified. This is achieved with another dual converter circuit, shown in Fig. 9.5 a, where both two-quadrant converters are conducting simultaneously. This calls for separate transformer secondaries as well as decoupling reactors because of the instantaneous voltage differences existing between the two converters C_1, C_2. As seen from Fig. 9.5 a the armature current can assume either sign,

$$i_a = i_1 - i_2 \tag{9.2}$$

depending on which of the currents prevails. The smaller of the currents represents a circulating current which flows through the converters but bypasses the load branch,

$$\min(i_1, i_2) = i_c ; \tag{9.3}$$

it should be kept as small as possible to reduce the associated active and reactive power losses. On the other hand, the circulating current should be

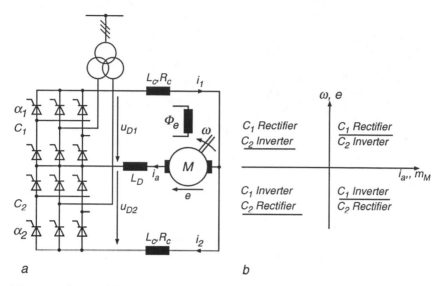

Fig. 9.5. Reversible converter with circulating current supplying DC machine

continuous so that there are no unnecessary waiting intervals delaying possible control action.

The operating states of the two converters are indicated in the four quadrants of the torque/speed- plane in Fig. 9.5 b; for $m_M > 0$, the armature- and circulating currents flow through C_1 whereas the auxiliary converter C_2 carries only the circulating current; the inverse is true for $m_M < 0$. To keep the circulating current at a sufficiently low level, such as 10% of nominal current, the mean voltage of the auxiliary converter must closely track the mean voltage of the main converter which is determined by the armature of the motor. If this condition is maintained all the time, both converters are active and ready to accept the motor current in a continuous transient, without any waiting interval. This two-variable control is effected with the help of the two firing circuits.

The auxiliary converter must be controlled to satisfy the condition

$$\overline{u_{D1}} + \overline{u_{D2}} = R_c(\overline{i_1} + \overline{i_2}) \approx 0 , \tag{9.4}$$

i.e. the sum of the mean voltages should be approximately zero; however, this is not true for the instantaneous voltages $u_{D1}(\tau)$, $u_{D2}(\tau)$, because one of the converters operates as rectifier, the other as inverter. The two voltages and their sum are plotted in Fig. 9.6 for three different conditions, neglecting commutation and internal voltage drops. Clearly at $\alpha_1 = \alpha_2 = \frac{\pi}{2}$ the sum of the instantaneous voltages assumes very large values eventhough the sum of the mean values is zero. To prevent these large alternating voltages from causing excessive currents, the circulating current reactors L_C seen in Fig. 9.5 a are required. Since there is always one which carries the small continuous

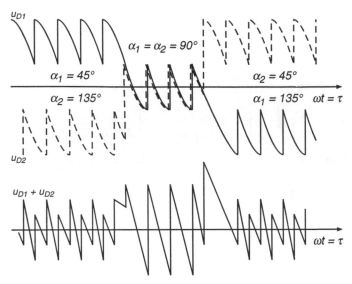

Fig. 9.6. Voltages in reversible converter with circulating current

circulating current, these reactors need not be entirely linear; in fact, by allowing the one carrying the armature current to saturate, their physical size is reduced. For smoothing the armature current i_a, a separate filter reactor L_D may be included in the armature circuit.

For obtaining the desired current distribution for the respective main- and auxiliary-converters, given a certain load current i_a, one might think first of an open loop control scheme, where both firing angles are controlled jointly according to $\alpha_2 \approx \pi - \alpha_1$, which approximately fulfills condition (9.4). However, the result would be far from satisfactory since the characteristics $\alpha(y)$ of the two firing circuits are likely to be somewhat different; also the current- dependent voltage drops would have to be taken into account. Thus the circulating current would either be too large causing unnecessary losses in the transformer, converters and reactors, or it would become discontinuous, preventing fast response, should the need for a reversal of armature current arise. The closed loop control scheme shown in Fig. 9.7 a promises a very simple and effective solution to this problem; it is proven in numerous applications over many years [K24].

The idea behind this scheme is as follows, Fig. 9.7: Based on the current reference $i_{a\,Ref}$ prescribed by the speed controller, separate current references $i_{1\,Ref}$, $i_{2\,Ref}$ are derived for each of the converters. This is done by an electronic function generator, the characteristics of which are shown in Fig. 9.7 b; they are so chosen that the conditions

$$i_{1\,Ref} - i_{2\,Ref} = i_{a\,Ref} \,,$$
$$\min(i_{1\,Ref}, i_{2\,Ref}) = i_{c\,Ref} = \text{const.}$$

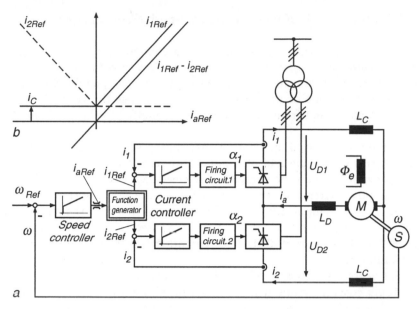

Fig. 9.7. Control of reversible converter with controlled circulating current

are fulfilled, with the circulating current reference assuring continuous current flow. Thus the task of controlling armature current and circulating current is assigned to two separate current control loops as described in Sect. 8.5. The circuit shown in Fig. 9.7 a can also be extended to include field weakening beyond base speed.

With this control method the drive can operate in all quadrants of the torque/speed plane allowing smooth transitions between the quadrants. However, this is achieved with considerable additional cost and complexity (transformer, reactors). Some savings may be possible, for example with continuous rolling mills, when the converter for reverse torque is only used for occasional braking and can be designed for a lower current rating; more substantial economies are made by turning to the reversible converter without circulating current that was discussed before. This circuit permits also the use of the full rectifier voltage ($\alpha = 0$), whereas with the circuit in Fig. 9.5, the rectifier voltage must be limited to the maximum value that can be supported by the inverter operating at maximum firing angle $\alpha_{\max}$. With today's state of control the loss in dynamic performance caused by the discontinuity at current zero is very slight, rendering the circuit in Fig. 9.4 the best solution for most applications.

9.2 DC Drives with Force-commutated Converters

The converter circuits dealt with so far have in common that they are fed from single or three-phase alternating voltages; this is a prerequisite for phase control employing natural commutation. As soon as a valve, having forward bias voltage, is fired, one of the supply voltages assumes the task of commutating the load current from the previously conducting to the newly fired valve.

In a number of potential applications AC supply voltages are not available, so that this simple method of commutation is not feasible. Because a conducting thyristor cannot be turned off by electronic control, special provisions in the power circuit are then required to ensure that the thyristor to be fired has, at least temporarily, forward bias voltage and that the outgoing thyristor becomes nonconducting and is briefly exposed to reverse voltage as a condition for blocking. A multitude of circuits exists which perform this "forced commutation".

In recent years thyristors have been developed that can also be switched off by suitable control signals while conducting (Gate-Turn-Off thyristors, GTO); this is a major improvement, even though the current gain for turn-off is very low, requiring a short high pulse of inverse control current. These devices have in the last 10 years been developed to the point, where they can replace normal thyristors with several kV and kA rating in high power applications, such as traction drives.

The most common switching device at lower power permitting electronic turn- off is the bipolar transistor. Since it can only conduct, when it is activated by a base current, it may be switched on and off at will, as long as the switching losses are not too high and the transistor is protected against over-voltages caused, for example, by an inductive load. Bipolar transistors in cascaded form (Darlington circuit) are now available for voltages exceeding 1ooo V and currents of several hundred A, so that converters for medium size drives up to about 500 kW can been built, but the design of the base control circuits remains cumbersome.

Another area where promising developments are under way are field effect transistors (MOSFET) for low power applications. These transistors are characterised by very short switching time and, consequently, high switching frequency beyond the audible range; being inherently voltage controlled, they require practically no steady state gate current, similar to vacuum tubes. However, caused by the relatively large gate capacity there is a substantial charging current which must be supplied by the control device for switching; also, the on- state resistance is larger than with bipolar transistors.

More recently Insulated Gate Bipolar Transistors (IGBT) have been developed as high power devices, combining the high current carrying capacity of bipolar transistors with the simple control of field effect transistors as well as high switching speed. IGBT's are presently the most promising switching devices for higher power applications, reaching into the kV, kA region. A further possibility, but still at an early stage of development, are MOSFET-controlled thyristors (MCT) but their potential cannot yet be assessed.

Clearly, the work on new and improved high power electronic switching devices is continuing at a rapid pace and a state of consolidation is not in sight. This strongly affects the development of the circuitry and control methods for converter equipment. In this section it is shown with the example of a simple force-commutated chopper circuits for DC drives, how the new devices may result in better dynamic performance, simplifications and, possibly, cost reductions.

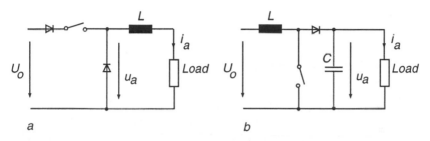

Fig. 9.8. Principle of single phase chopper circuits
(a) Step-down chopper (b) Step-up chopper

In Fig. 9.8 a,b two basic circuits for DC/DC choppers are seen for generating from a constant direct bus voltage with low losses a controllable direct voltage for supplying a DC load. The electronic switches in conjunction with an inductive energy storage are operated with a controllable On/Off duty cycle to produce a desired mean voltage $\overline{u_a}$ across the output terminals; the circuits could be used as single quadrant supply for the armature of a DC motor. The electronic switch is carrying only forward current, reverse blocking is not necessary with the diodes included.

As an example that is mainly of historical interest, a typical converter circuit of the type of Fig. 9.8 a, employing forced commutation with an ordinary thyristor is depicted in Fig. 9.9. Its purpose is to convert the constant direct voltage U_0 to a lower adjustable voltage $\overline{u_a} < U_0$ for controlling the speed of a DC series wound motor. This problem arises with vehicles, supplied through a DC rail or catenary or from a storage battery. As the thyristor T_1 which corresponds to the switch in Fig. 9.8 a cannot be turned off by electronic control, a resonant circuit with the auxiliary thyristor T_2 is provided with the purpose of creating a zero intersection of the current i_{T1}, thus facilitating the blocking of thyristor T_1. When using a GTO which can also be switched off electronically, the same effect could be achieved more directly.

The DC/DC converter operates as a chopper, which, by means of an electronic switch, connects the inductive load with the parallel shunting diode D_1 periodically to the constant supply voltage. By varying the duty cycle, the

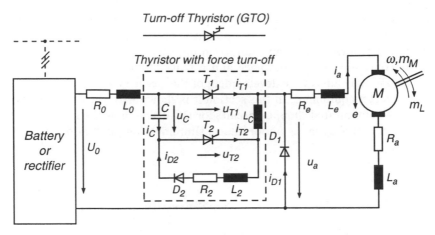

Fig. 9.9. DC/DC chopper supplying a series-wound motor

mean voltage $\overline{u_a}$ is changed. The switching frequency is chosen sufficiently high (> 100 Hz) so that a relatively smooth continuous current i_a flows in the load circuit. The converter shown in Fig. 9.9 is of the single quadrant variety, which is adequate for a series motor but with additional components it could be extended for two- or four-quadrant operation as well. The circuit should only be considered as an example; there are many different versions having similar properties. The operation of the DC/DC converter will be explained in steady-state where all variables are periodic functions with period T. The curves in Fig. 9.10 have been obtained by a detailed simulation of the circuit on a digital computer.

The mode of operation is briefly as follows: The main thyristor T_1 is periodically fired at time $t_1 + \nu T$, $\nu = 0, 1, \ldots$ connecting the supply voltage U_0 to the inductive load circuit, $u_a(t) = U_0$, at the same time blocking the shunting diode D_1 which had carried the continuous load current i_a. This causes the positively charged capacitor, $u_c(t_1) > 0$, to be discharged through the resonant commutation circuit consisting of L_c, L_2, R_2, D_2; after the first half wave this transient is interrupted by D_2 as the current is about to change its sign, leaving the capacitor with negative voltage, $u_c(t) < 0$. This voltage now serves as forward bias voltage for the auxiliary thyristor T_2 while the main thyristor T_1 is conducting; as T_1 cannot be turned off by a control signal, it must be blocked by temporarily cutting off the current i_{T1} and applying a negative bias voltage to the thyristor. This is achieved by firing the auxiliary thyristor T_2 at time t_2, thus connecting the negatively charged capacitor across T_1 and temporarily increasing the output voltage u_a. The load current i_a transfers quickly into the auxiliary thyristor, at the same time recharging the capacitor at a rate proportional to the load current

$$C \frac{du_c}{dt} = i_c = i_a \ . \tag{9.5}$$

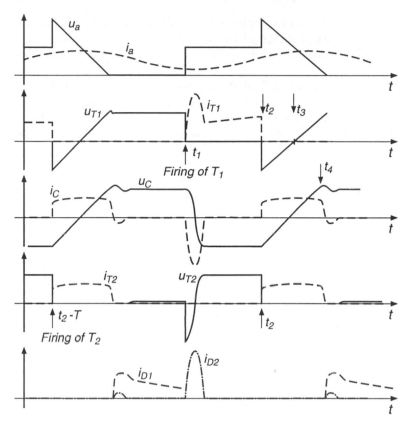

Fig. 9.10. Simulation results for DC/DC converter circuit

The bias voltage across thyristor T_1 eventually becomes positive at time t_3, when u_c passes through zero; in the interval $t_3 - t_2$ the recovery of T_1 must have been completed. The recharging of the capacitor continues until $u_c = U_0$, when the freewheeling diode D_1 takes over the load current and T_2 reverts to blocking condition. The electronic switch then remains in this state, with both thyristors blocked and $u_c > 0$, until the next switching cycle begins. In Fig. 9.10 the commutation interval has been enlarged for better visibility.

Clearly it is critical for the functioning of the circuit that the alternating voltage of the commutating capacitor is of sufficient magnitude to ensure an adequate hold-off interval $t_3 - t_2$ when the main thyristor is reversely biased. If this time becomes too short because the charge on the capacitor was insufficient or there is excessive load current, T_1 will refire and a short circuit condition develops which can only be cleared by a fuse link or a breaker. In this respect a commutation failure of this circuit is more serious than with a line-commutated converter where the situation could be resolved by advancing the next firing instant. This danger of losing control over the converter exists with all thyristor circuits employing forced commutation; it can only be avoided through careful design of the circuit and by having a fast current control which prevents excessive load current before it approaches the commutation limit.

Another detail is seen in Fig. 9.10 at time t_4 where the capacitor, having been charged beyond U_0 due to the effect of the source inductance L_0, returns some of its charge to the battery through diode D_2; since it is desirable to have the capacitor well charged for added commutation capability, this loss of charge should be prevented by using instead of D_2 another thyristor that is fired together with T_1.

The switching sequence is repeated at a frequency f between 50 Hz and perhaps 1 kHz, depending on the type and power rating of the thyristors. The control effect is achieved by pulse-width modulation; for example, if thyristor T_1 is fired at the frequency f with a fixed phase, the On-Off ratio is altered by advancing or retarding the firing instant t_2 of thyristor T_2. When neglecting the commutating transients the voltage u_a can be approximated by a periodical square wave having constant amplitude U_0 and variable pulse-width $t_2 - t_1$, so that the mean output voltage may be continuously changed,

$$0.1 < \frac{\overline{u_a}}{U_0} = \frac{t_2 - t_1}{T} < 0.9 ; \tag{9.6}$$

this avoids the large power losses that would occur with a series resistor. Of course, all components in the converter circuit involve some losses which have to be examined very carefully at the design stage; still, the over-all efficiency of a typical DC/DC converter may be better than 95%. The limits $(t_2 - t_1) \rightarrow 0$ and $(t_2 - t_1) \rightarrow T$ cannot be reached by continuous pulse-width control because some time is needed for the commutating transients.

From this simplified description it is apparent that the analysis and, consequently, the design of a converter with forced commutation is much more complex than that of a line-commutated converter. The voltages and currents, for example those directly affecting the thyristors, can only be determined by tedious hand calculations in the successive intervals, by measurement on a laboratory unit or by simulating the complete circuit on a digital computer. Very efficient programs have been developed for this purpose which require only a topological description of the circuit; from this information the differential equations are automatically generated. With given initial conditions, the numerical integration proceeds until either a control input occurs (e.g. firing of a thyristor) or one of the components reaches a discontinuity, such as a diode blocking, at which time new equations are defined (or new parameters entered into the former equations) to continue the integration, starting from the boundary conditions reached before [E6]. In complex converter circuits there may be a multitude of intervals, corresponding to different switching states of the circuit, before the operation repeats itself. Computations of this type are time consuming because the integration must be performed in small time steps as dictated by the rapid transients, for example in the commutating branches; nevertheless, digital simulation is an indispensible tool at the design stage. Naturally, after the converter design is completed, much simpler models can be used for designing the control system. An extensive specialised literature exists on DC/DC converters describing numerous details, particularly with regard to commutation circuits [3, 48, 51].

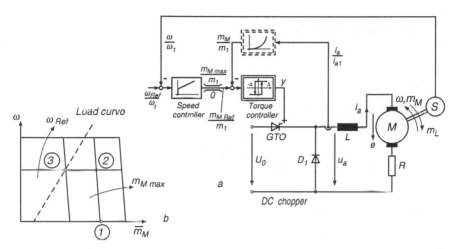

Fig. 9.11. (a) Simplified drive control scheme of drive with DC/DC converter, **(b)** steady-state characteristics

The DC/DC converter, like the line-commutated converter, is a power amplifier with fast response having limited overload capacity; hence it calls for effective protection which is best provided by closed loop control. With vehicle drives which are often coupled to a large load inertia this is of particular importance because acceleration should often take place at a chosen current limit. Therefore cascade control having an inner current loop is well suited also to DC/DC converters.

As was shown in Sect. 8.5, very simple dynamic models are quite adequate for the design of the control system, provided the innermost loop contains – besides the switching converter – a simple control plant with low-pass characteristics. When neglecting the commutation transients and assuming continuous load current i_a, the converter in Fig. 9.9 looks from the load side like an impressed voltage source, where u_a alternates between U_0 and zero with variable duty cycle. Hence the circuit can be represented by a mechanical switch as seen in Fig. 9.11 a that may be operated by an On-Off-controller having a narrow hysteresis loop of width 2Δ. With the real converter, the controller would issue firing pulses to the main- and auxiliary thyristors.

The block diagram of a series wound motor, Fig. 6.3, is quite nonlinear, but with an On-Off-controller for the inner current control loop no problems are encountered. The system results in a quasi-linear inner closed loop

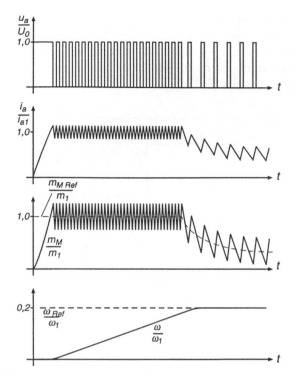

Fig. 9.12. Simulated starting transient of drive with DC/DC converter

with approximately unity gain. In the superimposed speed control loop the mechanical integration is added (neglecting friction), however with variable gain, because the torque of a series motor is determined by $\Phi_e i_a \approx i_a^2$. The result would be different speed transients depending on the operating point of the motor. To avoid this effect, it may be advantageous to add a nonlinear function generator producing a torque signal,

$$\frac{m_M}{m_1} = f\left(\frac{i_a}{i_{a1}}\right) , \qquad i_a \geq 0 , \tag{9.7}$$

and to form an inner torque control loop instead of a current loop; no particular accuracy of the function generator is required. The result is seen in the block diagram Fig. 9.11 a, where all nonlinearities are now contained in the inner loop. The pulse-width modulated waveform is generated by the closed torque control loop without the need for an external clock; the frequency can be adjusted by the choice of the hysteresis width. Again, with a simple On/Off-controller, the switching frequency of the converter is not constant but exhibits a maximum in the centre of the control range [38], but this causes no harm as long as the minimum frequency is sufficiently high. However, when a constant switching frequency is specified to avoid interference with signalling circuits laid along the railway tracks, a constant frequency pulse-width modulator is preferable. If the vehicle contains several independently fed drive motors, the firing of the converters may be synchronised out of phase to increase the ripple frequency of the current in the supply line.

Figure 9.12 shows a simulated starting transient with speed-dependent load torque; the operating point in Fig.9.11 b moves from 1 via 2 to 3. The inertia of the drive was chosen very small to clarify the details of the switching operation; with an actual traction drive, the acceleration phase may last many seconds while the switching takes place at several hundred Hz. When selecting the parameters of the controllers, simple design methods are again applicable, for example as discussed in [38]. As mentioned in Sect. 8.5, On/Off control can be treated more accurately as a nonlinear discrete control system which however would involve considerable complexity; in view of the good results with the approximate methods this is normally not justified.

As indicated in Chap. 6 the use of series connected motors is restricted to only a few applications; for high dynamic performance drives, such as reversible servo drives, DC motors with separate excitation or permanent magnets are normally preferred; they require converters with four quadrant capability in the i_a/u_a-plane.

The bridge type converter shown in Fig. 9.13 a is connected to the armature circuit of a DC motor; it may be supplied with constant voltage u_D from a DC bus or a battery. The converter contains four electronic switches where two in each half-bridge are drawn in the form of a transfer switch (at the same time excluding accidental short circuits of the DC bus); the diodes which can be part of the electronic switches allow an inductive load current to continue during the short protective intervals, when all contacts are open (similar to the red-light-overlap on a signal crossing).

By assigning logic symbols S_1, S_2 to the otherwise ideally assumed switches, the voltage equation of the load circuit is

$$L_a \frac{i_a}{dt} + R_a i_a + e = u_a , \tag{9.8}$$

where, depending on the switching state

$$u_a = \frac{1}{2}(S_1 - S_2)\, u_D \text{ and } i_D = \frac{1}{2}(S_1 - S_2)\, i_a \tag{9.9}$$

holds.

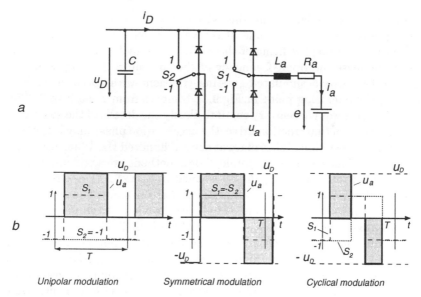

Fig. 9.13. Four-quadrant DC/DC converter with inductive load,
(a) Circuit, (b)Different modulation patterns

The pulse-width-modulation (PWM) of the converter at the frequency $f = 1/T$ can follow different switching strategies, as illustrated in Fig. 9.13 b with the output voltage u_a during a switching period:

- Unipolar modulation
  One of the switches is assumed to be stationary, e. g. S_2 = -1 = const., whereas the other half-bridge is pulse-width-modulated, $S_2 = \pm 1$, so that the output voltage u_a assumes the values u_D or zero; the same applies with $S_1 = -1$ = const. for negative output voltages.
- Symmetrical modulation
  With this modulation pattern the switches are operated in diagonal pairs, $S_1 = - S_2$, so that the short circuit interval is omitted and the output voltage alternates between the values u_D and $-u_D$. During the unavoidable (but in Fig. 9.13 neglected) protective intervals the diodes are carrying the load current.
- Cyclical modulation
  With this modulation scheme the two transfer switches are operated sequentially, so that the output is alternatively short circuited at the upper or lower supply bus. Hence the output voltage u_a assumes a ternary waveform, $u_a = u_D, 0, -u_D$. Whereas with symmetrical switching only the mean of the output voltage can be controlled in steady state, the cyclical modulation offers an additional degree of freedom that may for instance be used for eliminating harmonics of the output voltage u_a.

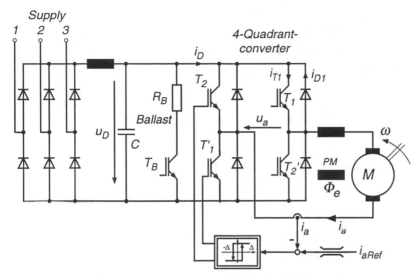

Fig. 9.14. DC Servo drive with voltage source DC/DC converter

A four-quadrant converter with IGBT-switches is depicted in Fig. 9.14 as frequently used for DC servo drives; protective circuitry is again omitted. For simplicity an On/Off device is drawn for current control but a linear current controller with constant frequency PWM would normally be preferred.

A diode rectifier followed by a smoothing filter, whose capacitor C absorbs also the modulation-induced ripple components of the link current i_D serves as the supply of the DC link with constant voltage u_D. For instance, when rapidly braking the drive, power released from the kinetic energy flows back into the DC-link causing negative current i_D and, because of the unidirectional line-side rectifiers, could result in an overcharge of the capacitor; this is prevented by dissipating the energy in a resistive ballast circuit that can also be pulse-width-modulated, depending on the link voltage. In view of the losses this is only practical with small drives or when it happens only occasionally; otherwise a reversible line-side supply (an active front-end converter) is preferable as will be shown in Fig. 9.18 and further discussed in Sect. 13.2.

Some of the steady state waveforms in a converter like the one in Fig. 9.14 are indicated in Fig. 9.15, showing the output voltage u_a alternating between u_D and $-u_D$ and the alternating current components of i_D, which must be absorbed by the capacitor. The control can be arranged as before; an inner current loop controlling the converter via a pulse-width modulator is important for safe operation. The current controller in Fig. 9.14 is again drawn as an On/Off switch, but this is only an illustrative example.

The typical response time of the current loop, employing a switched transistor converter in combination with a DC disk motor, is 1 or 2 ms. For many

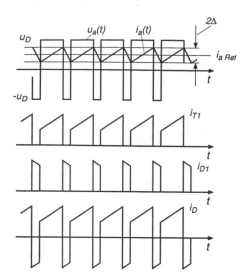

Fig. 9.15. Waveforms of DC/DC converter with symmetrical modulation

applications this justifies the assumption that the current control loop acts as controllable current source having instantaneous response. Current limit is achieved by limiting the current reference produced by the superimposed speed controller. The next higher level of control could be a position control loop as shown in Fig. 15.9, where the response may be further improved by feed-forward signals from a reference generator.

Transistor converters have the important advantage that they can be switched at frequencies > 5 kHz, thus enlarging the control bandwidth as compared to line-commutated converters. With field effect transistors or IGBT's, the frequency can even be increased beyond the audible threshold > 16 kHz, so that the drive is no longer emitting objectionable acoustic noise.

The four quadrant converter shown in Fig. 9.14 is also suitable for producing AC in the form of pulse-with-modulated voltages and currents, provided its frequency is sufficiently below the switching frequency of the converter; this is seen in Fig. 9.16 with the example of approximately sinusoidal output current i_a in a passive ($e = 0$) inductive load circuit. With the assumed On/Off controller the modulation is symmetrical and the switching frequency varies but this could be remedied with a PWM modulator and fixed clock frequency. An inspection of the curves reveals that the converter passes cyclically though all quadrants of the i_a/u_a - plane, where energy is alternatively flowing from or into the DC link; naturally, in view of the losses, the reverse flow is reduced, as seen from the trace of i_D. We will meet this kind of converter again in connection with AC drives.

DC servo motors with permanent magnets are usually built in one of two forms:

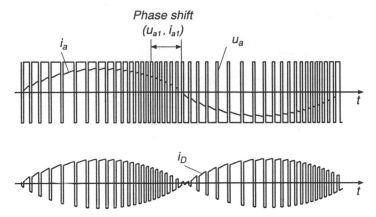

Fig. 9.16. Waveforms of symmetrically modulated four-quadrant converter with AC output

- Cylindrical motors with a slim rotor hence low inertia, but otherwise conventional design and
- Disk motors with axial magnetic field and radial armature winding, which is directly printed on a fiber reinforced disk; the brushes are sliding on the printed conductors of the commutator. These motors are charakterised by a very small armature time constant and low inertia; typical data are given in Fig. 14.1.

Because of the progress with controlled AC drives, servo motors are today mostly synchronous motors with permanent magnet excitation, Sect. 14.1.

When discussing phase controlled converters in Chap. 8, the problem of line-side interactions by harmonics and reactive current was mentioned. This applies also to uncontrolled rectifiers, such as in Fig. 9.14, particularly when a simple capacitive filter without smoothing choke is employed which causes line currents in the form of short spikes at the voltage peaks. With the growing use of power electronic equipment these effects are now receiving increased attention in technical standards and regulations.

With force controlled converters the reactive line currents can be greatly reduced while the harmonics are shifted to a higher frequency range, where it is easier to suppress them by filters; at the same time the problem of reverse power flow can be solved.

This is shown with the schematic diagram in Fig. 9.17 of a single phase converter which is of interest, apart from low power drives, on AC traction drives with intermediate DC link converter (an extension to three phases is discussed in Sect. 13.2). The circuit is essentially an inverted four-quadrant converter in Fig. 9.13, where the inductive impedance L_N is now represented by the leakage of the line-side transformer. The simplified equations are

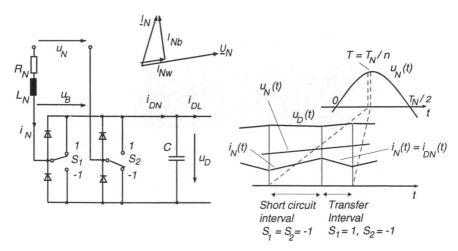

Fig. 9.17. Principle of a single phase line-fed four-quadrant converter for supplying a DC link

$$L_N \frac{i_N}{dt} + R_N i_N = u_N - u_B , \qquad (9.10)$$

and

$$C \frac{u_D}{dt} = i_{DN} - i_{DL} , \qquad (9.11)$$

where i_{DL} is the load-side link current; furthermore

$$u_B = \frac{1}{2}(S_1 - S_2)\, u_D \text{ und } i_{DN} = \frac{1}{2}(S_1 - S_2)\, i_N \qquad (9.12)$$

holds. Approximately sinusoidal line current i_N can only be achieved if the DC link, usually operated at constant voltage u_D, can also be supplied with energy when the line voltage is much lower than u_D, i.e. near the zero intersections of u_N; hence it is necessary, as shown in Fig. 9.17 b, to create zero intervals of the voltage u_B for storing in the line-side inductance energy that is subsequently transferred in the form of a current pulse to the link capacitor. Thus it is necessary to employ a unipolar or cyclical modulation, where the converter bridge is temporarily short circuited at a DC bus.

The switching frequency of the converter is best chosen as an integer multiple n of the line frequency to avoid in steady state subharmonics of the line current; with a 50 Hz supply and a clock frequency for instance of 2 kHz, $n = 40$ switching cycles would fit into one line voltage period, resulting in a fairly smooth waveform of the current.

A possible control scheme is shown in Fig. 9.18, where a voltage control loop for u_D is superimposed on an inner current loop for i_N. The current reference $i_{N\,\text{Ref}}$ is derived from the line voltage for obtaining a sinusoidal line current reference, the magnitude of which is determined by the voltage controller. When shifting the current reference by t_0 with respect to the line voltage u_N, a reactive component of the line current results.

Caused by the single phase supply, periodic fluctuations of the link voltage u_D at twice line frequency cannot be avoided but are greatly diminished by inserting a resonant circuit tuned to double line frequency, as is indicated in Fig. 9.18; this is the usual practice on traction drives. The load-side DC link current i_{DL} acts as a disturbance to the voltage control, its effect may be reduced by feed-forward to the current reference.

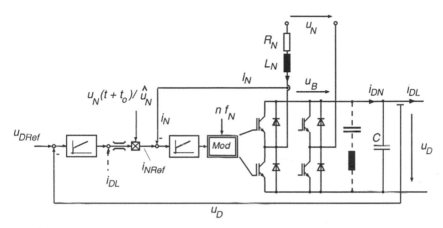

Fig. 9.18. Control of a single phase line-fed four-quadrant converter

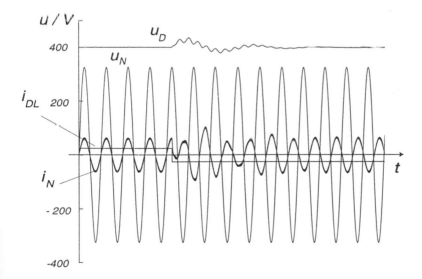

Fig. 9.19. Simulated load reversal of a line-fed four-quadrant converter

A simulated load reversal is plotted in Fig. 9.19, initiated by inverting the current i_{DL}, which results in a phase reversal of the line current in steady state, i. e. feeding power back to the line [I3]. The resonant circuit shown in Fig. 9.18 is included in the simulation model, thus reducing the harmonics content of the link voltage.

10. Symmetrical Three–Phase AC Machines

The asynchronous or induction motor is the most widely used electrical drive motor; its invention at the end of the last century has given a strong impetus to the transition from DC to AC in the field of generation, transmission and distribution of electrical energy. Its main advantage is the elimination of all sliding contacts, resulting in an exceedingly simple and rugged construction. Induction machines are built in a variety of designs with ratings from a few watts to many megawatts.

Unfortunately, the speed of line- fed induction motors cannot be continuously varied without additional equipment or without incurring heavy power losses. Even though the problems of efficiently controlling the speed of induction motors have been investigated for decades, all solutions realisable until some years ago were unsatisfactory with regard to complexity, efficiency, dynamic performance or cost. It is only due to the progress of semiconductor technology in the last 30 years that suitable static frequency converters can now be built at acceptable cost, making the induction machine a versatile adjustable speed drive for many applications.

The substitution of the mechanical commutator, by a power electronic, solid state converter reduces the AC machine solely to the task of electromechanical energy conversion, resulting in much higher power density per volume or mass. This is exemplified by two main line traction motors, the outlines of which are shown in Fig. 10.1, together with some characteristic data. The motor at left is a proven design, operated for many years in Intercity locomotives; it is a single phase series-wound commutator motor as mentioned in Chap. 6, voltage controlled by transformer tap changer. At the right is a modern AC induction motor, fed and controlled by a GTO voltage source converter. Clearly, the new motor is superior in every respect; it is also suitable for high dynamic performance torque control, thus allowing full utilisation of the adhesive frictional forces between wheel and rail. As a result, a locomotive containing four of the new motors can produce a similar traction force as the former much heavier locomotive with six single phase motors, the torque of which is pulsating at double the supply frequency.

The theory of the induction motor under dynamic conditions is somewhat involved because of the rotating magnetic fields, the spatial relationships of which depend on speed and load; it can be treated here only in simplified form. On the other hand, the equivalent circuits, derived for steady state operation with sinusoidal voltages and currents, are inadequate when dealing with transients or when the motor is supplied from a switching converter. In the following, the steady state condition will be treated as a special case of the dynamic case.

The mathematical model to be used is tailored to the needs of controlled drives. It incorporates most of the qualitative features of an actual motor but would not, of course, be accurate enough for the purpose of designing the machine.

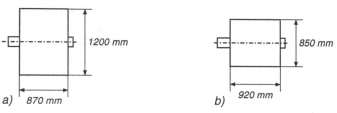

Series-wound commutator motor		Three-phase AC - induction motor
1230 kW, 7730 Nm at 1520 min^{-1}	Cont. rating	1428 kW, 9155 Nm at 1490 min^{-1}
8530 Nm	Torque (5 min)	11600 Nm
1600 min^{-1}	Max. speed	4200 min^{-1}
3550 kg	Mass	2660 kg
120 kg m^2	Inertia	22 kg m^2

Fig. 10.1. Comparison of traction motors (Siemens).
(a) Single phase series-wound AC commutator motor,
(b) Three-phase AC induction motor

10.1 Mathematical Model of a General AC Machine

It is assumed that the stator S of the machine is a hollow iron cylinder with circular cross section, containing a concentric rotor R so that a narrow airgap of constant radial length h exists between the smooth cylindrical surfaces to which symmetrical three- phase windings of negligible height are assumed to be attached; their spatial ampereturns distributions may be thought of as being produced by suitably placed thin strands of axial conductors or conductive sheets. Both neutrals of the star-connected windings are isolated, the terminals of the rotor winding are either connected to sliprings or short circuited internally. N_S, N_R are the numbers of full pitch turns in each phase winding.The permeability of the fully laminated stator and rotor iron is assumed to be infinite; saturation, iron losses, end-windings and slot-effects are ignored. The following discussion applies to two-pole motors; with multi-pole machines the synchronous speed is reduced correspondingly. This effect is exploited with pole-changing windings, however at some loss in utilisation of the motor.

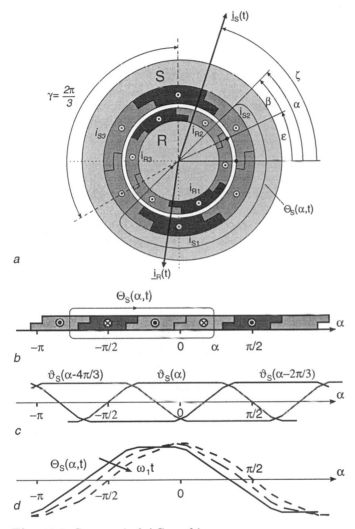

Fig. 10.2. Symmetrical AC machine
(a) Cross section with two-layer airgap windings,
(b) Unwrapped stator windings,
(c) Ampereturns distribution of stator windings,
(d) Travelling ampereturns wave caused by three-phase currents

The variables are defined in Fig. 10.2 a. α is the angular coordinate in the stator with reference to the axis of stator winding 1 which is indicated by its central turn. The centres of the identical windings 2 and 3 are positioned at $\alpha = \gamma = 120°$ and $\alpha = 2\gamma = 240°$ respectively. Corresponding definitions hold for the rotor windings, where β is the angular coordinate, again with reference to the centre of winding 1. $\varepsilon(t)$ is the angle of rotation of the rotor, measured in the stator frame of coordinates, $\omega(t) = d\varepsilon/dt$ is the instantaneous angular velocity of the rotor.

The magnetic field in the airgap of the machine has radial direction because of the smooth parallel stator and rotor surfaces and the postulated infinite permeability of the iron; since end-effects are neglected, we are dealing with a two-dimensional magnetic field problem.

The three stator currents $i_{S1}(t)$, $i_{S2}(t)$, $i_{S3}(t)$ may exhibit any waveforms; for reasons of symmetry all currents are retained throughout the calculations, even though one of the currents is redundant because, due to the isolated neutral,

$$i_{S1}(t) + i_{S2}(t) + i_{S3}(t) = 0 \tag{10.1}$$

is valid at any instant, i.e. the currents are balanced; this applies also to Δ - connected motors.

With these simplifications and definitions, the radial ampereturns wave of the stator windings consists of three spatially sinusoidal, stationary terms, which are modulated by the currents

$$\Theta_S(\alpha,\,t) = N_S\left[i_{S1}(t)\,\vartheta(\alpha) + i_{S2}(t)\,\vartheta(\alpha - \gamma) + i_{S3}(t)\,\vartheta(\alpha - 2\gamma)\right],\gamma = \frac{2\,\pi}{3}\,. \tag{10.2}$$

According to Fig. 10.2, $\Theta_S(\alpha, t)$ corresponds to the stator ampereturns enclosed by a radial magnetic field line crossing the motor at angle α; because of the assumed high permeability of the iron these ampere turns appear as magnetomotive forces at the two airgap crossings. The nondimensional spatial function $-1 \leq \vartheta(\alpha) \leq 1$ in Fig. 10.2 c characterises the ampereturns distribution of one phase of the stator winding, it could be written as a Fourier series. When the windings are fed with alternating currents, each term in Eq. (10.2) oscillates as a standing wave, fixed in space; their superposition results in a travelling wave $\Theta_S(\alpha,\,t)$, as seen in Fig. 10.2 d. The spatial harmonics may be diminished by reduced pitch and skew of the windings, resulting in $\vartheta(\alpha) = \cos\alpha$.

By introducing complex notation,

$$\cos\alpha = \tfrac{1}{2}\left(e^{j\alpha} + e^{-j\alpha}\right), \quad \text{etc.,} \tag{10.3}$$

and rearranging, Eq. (10.2) becomes

$$\Theta_S(\alpha,\,t) = \tfrac{1}{2}\,N_S\left[\underline{i}_S(t)\,e^{-j\alpha} + \underline{i}_S^*(t)\,e^{j\alpha}\right]\,,\ \text{real}\,, \tag{10.4}$$

where

$$i_S(t) = i_{S1}(t) + i_{S2}(t)\, e^{j\gamma} + i_{S3}(t)\, e^{j\,2\,\gamma} \tag{10.5}$$

is a time-dependent current vector in the complex plane perpendicular to the motor axis and

$$i_S^*(t) = i_{S1}(t) + i_{S2}(t)\, e^{-j\gamma} + i_{S3}(t)\, e^{-j\,2\,\gamma} \tag{10.6}$$

is the corresponding conjugate complex vector. The ampereturns $\Theta_S(\alpha, t)$ are of course real, being a measurable physical quantity depending on α and t.

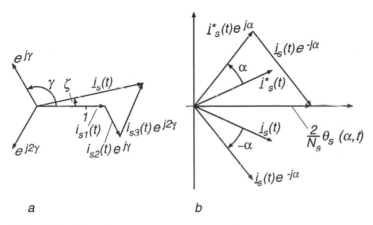

a   b

Fig. 10.3. Complex current vector

The construction of the current vector $i_S(t)$ is shown in Fig. 10.3 for assumed values of $i_{S1} > 0$, $i_{S2}, i_{S3} < 0$. Magnitude and angle each vary with time according to

$$i_S(t) = i_S(t)\, e^{j\,\zeta(t)} \,. \tag{10.7}$$

The current vector determines the instantaneous magnitude and angular position of the peak of the sinusoidally distributed ampereturns wave produced by the three spatially displaced stator windings. By combining Eqs. (10.4, 10.7) the ampereturns may be expressed as a travelling wave, the peak of which follows the angle $\alpha = \zeta(t)$ of the current vector,

$$\Theta_S(\alpha,\, t) = N_S i_S(t)\, \cos(\zeta(t) - \alpha) \,. \tag{10.8}$$

If the stator currents are sinusoidal and form a symmetrical three phase system, the ampereturns wave revolves with constant magnitude and velocity $\omega_1 = d\zeta/dt$ in the airgap of the motor.

Since the complex vectors defined in Eqs. (10.5 – 10.8) describe the spatial distribution of a magnetic field in a plane perpendicular to the motor axis, they are also called space-vectors or space-phasors [N7, S73, 31]. However, they are not to be confused with the constant complex phasors for describing steady state sinusoidal alternating quantities.

The same reasoning as before holds for the ampereturns wave produced by the moving three-phase rotor windings,

$$\Theta_R(\beta, t) = N_R \left[i_{R1}(t) \cos\beta + i_{R2}(t) \cos(\beta - \gamma) + i_{R3}(t) \cos(\beta - 2\gamma) \right].$$

(10.9)

When defining a rotor current vector,

$$\underline{i}_R(t) = i_{R1}(t) + i_{R2}(t)\, e^{j\gamma} + i_{R3}(t)\, e^{j\,2\gamma} = i_R(t)\, e^{j\,\xi(t)},$$

(10.10)

$$\underline{i}_R^*(t) = i_{R1}(t) + i_{R2}(t)\, e^{-j\gamma} + i_{R3}(t)\, e^{-j\,2\gamma} = i_R(t)\, e^{-j\,\xi(t)},$$

(10.11)

the ampereturns-wave, excited by the rotor currents and moving with the rotor, assumes the form

$$\Theta_R(\beta, t) = \tfrac{1}{2} N_R \left[\underline{i}_R(t)\, e^{-j\beta} + \underline{i}_R^*(t)\, e^{j\beta} \right], \text{ real};$$

(10.12)

its effect on the stator is obtained by substituting

$$\beta = \alpha - \varepsilon,$$

(10.13)

$$\Theta_R(\alpha, \varepsilon, t) = \tfrac{1}{2} N_R \left[\underline{i}_R(t)\, e^{-j\,(\alpha-\varepsilon)} + \underline{i}_R^*(t)\, e^{j\,(\alpha-\varepsilon)} \right].$$

(10.14)

The resultant wave is a superposition of the stator- and rotor-ampere turns

$$\Theta(\alpha, \varepsilon, t) = \Theta_S(\alpha, t) + \Theta_R(\alpha, \varepsilon, t).$$

(10.15)

Since the permeability of the iron is assumed infinite, the combined magnetomotive force becomes effective at the two airgap crossings, causing a local flux density at the stator side of the airgap,

$$B_S(\alpha, \varepsilon, t) = \frac{1}{2\,h}\, \mu_0 \left[(1+\sigma_S)\Theta_S(\alpha, t) + \Theta_R(\alpha, \varepsilon, t) \right],$$

(10.16)

where σ_S accounts for magnetic leakage; μ_0 is the permeability constant. A detailed modelling of leakage effects is not possible with the simplifications made.

When calculating flux linkages, the spatial distribution of the conductors must be considered as indicated in Fig. 10.4 for the example of one stator winding. By assuming a quasi-continuous distribution of turns with an "incremental density" $\tfrac{1}{2} N_S \cos\lambda$, the postulated sinusoidal distribution results; at the same time the total number of turns is N_S, as proved by integration

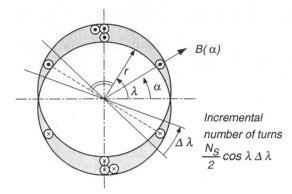

Fig. 10.4. Distributed winding

in the interval $-\frac{\pi}{2} < \lambda < \frac{\pi}{2}$. Thus the flux linkage in stator winding 1 is obtained by a double integration,

$$\psi_{S1}(t) = \tfrac{1}{2} N_S \int\limits_{\lambda=-\frac{\pi}{2}}^{\frac{\pi}{2}} \cos\lambda \left[\int\limits_{\alpha=\lambda-\frac{\pi}{2}}^{\lambda+\frac{\pi}{2}} l\,r\,B_S(\alpha,\varepsilon,t)\,d\alpha \right] d\lambda , \qquad (10.17)$$

where l is the effective axial length and r the radius of the rotor. The integration over α is a consequence of the non-uniform field in the airgap while the integration over λ is required by the nonuniform distribution of the winding. Inserting Eqs. (10.4, 10.14, 10.16) results in

$$\psi_{S1}(t) = \frac{N_S^2\,l\,r}{16\,h} (1+\sigma_S)\mu_0 \int\limits_{-\frac{\pi}{2}}^{\frac{\pi}{2}} \left[e^{j\lambda} + e^{-j\lambda} \right] \int\limits_{\lambda-\frac{\pi}{2}}^{\lambda+\frac{\pi}{2}} \left[\underline{i}_S(t)\,e^{-j\alpha} + \underline{i}_S^*(t)\,c^{j\alpha} \right] d\alpha d\lambda$$

$$+ \frac{N_S\,N_R\,l\,r}{16\,h} \mu_0 \int\limits_{-\frac{\pi}{2}}^{\frac{\pi}{2}} \left[e^{j\lambda} + e^{-j\lambda} \right] \int\limits_{\lambda-\frac{\pi}{2}}^{\lambda+\frac{\pi}{2}} \left[\underline{i}_R(t)\,e^{-j(\alpha-\varepsilon)} + \underline{i}_R^*(t)\,e^{j(\alpha-\varepsilon)} \right] d\alpha d\lambda .$$

$$(10.18)$$

The evaluation of the integral is greatly simplified by the complex notation due to the periodicity of the integrand. With the abbreviations for self and mutual inductances

$$(1+\sigma_S)\frac{N_S^2\,l\,r}{8\,h}\,\pi\,\mu_0 = \tfrac{1}{3}\,L_S , \qquad (10.19)$$

$$\frac{N_S\,N_R\,l\,r}{8\,h}\,\pi\,\mu_0 = \tfrac{1}{3}\,M , \qquad (10.20)$$

a simple result is obtained

$$\psi_{S1}(t) = \tfrac{1}{3} L_S \left[\underline{i}_S(t) + \underline{i}_S^*(t)\right] + \tfrac{1}{3} M \left[\underline{i}_R(t)\, e^{j\,\varepsilon} + \underline{i}_R^*(t)\, e^{-j\,\varepsilon}\right] . \qquad (10.21)$$

The flux linkage in the stator winding 1 is a physical integral quantity and must, of course, be real. The flux in the other stator windings is computed likewise, with the exception that the integration over λ is shifted to $\gamma \pm \frac{\pi}{2}$ and $2\gamma \pm \frac{\pi}{2}$ respectively, reflecting the position of the windings. This results in

$$\psi_{S2}(t) = \tfrac{1}{3} L_S \left[\underline{i}_S(t)\, e^{-j\,\gamma} + \underline{i}_S^*(t)\, e^{j\,\gamma}\right]$$
$$+ \tfrac{1}{3} M \left[\underline{i}_R(t)\, e^{j\,(\varepsilon-\gamma)} + \underline{i}_R^*(t)\, e^{-j\,(\varepsilon-\gamma)}\right] , \qquad (10.22)$$

$$\psi_{S3}(t) = \tfrac{1}{3} L_S \left[\underline{i}_S(t)\, e^{-j\,2\,\gamma} + \underline{i}_S^*(t)\, e^{j\,2\,\gamma}\right]$$
$$+ \tfrac{1}{3} M \left[\underline{i}_R(t)\, e^{j\,(\varepsilon-2\,\gamma)} + \underline{i}_R^*(t)\, e^{-j\,(\varepsilon-2\,\gamma)}\right] . \qquad (10.23)$$

The symmetry of these equations gives rise to the definition of a complex vector of flux linkages

$$\underline{\psi}_S(t) = \psi_{S1}(t) + \psi_{S2}(t)\, e^{j\,\gamma} + \psi_{S3}(t)\, e^{j\,2\,\gamma} , \qquad (10.24)$$

allowing Eqs. (10.21)–(10.23) to be combined and disposing of the conjugate terms

$$\underline{\psi}_S(t) = L_S\, \underline{i}_S(t) + M\, \underline{i}_R(t)\, e^{j\,\varepsilon(t)} . \qquad (10.25)$$

This flux vector describes the magnitude and angular position of the peak of the sinusoidal flux density in the airgap of the machine. The exponential term attached to $\underline{i}_R(t)$ indicates that the rotor current vector must be rotated by the angle of mechanical rotation before its effect can be superimposed upon that of the stator current vector $\underline{i}_S(t)$.

The flux linkage of the moving rotor windings is computed in exactly the same way. Conversion of the stator current vector into rotor coordinates by means of Eq. (10.13) yields

$$\Theta_S(\beta,\,\varepsilon,\,t) = \tfrac{1}{2} N_S \left[\underline{i}_S(t)\, e^{-j\,(\beta+\varepsilon)} + \underline{i}_S^*(t)\, e^{j\,(\beta+\varepsilon)}\right] , \qquad (10.26)$$

which produces a flux density on the rotor surface, corresponding to Eq. (10.16)

$$B_R(\beta,\,\varepsilon,\,t) = \frac{1}{2\,h}\, \mu_0 \left[(1+\sigma_R)\Theta_R(\beta,\,t) + \Theta_S(\beta,\,\varepsilon,\,t)\right] . \qquad (10.27)$$

σ_R again stands for the unavoidable magnetic leakage.

Integration around the circumference of the rotor, again presupposing a finely distributed rotor winding, leads to an expression for the flux linkage in rotor winding 1,

$$\psi_{R1}(t) = \tfrac{1}{3} L_R \left[\underline{i}_R(t) + \underline{i}_R^*(t) \right] + \tfrac{1}{3} M \left[\underline{i}_S(t) e^{-j\,\varepsilon} + \underline{i}_S^*(t) e^{j\,\varepsilon} \right] , \qquad (10.28)$$

which has a similar form as Eq. (10.21). However, as this flux is defined in rotor coordinates, the stator current vector is now turned backwards relative to the position of the rotor.

$$(1 + \sigma_R) \frac{N_R^2 \, l \, r}{8 \, h} \pi \, \mu_0 = \tfrac{1}{3} L_R \qquad (10.29)$$

defines the self inductance of a rotor winding.

Likewise the flux linkages of the other rotor windings are obtained,

$$\psi_{R2}(t) = \tfrac{1}{3} L_R \left[\underline{i}_R(t) e^{-j\,\gamma} + \underline{i}_R^*(t) e^{j\,\gamma} \right]$$
$$+ \tfrac{1}{3} M \left[\underline{i}_S(t) e^{-j\,(\varepsilon+\gamma)} + \underline{i}_S^*(t) e^{j\,(\varepsilon+\gamma)} \right] , \qquad (10.30)$$

$$\psi_{R3}(t) = \tfrac{1}{3} L_R \left[\underline{i}_R(t) e^{-j\,2\gamma} + \underline{i}_R^*(t) e^{j\,2\gamma} \right]$$
$$+ \tfrac{1}{3} M \left[\underline{i}_S(t) e^{-j\,(\varepsilon+2\,\gamma)} + \underline{i}_S^*(t) e^{j\,(\varepsilon+2\,\gamma)} \right] . \qquad (10.31)$$

These expressions are again simplified by forming a complex vector for the rotor flux

$$\underline{\psi}_R(t) = \psi_{R1}(t) + \psi_{R2}(t) e^{j\,\gamma} + \psi_{R3}(t) e^{j\,2\gamma}$$
$$= L_R \, \underline{i}_R(t) + M \, \underline{i}_S(t) e^{-j\,\varepsilon} . \qquad (10.32)$$

The magnetic linkages described by Eqs. (10.21)–(10.23) are now used to derive the voltage equations for the stator and rotor circuits, depicted in Fig. 10.5.

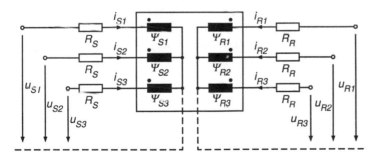

Fig. 10.5. Magnetic linkages and voltages

The line-to-neutral voltages in the stator circuit are

$$R_S \, i_{S1} + \frac{d\psi_{S1}}{dt} = u_{S1}(t) ,$$

$$R_S \, i_{S2} + \frac{d\psi_{S2}}{dt} = u_{S2}(t) , \qquad (10.33)$$

$$R_S \, i_{S3} + \frac{d\psi_{S3}}{dt} = u_{S3}(t) ,$$

where R_S is the stator resistance per phase and u_{S1}, u_{S2}, u_{S3} are voltages of arbitrary waveforms. These equations may again be combined by introducing complex vectors. By formal definition of a voltage vector, eventhough there is no immediate physical interpretation as with the currents,

$$\underline{u}_S(t) = u_{S1}(t) + u_{S2}(t)\, e^{j\,\gamma} + u_{S3}(t)\, e^{j\,2\,\gamma}, \tag{10.34}$$

and with Eqs. (10.5, 10.10, 10.25), this results in

$$R_S\, \underline{i}_S + \frac{d\underline{\psi}_S}{dt} = R_S\, \underline{i}_S + L_S\, \frac{d\underline{i}_S}{dt} + M\, \frac{d}{dt}\, (\underline{i}_R\, e^{j\,\varepsilon}) = \underline{u}_S(t)\,. \tag{10.35}$$

Employing the rules of differentiation, this equation assumes the form

$$R_S\, \underline{i}_S + L_S\, \frac{d\underline{i}_S}{dt} + M\, \frac{d\underline{i}_R}{dt}\, e^{j\,\varepsilon} + j\,\omega\, M\, \underline{i}_R\, e^{j\,\varepsilon} = \underline{u}_S(t)\,, \tag{10.36}$$

where $\omega = d\varepsilon/dt$ is the angular velocity of the rotor. The two terms containing the rotor current may be interpreted as voltages due to mutual induction and rotation, respectively.

By inserting Eqs. (10.21–10.23) into Eq. (10.33) and considering Eq. (10.1) which is due to the isolated neutral, it is found that the line-to-neutral voltages of the symmetrical machine are also balanced at any instant,

$$u_{S1}(t) + u_{S2}(t) + u_{S3}(t) \equiv 0\,. \tag{10.37}$$

On most motors the neutral point of the stator winding is not accessible, so that only the terminal voltages $u_{S12} = u_{S1} - u_{S2}$ etc. are available for external use, while the line- to neutral voltages are dependent quantities. This corresponds to an actual situation, where the motor may in fact be Δ – connected. With the terminal voltages $u_{12} = u_{S1} - u_{S2}$ etc. the voltage vector in Eq. (10.34) can also be written as

$$\underline{u}_S(t) = u_{12}(t) - u_{23}(t)\, e^{-j\,\gamma}. \tag{10.38}$$

The same arguments are valid for the rotor winding, where the currents are also balanced, due to the assumed fictitious neutral point

$$i_{R1}(t) + i_{R2}(t) + i_{R3}(t) \equiv 0\,; \tag{10.39}$$

the dashed lines connecting the neutrals in Fig. 10.5 are redundant in reality. Hence the voltage equations are

$$R_R\, i_{R1} + \frac{d\psi_{R1}}{dt} = u_{R1}(t)\,,$$

$$R_R\, i_{R2} + \frac{d\psi_{R2}}{dt} = u_{R2}(t)\,, \tag{10.40}$$

$$R_R\, i_{R3} + \frac{d\psi_{R3}}{dt} = u_{R3}(t)\,.$$

With the help of Eqs. (10.10, 10.32) they are again combined to form a vectorial equation,

$$R_R \, \underline{i}_R + \frac{d\underline{\psi}_R}{dt} = R_R \, \underline{i}_R + L_R \frac{d\underline{i}_R}{dt} + M \frac{d}{dt} \left(\underline{i}_S \, e^{-j\varepsilon} \right) = \underline{u}_R(t) , \qquad (10.41)$$

where all variables are defined in rotor coordinates. The vector of the rotor voltages,

$$\underline{u}_R(t) = u_{R1}(t) + u_{R2}(t) \, e^{j\,\gamma} + u_{R3}(t) \, e^{j\,2\gamma} \qquad (10.42)$$

may be impressed by an external voltage source. In case of a cage rotor, the voltages are zero and the rotor circuit is internally short circuited.

The vector differential equations (10.35, 10.41) describing the electromagnetic interactions of the symmetrical AC motor in steady state and transient condition are now to be supplemented by equations for the electrical torque and the mechanical transients. On a real motor the tangential forces are acting at the sides of the slots, where the magnetic field enters the iron. Since there are no slots in our simplified model with airgap windings, the torque is computed with the help of the tangential Lorentz- forces exerted on the axial current carrying conductors orthogonally crossed by the radial magnetic field.

The component of the flux density on the rotor surface which is caused by the stator currents follows from Eq. (10.27)

$$B_{RS}(\beta, \, \varepsilon, \, t) = \frac{N_S \, \mu_0}{4 \, h} \left[\underline{i}_S(t) \, e^{-j \, (\beta+\varepsilon)} + \underline{i}_S^*(t) \, e^{j \, (\beta+\varepsilon)} \right] . \qquad (10.43)$$

The magnetic field produced by the rotor currents themselves does not generate any net tangential forces with the rotor currents since there is no reluctance torque with the uniform air-gap; this could be shown by carrying out the calculations with B_R instead of B_{RS}.

The current distribution $a_R(\beta, t)$ along the circumference of the rotor is also, with the assumptions made, sinusoidal. According to Fig. 10.6, it is defined as the spatial derivative of the rotor ampereturns,

$$a_R(\beta, \, t) = \frac{1}{2} \frac{\partial \Theta_R(\beta, \, t)}{\partial (r \, \beta)} = -j \, \frac{N_R}{4 \, r} \left[\underline{i}_R \, e^{-j\beta} - \underline{i}_R^* \, e^{j\beta} \right] . \qquad (10.44)$$

The tangential force df acting on an axial strip of width $r \, d\beta$ of the rotor surface is the product of radial flux density and axial rotor current distribution

$$df = -B_{RS}(\beta, \, \varepsilon, \, t) \, a_R(\beta, \, t) \, l \, r \, d\beta , \qquad (10.45)$$

which upon integration along the circumference yields the electric torque in the direction of rotation,

$$m_M(t) = r \int_F df = -r^2 \, l \int_0^{2\pi} B_{RS}(\beta, \, \varepsilon, \, t) \, a_R(\beta, \, t) \, d\beta . \qquad (10.46)$$

Inserting Eqs. (10.43, 10.44) results in

$$m_M(t) = -\frac{M}{6\,\pi\,j} \int\limits_0^{2\,\pi} \left[\underline{i}_S\, e^{-j(\beta+\varepsilon)} + \underline{i}_S^*\, e^{j(\beta+\varepsilon)}\right] \left[\underline{i}_R\, e^{-j\beta} - \underline{i}_R^*\, e^{j\,\beta}\right]\, d\beta \;,$$

$$(10.47)$$

where M is the coefficient of mutual inductivity defined in Eq. (10.20). When integrating over the full circumference of the rotor, the exponential terms containing β cancel. Hence we find

$$m_M(t) = \frac{M}{3\,\pi} \int\limits_0^{2\,\pi} \frac{\underline{i}_S\,\underline{i}_R^*\, e^{-j\,\varepsilon} - \underline{i}_S^*\,\underline{i}_R\, e^{j\,\varepsilon}}{2\,j}\, d\beta = \tfrac{2}{3}\, M\, \mathrm{Im}\left[\underline{i}_S(t)\,(\underline{i}_R\, e^{j\,\varepsilon})^*\right]\;;$$

$$(10.48)$$

the imaginary part of the bracket is equivalent to a vector-product, being proportional to the product of the two current vectors and the sine of the angular displacement χ, as seen in Fig. 10.6 b.

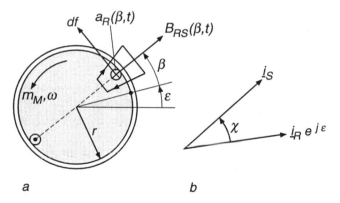

Fig. 10.6. Rotor current distribution and torque

When assuming that the stator and rotor windings have an equal number of turns, $N_R' = N_S$, and introducing the usual leakage factors,

$$L_S = (1 + \sigma_S)\, L_0\,, \qquad L_R = (1 + \sigma_R)\, L_0\,, \qquad M = L_0 \qquad (10.49)$$

the complete mathematical model of the symmetrical general AC machine with the lumped inertia J then assumes the following form, with m_L being the net load torque at the coupling of the motor,

$$R_S \underline{i}_S + L_S \frac{d\underline{i}_S}{dt} + L_0 \frac{d}{dt}(\underline{i}_R e^{j\,\varepsilon}) = \underline{u}_S(t)\,, \tag{10.50}$$

$$R_R \underline{i}_R + L_R \frac{d\underline{i}_R}{dt} + L_0 \frac{d}{dt}(\underline{i}_S e^{-j\,\varepsilon}) = \underline{u}_R(t)\,, \tag{10.51}$$

$$J \frac{d\omega}{dt} = m_M(t) - m_L(t) = \tfrac{2}{3} L_0 \operatorname{Im}\left[\underline{i}_S (\underline{i}_R e^{j\,\varepsilon})^*\right] - m_L(\varepsilon, \omega, t), \tag{10.52}$$

$$\frac{d\varepsilon}{dt} = \omega\,. \tag{10.53}$$

Since the first two equations can be split into real and imaginary parts, this represents a set of 6 scalar nonlinear differential equations. They are valid for any waveforms of voltages, currents and at varying load torque and speed.

The physical currents, for example in the stator windings, are easily derived from the vectorial representation. With Eq. (10.1), the current vector, defined in Eq. (10.5), is

$$\underline{i}_S(t) = \tfrac{3}{2} i_{S1}(t) + j \frac{\sqrt{3}}{2} [i_{S2}(t) - i_{S3}(t)]\,; \tag{10.54}$$

hence

$$i_{S1}(t) = \tfrac{2}{3} \operatorname{Re}[\underline{i}_S(t)]\,, \tag{10.55}$$

and correspondingly,

$$i_{S2}(t) = \tfrac{2}{3} \operatorname{Re}\left[\underline{i}_S e^{-j\,\gamma}\right], \text{ etc.} \tag{10.56}$$

Likewise the currents in the rotor winding are obtained; of course, in case of a squirrel cage rotor, the notion of phase currents loses its physical meaning, even though the simplified theory remains applicable.

The Eqs. (10.50–10.53) are the starting point for all subsequent discussions of transient and steady state phenomena. They describe the electromechanical interactions of a symmetrical AC machine and can be adapted to various constraints, imposed by the type of machine used, as long as the basic assumptions on which the model was based, remain valid. This mainly refers to radial symmetry and constant airgap, excluding saliency of the stator or rotor.

Table 10.1 contains the rotor circuits of different nonsalient AC machines used in controlled drives that may be represented by the model equations. The details of the constraints are discussed in later chapters.

The model equations contain, of course, also the special case of steady state operation, when the AC motor is fed with sinusoidal impressed voltages by a symmetrical three–phase supply and is loaded with constant torque. To form a link to this commonly used presentation and to gain a better understanding of the model equations, this case will be discussed first. With $\underline{u}_R = 0$, a cage motor or a wound rotor results, where external symmetrical rotor resistors may have been added.

Table 10.1. Mathematical models for nonsalient rotors of different AC machines used in controlled electrical drives

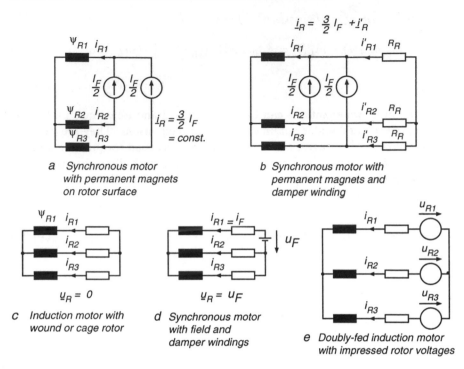

a   Synchronous motor
   with permanent magnets
   on rotor surface

b Synchronous motor with
   permanent magnets and
   damper winding

c   Induction motor with
   wound or cage rotor

d Synchronous motor
   with field and
   damper windings

e Doubly-fed induction motor
   with impressed rotor voltages

10.2 Induction Motor with Sinusoidal Symmetrical Voltages in Steady State

10.2.1 Stator Current, Current Locus

A symmetrical three-phase system of sinusoidal voltages having the angular frequency ω_1 may be defined as follows, using complex notation,

$$u_{S1}(t) = \sqrt{2}\,U_S\,\cos(\omega_1 t + \tau_1) = \mathrm{Re}\left[\sqrt{2}\,U_S\,e^{j\,\tau_1}\,e^{j\,\omega_1 t}\right]$$

$$= \frac{\sqrt{2}}{2}\left[\underline{U}_S\,e^{j\,\omega_1 t} + \underline{U}_S^*\,e^{-j\,\omega_1 t}\right], \tag{10.57}$$

where $\underline{U}_S = U_S\,e^{j\,\tau_1}$ is the constant voltage phasor; corresponding expressions hold for phases 2 und 3,

$$u_{S2}(t) = u_{S1}\left(t - \frac{\gamma}{\omega_1}\right)$$

$$= \frac{\sqrt{2}}{2}\left[\underline{U}_S\, e^{j\,(\omega_1 t - \gamma)} + \underline{U}_S^*\, e^{-j\,(\omega_1 t - \gamma)}\right], \qquad \gamma = \frac{2\,\pi}{3} \tag{10.58}$$

$$u_{S3}(t) = u_{S1}\left(t - \frac{2\,\gamma}{\omega_1}\right)$$

$$= \frac{\sqrt{2}}{2}\left[\underline{U}_S\, e^{j\,(\omega_1 t - 2\,\gamma)} + \underline{U}_S^*\, e^{-j\,(\omega_1 t - 2\,\gamma)}\right]. \tag{10.59}$$

The voltage phasor $\underline{U}_S$ is a complex constant, having a magnitude equal to the RMS line-to-neutral voltage and an arbitrary phase angle τ_1 which may be set at will because it is defined by the chosen instant for time zero.

When the complex voltage vector according to Eq. (10.34) is formed, a very simple result is obtained,

$$\underline{u}_S(t) = u_{S1} + u_{S2}\, e^{j\,\gamma} + u_{S3}\, e^{j\,2\,\gamma} = \frac{3\sqrt{2}}{2}\,\underline{U}_S\, e^{j\,\omega_1 t}\,; \tag{10.60}$$

the conjugate complex part vanishes, leaving a vector with constant magnitude and constant angular velocity describing a circular path around the origin.

The stator and rotor currents form also symmetrical three-phase systems in steady state. With

$$i_{S1}(t) = \frac{\sqrt{2}}{2}\left[\underline{I}_S\, e^{j\,\omega_1 t} + \underline{I}_S^*\, e^{-j\,\omega_1 t}\right], \quad \text{etc.} \tag{10.61}$$

we find

$$\underline{i}_S(t) = \frac{3\sqrt{2}}{2}\,\underline{I}_S\, e^{j\,\omega_1 t}\,, \tag{10.62}$$

where the current phasor $\underline{I}_S$ is still to be determined. A similar definition exists for the rotor currents oscillating with slip frequency $\omega_2 = \omega_1 - \omega$

$$i_{R1}(t) = \frac{\sqrt{2}}{2}\left[\underline{I}_R\, e^{j\,(\omega_1 - \omega)\,t} + \underline{I}_R^*\, e^{-j\,(\omega_1 - \omega)\,t}\right], \quad \text{etc.} \tag{10.63}$$

or

$$\underline{i}_R(t) = \frac{3\sqrt{2}}{2}\,\underline{I}_R\, e^{j\,(\omega_1 - \omega)\,t}\,. \tag{10.64}$$

The rotor current in stator coordinates, i.e. as seen by a stationary observer, is with $\varepsilon = \omega t$ because of constant speed,

$$\underline{i}_R(t)\, e^{j\,\varepsilon(t)} = \frac{3\sqrt{2}}{2}\,\underline{I}_R\, e^{j\,\omega_1 t}\,. \tag{10.65}$$

Hence the rotor ampereturns wave, moving with slip frequency across the rotor surface, rotates in synchronism with the stator current wave; this is a precondition for constant electrical torque. By introducing these expressions into Eqs. (10.50, 10.51) and omitting the redundant terms, the differential equations are converted with $u_R = 0$ to algebraic equations for the phasors,

$$(R_S + j\,\omega_1\,\sigma_S\,L_0)\,\underline{I}_S + j\,\omega_1\,L_0\,(\underline{I}_S + \underline{I}_R) = \underline{U}_S$$
$$(R_R + j\,\omega_2\,\sigma_R\,L_0)\,\underline{I}_R + j\,\omega_2\,L_0\,(\underline{I}_S + \underline{I}_R) = 0\,. \tag{10.66}$$

Normalising the second equation with the slip frequency

$$S = \frac{\omega_2}{\omega_1} = \frac{\omega_1 - \omega}{\omega_1} \tag{10.67}$$

results in

$$\left(\frac{R_R}{S} + j\,\omega_1\,\sigma_R\,L_0\right)\,\underline{I}_R + j\,\omega_1\,L_0\,(\underline{I}_S + \underline{I}_R) = 0\,. \tag{10.68}$$

Equations (10.66, 10.68) lead immediately to the familiar single phase equivalent circuit seen in Fig. 10.7 a; its validity is restricted to steady state, assuming sinusoidal symmetrical currents and constant load torque.

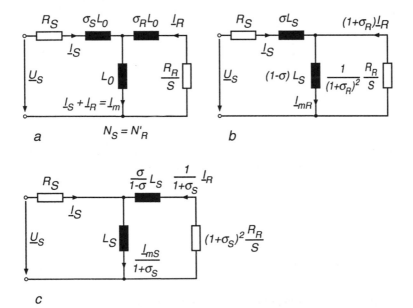

Fig. 10.7. Single phase equivalent circuits of induction motor in steady state
(**a**) with leakage inductances on stator and rotor side,
(**b**) with leakage inductance on stator side,
(**c**) with leakage inductance on rotor side

By inspecting the equivalent circuit in Fig. 10.7 a, the stator impedance per phase is derived. It may be written in the form

$$Z_S = R_S + j\,\omega_1\,L_S \frac{1 + j\,\frac{S\,\omega_1\,\sigma\,L_R}{R_R}}{1 + j\,\frac{S\,\omega_1\,L_R}{R_R}} \;,\qquad (10.69)$$

where

$$\sigma = 1 - \frac{1}{(1 + \sigma_S)(1 + \sigma_R)} \qquad\qquad (10.70)$$

is the total leakage factor of the motor, which has an important influence on the characteristics of the machine; it is chosen by suitable design parameters such as the shape of the slots and the length of the airgap. Normal values of σ range between 0.05 for low leakage machines to about 0.20.

The expression in Eq. (10.69) gives rise to the definition of yet another important parameter

$$S_p = \frac{R_R}{\omega_1\,\sigma\,L_R} = \frac{1}{\omega_1\,\sigma\,T_R} \;,\; \text{with} \quad T_R = \frac{L_R}{R_R}\;, \qquad (10.71)$$

which is called pull-out slip because a motor with zero stator resistance produces maximum torque at $S = S_p$, as will be shown later. The pull-out slip of normal medium size motors is $S_p < 0.25$.

The steady state equivalent circuit in Fig. 10.7 a contains three synthetic inductances which, because of $\underline{I}_m = \underline{I}_S + \underline{I}_R$, are not representing independent energy storages; hence it must be possible to redraw the circuit diagram with only two inductances. One possibility is shown in Fig. 10.7 b, where all the leakage is concentrated at the stator side. There is a different fictitious rotor resistance leading to a different current in the right hand side branch of the circuit but the stator impedances are identical. The new synthetic magnetising current I_{mR} represents the rotor flux, as will be shown in a later chapter. A third equivalent circuit is seen in Fig. 10.7 c, where all the leakage is referred to the rotor side and where I_{mS} stands for the stator flux. This arrangement can also be used for designing controls.

The stator current is now computed with the added simplification $R_S = 0$ which is of little significance with larger machines operating with low slip at line frequency. Thus the normalised expression for the stator current phasor becomes

$$\underline{I}_S = \frac{\underline{U}_S}{j\,\omega_1\,L_S} \frac{1 + j\,\frac{1}{\sigma}\,\frac{S}{S_p}}{1 + j\,\frac{S}{S_p}} \;; \qquad (10.72)$$

which is a linear complex function mapping the real S/S_p-axis onto the $\underline{I}_S$-plane. According to the rules of conformal mapping, the result is a circular locus, the circle diagram, also called Ossanna's or Heyland's circle. This is confirmed by the following transformation

$$\underline{I}_S = \underline{I}_{S0} \left[\frac{1+\sigma}{2\,\sigma} - \frac{1-\sigma}{2\,\sigma} \frac{1-j\frac{S}{S_p}}{1+j\frac{S}{S_p}} \right]$$

$$= \underline{I}_{S0} \left[\frac{1+\sigma}{2\,\sigma} - \frac{1-\sigma}{2\,\sigma} e^{-j\,2\,\text{arctan}\,S/S_p} \right] , \qquad (10.73)$$

where

$$\underline{I}_{S0} = \frac{U_S}{j\,\omega_1\,L_S} \qquad (10.74)$$

is the ideal no load current valid for $S = 0$, i.e. when the rotor moves synchronously with the revolving stator field. When no speed difference exists, no rotor current will be induced and there is no torque; hence the no-load current is a purely reactive magnetising current. This is in agreement with the equivalent circuits in Fig. 10.7 for $S = 0$.

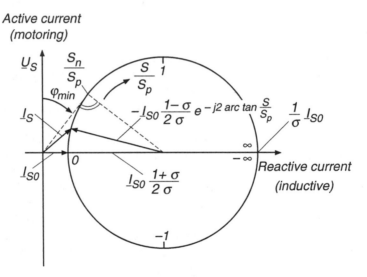

Fig. 10.8. Locus of stator current with stator resistance neglected, circle diagram for an induction motor

The complex locus $\underline{I}_S(S/S_p)$ is depicted in Fig. 10.8 with the voltage phasor assumed imaginary, $\underline{U}_S = j\,U_S$, in order to place the circle in the customary orientation. The curve covers the full range $-\infty < S/S_p < \infty$; for $S > 0$ the motor runs below synchronism, drawing active power, while for $S < 0$ it operates as a generator and supplies active power to the line. Reactive power for magnetisation is always needed which is a characteristic of induction machines having no internal voltage source which is present in synchronous machines. At $S = 0$ the current assumes its minimum value representing (for $R_S = 0$) the no-load magnetising current as mentioned before. Minimum phase shift $\varphi = \arg(\underline{U}_S, \underline{I}_S)$, i.e. maximum power factor is reached when the current phasor forms a tangent to the circle. The optimal power factor then follows from the geometrical relations of Fig. 10.8,

$$\cos\varphi_n = \frac{1-\sigma}{1+\sigma} \; ; \tag{10.75}$$

this is usually defined as the rated or nominal operating point. The pertinent value of S_n/S_p is obtained from the phase angle, Eq. (10.72),

$$\varphi = -\frac{\pi}{2} + \arctan\frac{1}{\sigma}\frac{S}{S_p} - \arctan\frac{S}{S_p} \tag{10.76}$$

by differentiation, $\dfrac{d\varphi}{d(S/S_p)} = 0$; the result is

$$\frac{S_n}{S_p} = \sqrt{\sigma} \, . \tag{10.77}$$

Hence, not only the geometry of the current locus, but also the maximum power factor, the pull-out slip and the normalised slip at the nominal operating point are all determined by the leakage factor σ. The magnitude of the normalised stator current,

$$\frac{I_S}{I_{S0}} = \frac{\sqrt{1 + \left(\frac{1}{\sigma}\frac{S}{S_p}\right)^2}}{\sqrt{1 + \left(\frac{S}{S_p}\right)^2}} \tag{10.78}$$

is an even function of slip, as seen in Fig. 10.9 a. The normal operating region comprises only the central portion of the curve. At larger values of slip, for instance when starting the motor at $S = 1$, the current rises sharply so that the simplifications employed become questionable; this is particularly true for the iron saturation that is not negligible at high currents.

The stator current at nominal slip, $S_n/S_p = \sqrt{\sigma}$, is

$$\frac{I_{Sn}}{I_{S0}} = \frac{1}{\sqrt{\sigma}} \, , \tag{10.79}$$

while the asymptotic value of the current is

$$\lim_{S\to\infty} \frac{I_S}{I_{S0}} = \frac{1}{\sigma} \, . \tag{10.80}$$

This underlines the importance of the leakage factor σ as a design parameter.

The maximum and minimum values of the active current and active power, assuming $\underline{U}_S = $const., are reached at $S = \pm S_p$, i.e. in the vertices of the circle. In the next section it is shown that these are also the points where the torque assumes its peak values.

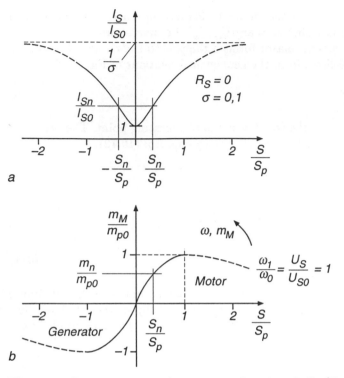

Fig. 10.9. Stator current and torque as a function of slip ($R_S = 0$)

10.2.2 Steady State Torque, Efficiency

An expression for the steady state torque is obtained by inserting the vectors for sinusoidal symmetrical currents, Eqs. (10.62, 10.65), into the general formula for the torque, Eq. (10.48),

$$m_M = 3\,L_0\,\mathrm{Im}\left(\underline{I}_S\,\underline{I}_R^*\right)\,, \tag{10.81}$$

yielding a constant value, as expected. From Fig. 10.7 a the following relation between stator- and rotor-current is derived,

$$\underline{I}_R = \frac{-j\,\omega_1\,L_0}{R_R/S + j\,\omega_1\,L_R}\,\underline{I}_S\,, \tag{10.82}$$

which leads to

$$m_M = 3\,L_0\,I_S^2\,\mathrm{Im}\left[\frac{j\,\omega_1\,L_0}{R_R/S - j\,\omega_1\,L_R}\right]\,. \tag{10.83}$$

With the help of Eq. (10.78) this can be rearranged to

$$m_M = \frac{3}{2} \frac{1-\sigma}{\sigma} \frac{U_S^2}{\omega_1^2 L_S} \frac{2}{S/S_p + S_p/S} = m_p \frac{2}{S/S_p + S_p/S} ,$$ (10.84)

where m_p represents the peak torque, available at the pull-out slip $S = \pm S_p$. The function of torque vs. normalised slip is plotted in Fig. 10.9 b.

When the induction motor is part of a variable speed drive, the stator voltage U_S and frequency ω_1 may differ from their nominal values U_{S0}, ω_0 shown on the nameplate; hence the following extension of Eq. (10.84) applies

$$m_M = m_{p0} \left(\frac{U_S/\omega_1}{U_{S0}/\omega_0} \right)^2 \frac{2}{S/S_p + S_p/S} ,$$ (10.85)

where

$$m_{p0} = \frac{3}{2} \frac{1-\sigma}{\sigma} \frac{U_{S0}^2}{\omega_0^2 L_S}$$ (10.86)

is the maximum torque (for $R_S = 0$) at nominal voltage and frequency. This indicates that the torque rises with the square of the flux; the reason is of course that the rotor current, being caused by magnetic induction, is also depending on flux.

The torque vs. slip curve seen in Fig. 10.9 b may be qualitatively interpreted as follows:
When the rotor speed deviates from the angular velocity ω_1 of the stator field, currents are induced in the rotor winding producing a torque with the tendency of reducing the speed difference. Below synchronism $(S > 0)$ the torque acts in the direction of the rotating field, i.e. the machine operates between standstill and synchronous speed as motor; above synchronous speed, the torque is opposed to the direction of rotation, hence the machine acts as generator, feeding electrical power to the line. Steady state supersynchronous speed can thus only occur if the machine is mechanically driven. Because of the reactive power problem, induction generators are rarely used except in small unattended hydro or wind power stations, but temporary regeneration is a normal feature of controlled AC drives.

The occurrence of a peak torque can be explained by a simplified model of the rotor circuit as shown in Fig. 10.10. The rotor may be represented by a R–L-circuit that is supplied from a voltage source, the amplitude U_2 of which is proportional to frequency ω_2. The current phasor

$$\underline{I}_2 = \frac{U_{20} \frac{\omega_2}{\omega_0}}{R_2 + j\,\omega_2\,L_2} = \frac{U_{20}}{j\,\omega_0\,L_2} \frac{1}{1 - j \frac{R_2}{\omega_2 L_2}}$$ (10.87)

as a function of frequency ω_2 is described by the semicircle shown in Fig. 10.10 b. While the magnitude I_2 rises asymptotically with ω_2, the active component

$$\mathrm{Re}\,\underline{I}_2 = \frac{U_{20}}{\omega_0\,L_2} \frac{1}{\frac{R_2}{\omega_2 L_2} + \frac{\omega_2 L_2}{R_2}}$$ (10.88)

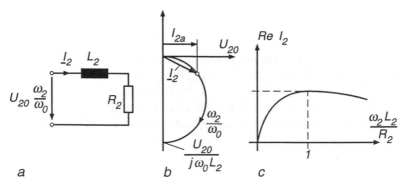

Fig. 10.10. Simplified model of rotor circuit.
(a) Circuit; (b) Locus of current phasor;
(c) Real current component

passes through a peak value; the curve, plotted in Fig. 10.10 c, has a similar shape as the torque characteristic in Fig. 10.9 b.

This points to the reason for the decreasing torque of the induction motor at higher slip frequency: Due to the rotor leakage, the rotor current is increasingly lagging, causing the rotor current wave to be shifted and eventually become orthogonal to the stator flux wave. Hence the torque is reduced whereas the stator and rotor currents continue to increase asymptotically.

In view of the stability of the drive and the rising losses, steady state operation of the motor makes only sense below pull-out slip.

When declaring the point, where the power factor is optimal, as the nominal operating point, $S_n/S_p = \sqrt{\sigma}$, the ratio of the peak to nominal torque is

$$\frac{m_{p0}}{m_n} = \frac{1+\sigma}{2\sqrt{\sigma}}, \tag{10.89}$$

which is a measure for the overload capacity of the motor. The pull-out torque, Eq. (10.86), shows that the rotor resistance has no effect on the maximum torque but only influences the pull-out slip. A wound rotor motor with external resistors in the rotor circuit then offers a possibility of modifying the torque/speed characteristics by changing the pull-out slip S_p through the rotor resistance. The resulting family of curves is depicted in Fig. 10.11 in the accustomed orientation. In combination with the dashed curves for inverse rotation of the field by interchanging two stator terminals a fairly complete coverage of the four quadrants in the torque/speed plane is achieved, albeit the shape of some curves is unfavourable. Lowering the stator voltages, for instance through a tap changing transformer, reduces the peak torque by the square of the voltage reduction, further compressing the curves in horizontal direction.

The main drawback of continuous speed control by rotor resistors or by reducing the stator voltages is the decrease in efficiency. Only for short

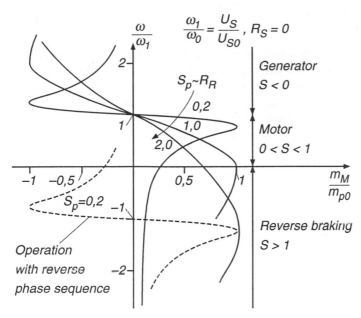

Fig. 10.11. Torque/speed curves of a wound rotor induction motor

time duty, for instance when starting the motor or with crane hoist applications, are these losses tolerable; simple control schemes of this type are in widespread use.

The efficiency of the simplified symmetrically fed induction machine in steady state operation may be assessed as follows: Because of the neglected iron and stator copper losses the active power P_S supplied to the stator windings is passed on to the rotor, where it is transformed into mechanical power and heat losses. In the steady state equivalent circuit (Fig. 10.7 a) all this power is dissipated in the synthetic resistance R_R/S,

$$P_R = 3 I_R^2 \frac{R_R}{S} . \tag{10.90}$$

In reality the resistance in each rotor phase is only R_R, hence the difference in power must correspond to mechanical power that cannot be modelled directly by a static circuit, Fig. 10.12,

$$P_m = 3 I_R^2 R_R \left[\frac{1}{S} - 1\right] = 3 I_R^2 R_R \frac{1-S}{S} . \tag{10.91}$$

This yields an estimate for the efficiency in the motoring region, $0 < S < 1$,

$$\eta_m = \frac{P_m}{P_R} = 1 - S = \frac{\omega}{\omega_1} < 1 , \tag{10.92}$$

as well as for operation as a generator, $S < 0$,

$$\eta_g = \frac{P_R}{P_m} = \frac{1}{1-S} = \frac{\omega_1}{\omega} < 1 \, , \tag{10.93}$$

which means that the efficiency even of an idealised motor is lowered in proportion to speed, whether the reduction in speed is caused by additional rotor resistors or by reduced stator voltage. A similar situation exists in the generation region, where the losses rise also with the magnitude of the slip. With a real motor, having primary copper losses as well as iron and friction losses, the efficiency will drop below this idealised case.

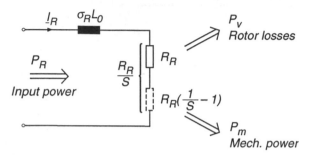

Fig. 10.12. Power balance in the rotor

The explanation is, of course, that the induction motor functions like a mechanical clutch, where the product of speed difference and torque causes an unavoidable power loss. The definition of efficiency in the braking region at reverse speed, $S > 1$, does not make sense, because electrical as well as mechanical power is supplied to the rotor and converted to heat.

With high slip control schemes for short duty drives the main problem is the dissipation of the rotor heat loss. In this respect a wound rotor motor with external rotor resistors is preferable, however at the cost of moving contacts.

These results give ample evidence that an efficient adjustable speed AC drive can only be realised when either the slip power is recovered by some means or the stator frequency ω_1 is changed to track the mechanical speed, thus restricting the rotor frequency to small values.

10.2.3 Comparison with Practical Motor Designs

The simplified theory of a symmetrical induction motor in steady state points to the total leakage factor σ as the most influential parameter of the machine. As was mentioned before, σ depends on the shape of the slots and increases with the airgap.

The scale of the circular current locus is determined by the ideal no-load current I_{So} which also depends to a large extent on the airgap. For mechanical reasons, the airgap is larger with multipole- than two pole-machines of equal size.

The scale of the normalised slip/torque curve, Fig. 10.9 b, is fixed by the pull-out slip S_p and the peak torque m_p; in the interest of good efficiency in steady state, S_p should be as small as possible. The peak torque is also affected by the leakage factor.

In Table 10.2 some of the characteristic parameters obtained with the simplified model are compared to those of standard motors, as taken from manufacturers lists on medium size (> 100 kW) 50 Hz motors, for a two pole- and an eight pole-machine. The comparison indicates a fairly close agreement, at least from the standpoint of the user.

Table 10.2. Comparison of idealised model with parameters of standard motors

	Model for $\sigma = 0.05$, $n_0 = 3000$ 1/min		Real	Model for $\sigma = 0.1$, $n_0 = 750$ 1/min	Real
$\frac{S_n}{S_p}$	$\sqrt{\sigma}$	$= 0.22$	0.20	0.32	0.30
$\frac{I_{so}}{I_{Sn}}$	$\sqrt{\sigma}$	$= 0.22$	0.30	0.32	0.40
$\cos \varphi_n$	$\frac{1-\sigma}{1+\sigma}$	$= 0.90$	0.90	0.82	0.84
$\frac{m_{p0}}{m_n}$	$\frac{1+\sigma}{2\sqrt{\sigma}}$	$= 2.35$	2.30	1.82	2.0

10.2.4 Starting of the Induction Motor

When selecting suitable values for efficiency, power factor and overload capacity at the nominal operating point, the starting performance of the motor is unsatisfactory; this is apparent from Fig. 10.13, where characteristic variables of a motor with a leakage factor $\sigma = 0.05$ are plotted against normalised slip. The conditions for acceptable performance at the nominal operating point are:

a) The normalised slip at the nominal operating point should be about $S_n/S_p \approx \sqrt{\sigma}$ to achieve a good power factor and adequate overload capacity.

b) The need for good efficiency $\eta_n < 1 - S_n$ calls for a minimal value of nominal and, hence, pull-out slip. With motors above 100 kW, a nominal slip $S_n \approx 0.02$ is realisable; the pull-out slip would then be about $S_p \approx 0.10$.

This means that an induction motor with a short circuited, constant resistance rotor would have to start from standstill ($S = 1$) at $S/S_p \approx 10$ which corresponds to an operating point at the far right of the circular diagram, Fig. 10.8. The starting performance at $S = 1$, characterised by

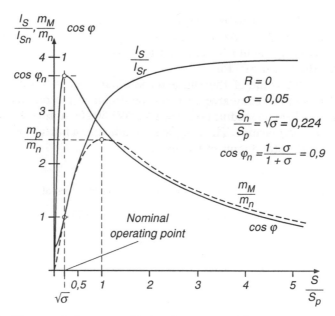

Fig. 10.13. Characteristic curves of induction motor with constant rotor resistance

$$\frac{I_S}{I_{Sn}} \approx \frac{1}{\sqrt{\sigma}} \approx 4.5 , \quad \cos\varphi \approx 0.095 , \quad \frac{m}{m_n} \approx 0.47 , \tag{10.94}$$

would be very poor, since low torque is combined with a high predominantly reactive current. When the motor is fed from a grid with a relatively large internal impedance, there could be an appreciable voltage drop, reducing the current linearly but the torque with the square of the voltage. This may even lead to a situation where the motor does not start under load; the protective devices would then have to disconnect it from the line to prevent damage due to overheating. The conditions are particularly severe in small grids such as on board ships, where the generators are not by orders of magnitude larger than the largest motors; this also calls for fast voltage control of the generators. Two options are available to improve the starting performance:

a) Use of a motor with a wound instead of a cage rotor which permits the rotor resistance and, hence, the pull-out slip S_p to be temporarily increased at will, creating optimal conditions for starting as well as at operational speed. The rotor resistance may be shorted in steps by contactors controlled in a fixed time sequence or depending on stator current or speed (see also Sect. 3.1.4). The larger part of the power losses during start occurs in the starting resistor, i.e. outside the motor, so that there are no additional cooling problems. When the operating speed has been reached, the slip-rings can be shorted internally with a special mechanism while the brushes are lifted to avoid the unnecessary friction and wear. A motor with $\sigma = 0.05$, starting with $S_p = 1.0$ would show the following starting performance at $S = 1$

$$\frac{I_S}{I_{Sn}} \approx 3.2 \,, \quad \cos\varphi \approx 0.66 \,, \quad \frac{m}{m_n} \approx 2.35 \,,$$

signifying a decisive improvement over the case with a short circuited rotor winding. With larger motors the starting resistor may have the form of a tank of water with enhanced conductivity into which the electrodes connected to the sliprings are progressively submerged.

b) A more elegant method requiring no moving contacts is the utilisation of eddy currents in the rotor bars leading to a concentration of the current in the outer portions of the conductors at higher rotor frequency (skin effect), thus increasing the effective rotor resistance at larger slip. As a consequence, the starting torque is improved with an accompanying reduction of current, while in normal operation at low slip the desirable features of the motor with short circuited rotor are maintained.

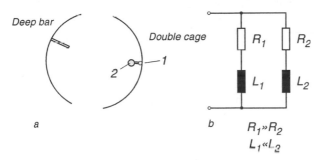

Fig. 10.14. Cross section of eddy current rotors and equivalent circuit

There are various designs which tend to emphasise the effect of eddy currents in the rotor bars. One solution employing two cage windings is shown in Fig. 10.14 a. The starting cage (1), being of higher resistivity, is placed in open slots immediately below the rotor surface and exhibits low magnetic leakage; as a consequence this winding produces a high pull-out slip S_{p1}. Because of the low leakage this winding dominates at high rotor frequency, i.e. at low speed.

The operational rotor winding (2), in contrast, is characterised by a larger cross section of the conductors, possibly employing also material of higher conductivity; the leakage reactance is increased because the bars are now embedded in the rotor iron. Clearly, this winding becomes mainly effective at low slip frequency. Figure 10.14 b shows the equivalent circuit with the two rotor windings connected in parallel.

Because of $R_1 \gg R_2$, $L_1 \ll L_2$ the two windings are complementary. The current locus of the machine is, of course, no longer a circle. Typical

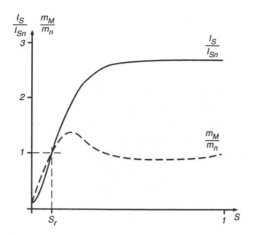

Fig. 10.15. Characteristic curves of induction motor with eddy current rotor

curves of an induction motor with double cage rotor are seen in Fig. 10.15; a comparison with Fig. 10.13 indicates the improvements achieved. Another design of a rotor winding with strong eddy-current effect is a cage with narrow bars extending deep into the rotor iron; this is also seen in Fig. 10.14 a.

When line-starting larger motors it is sometimes necessary to reduce the inrush current by lowering the stator voltages or by using only part of the stator winding, thus increasing the leakage factor. In both cases the starting torque is further reduced which limits this method to starts with light load. The simplest possibility for reducing the voltage during the start is, of course, a $\mathsf{Y}/\triangle$-reconnection of the stator windings. It entails a reduction of voltage and current by $1/\sqrt{3}$ and of input power and torque by $1/3$.

10.3 Induction Motor
Supplied by Impressed Voltages of Arbitrary Waveforms

After this discussion of the steady state with sinusoidal symmetrical three-phase voltages let us now return to the more general model of the symmetrical induction motor in dynamic condition, when torque and speed are varying.

We assume that the stator is supplied by impressed voltages having, in principle, variable frequency and arbitrary waveform, as long as Eq. (10.37) is fulfilled. In fact, due to the isolated neutral point of the motor, even a violation of this condition would be of no consequence; it would simply mean that the neutral of the windings assumes a different potential as the neutral of the supply voltages u_{S1}, u_{S2}, u_{S3}. Also, the results may be generalised to a supply having constant and equal internal impedances in each phase; they could be included in the stator impedances $(R_S, \sigma_S L_0)$ of the motor.

Should the motor be fed with sinusoidal symmetrical three-phase voltages of constant frequency and loaded with constant torque, the transients will eventually decay and the steady state condition discussed in Sect. 10.2 will emerge, where all quantities in the stator are periodic functions with the angular frequency ω_1. A better insight into the dynamic behaviour of the machine is obtained by choosing a moving system of coordinates, where these steady state oscillations disappear.

A suitable choice of coordinate system, independent of the waveform of the voltages, is that defined by the vector of stator voltages, Eqs. (10.34), (10.38),

$$\underline{u}_S(t) = u_{S1}(t) + u_{S2}(t)\, e^{j\,\gamma} + u_{S3}(t)\, e^{j\,2\,\gamma}$$

$$= u_{12}(t) - u_{23}(t)\, e^{-j\,\gamma} = u_S(t)\, e^{j\,\lambda(t)}\ ;$$

(10.95)

its instantaneous angular velocity

$$\omega_1(t) = \frac{d\lambda}{dt}$$

(10.96)

corresponds to the angular frequency of the stator voltages. Should the motor be supplied with sinusoidal symmetrical three-phase voltages of constant frequency and amplitude, Eq. (10.60) again becomes valid

$$\underline{u}_S(t) = \frac{3\sqrt{2}}{2} U_S\, e^{j\,(\omega_1 t + \tau_1)}\ ,$$

(10.97)

resulting in

$$u_S(t) = \frac{3\sqrt{2}}{2} U_S\ ,\quad \lambda(t) = \omega_1 t + \tau_1\ ,$$

(10.98)

where U_S is the RMS line-to-neutral voltage.

All electrical variables are now defined in this coordinate system; i.e. they are transformed as they appear to a moving observer positioned at angle $\lambda(t)$; by making use of Eqs. (10.7, 10.10) we define

$$\underline{i}_S(t)\, e^{-j\,\lambda} = i_S(t)\, e^{j\,(\zeta - \lambda)} = {}^1\underline{i}_S(t)$$

(10.99)

and

$$\underline{i}_R(t)\, e^{j\,\varepsilon}\, e^{-j\,\lambda} = i_R(t)\, e^{j\,(\xi + \varepsilon - \lambda)} = {}^1\underline{i}_R(t)$$

(10.100)

as the current vectors in moving coordinates. With sinusoidal symmetrical three-phase stator voltages of fixed frequency and constant load torque, these quantities are constant in steady state. When multiplying Eq. (10.50) by $e^{-j\,\lambda}$ we find

$$R_S\,\underline{i}_S\, e^{-j\,\lambda} + L_S\, \frac{d\underline{i}_S}{dt}\, e^{-j\,\lambda} + L_0\, \frac{d}{dt}(\underline{i}_R\, e^{j\,\varepsilon})\, e^{-j\,\lambda} = \underline{u}_S\, e^{-j\,\lambda}\ ,$$

(10.101)

which is transformed by substituting

$$\frac{d\,^1\underline{i}_S}{dt} = \frac{d}{dt}(\underline{i}_S\,e^{-j\lambda}) = \frac{d\underline{i}_S}{dt}e^{-j\lambda} - j\frac{d\lambda}{dt}\,\underline{i}_S\,e^{-j\lambda}\,. \tag{10.102}$$

Developing the expression

$$\frac{d\underline{i}_S}{dt}e^{-j\lambda} = \frac{d\,^1\underline{i}_S}{dt} + j\,\omega_1\,^1\underline{i}_S \tag{10.103}$$

correspondingly,

$$\frac{d(\underline{i}_R\,e^{j\,\varepsilon})}{dt}e^{-j\lambda} = \frac{d\,^1\underline{i}_R}{dt} + j\,\omega_1\,^1\underline{i}_R\,, \tag{10.104}$$

and inserting Eqs. (10.103), (10.104) into Eq. (10.101), the stator voltage equation in the moving reference frame results,

$$R_S\,^1\underline{i}_S + \frac{d}{dt}(L_S\,^1\underline{i}_S + L_0\,^1\underline{i}_R) + j\,\omega_1\,(L_S\,^1\underline{i}_S + L_0\,^1\underline{i}_R) = u_S(t)\,. \tag{10.105}$$

The rotor equation Eq.(10.51) is transformed accordingly, with $\underline{u}_R(t) \equiv 0$,

$$R_R\,^1\underline{i}_R + \frac{d}{dt}(L_R\,^1\underline{i}_R + L_0\,^1\underline{i}_S) + j\,(\omega_1 - \omega)\,(L_R\,^1\underline{i}_R + L_0\,^1\underline{i}_S) = 0\,. \tag{10.106}$$

These equations may be simplified by substituting the flux vectors, defined in Eqs. (10.25), (10.32) which are also transformed into moving coordinates,

$$^1\underline{\psi}_S = \underline{\psi}_S\,e^{-j\,\lambda} = L_S\,^1\underline{i}_S + L_0\,^1\underline{i}_R\,, \tag{10.107}$$

$$^1\underline{\psi}_R = \underline{\psi}_R\,e^{j\,(\varepsilon-\lambda)} = L_R\,^1\underline{i}_R + L_0\,^1\underline{i}_S\,. \tag{10.108}$$

Hence the current vectors are

$$^1\underline{i}_S = \frac{1}{\sigma\,L_S}\left(^1\underline{\psi}_S - \frac{1}{1+\sigma_R}\,^1\underline{\psi}_R\right)\,, \tag{10.109}$$

$$^1\underline{i}_R = \frac{1}{\sigma\,L_R}\left(^1\underline{\psi}_R - \frac{1}{1+\sigma_S}\,^1\underline{\psi}_S\right)\,. \tag{10.110}$$

Inserting Eqs. (10.107–10.110) into Eqs. (10.105, 10.106) results in

$$T_S'\,\frac{d\,^1\underline{\psi}_S}{dt} + [1 + j\,\omega_1\,T_S']\,^1\underline{\psi}_S - \frac{1}{1+\sigma_R}\,^1\underline{\psi}_R = T_S'\,u_S(t)\,, \tag{10.111}$$

$$T_R'\,\frac{d\,^1\underline{\psi}_R}{dt} + [1 + j\,(\omega_1 - \omega)\,T_R']\,^1\underline{\psi}_R - \frac{1}{1+\sigma_S}\,^1\underline{\psi}_S = 0\,, \tag{10.112}$$

where

$$T'_S = \frac{\sigma L_S}{R_S}, \quad T'_R = \frac{\sigma L_R}{R_R} \tag{10.113}$$

are transient time constants.

The expression for the electrical torque, Eq. 10.48, can be written in the form

$$m_M(t) = \tfrac{2}{3} L_0 \operatorname{Im}\left[\underline{i}_S \left(\underline{i}_R\, e^{j\varepsilon}\right)^*\right] = \tfrac{2}{3} L_0 \operatorname{Im}\left[{}^1\underline{i}_S \left({}^1\underline{i}_R\right)^*\right] ; \tag{10.114}$$

the torque is independent of the coordinate system because the inverse transformations with $e^{j\lambda}$ are cancelling. Hence, using Eqs. (10.109, 10.110) we have

$$m_M(t) = \frac{2}{3} \frac{L_0}{\sigma^2 L_S L_R} \operatorname{Im}\left[\left({}^1\underline{\psi}_S - \frac{1}{1+\sigma_R}\,{}^1\underline{\psi}_R\right)\left({}^1\underline{\psi}_R - \frac{1}{1+\sigma_S}\,{}^1\underline{\psi}_S\right)^*\right] . \tag{10.115}$$

Since only the mixed terms of the product contribute to the imaginary part, this reduces to

$$\begin{aligned} m_M(t) &= \frac{2}{3} \frac{L_0}{\sigma^2 L_S L_R} \operatorname{Im}\left[{}^1\underline{\psi}_S\,{}^1\underline{\psi}_R^* + \frac{1}{(1+\sigma_S)(1+\sigma_R)}\,{}^1\underline{\psi}_R\,{}^1\underline{\psi}_S^*\right] \\ &= \frac{2}{3} \frac{1-\sigma}{\sigma} \frac{1}{L_0} \operatorname{Im}\left[{}^1\underline{\psi}_S\,{}^1\underline{\psi}_R^*\right] . \end{aligned} \tag{10.116}$$

This is inserted into Eqs. (10.52, 10.53) for the mechanical motion.

To arrive at dimensionless real equations, the transformed flux vectors are now normalised and split in real and imaginary parts, representing the components that are parallel (x) and orthogonal (y) to the voltage vector. The flux linkages in steady state operation with sinusoidal voltages and nominal frequency, as indicated on the nameplate, and neglected stator resistance serve as flux references,

$$ {}^1\underline{\psi}_S(t) = \frac{3\sqrt{2}}{2} \frac{U_{S0}}{\omega_0} \left[x_S(t) + j\, y_S(t)\right] , \tag{10.117}$$

$$ {}^1\underline{\psi}_R(t) = \frac{3\sqrt{2}}{2} \frac{U_{S0}}{\omega_0} \left[x_R(t) + j\, y_R(t)\right] . \tag{10.118}$$

The magnitude of the voltage vector in Eq. (10.98) is also referred to the magnitude at nominal operating conditions,

$$u_S(t) = \frac{3\sqrt{2}}{2} U_{S0} \frac{U_S(t)}{U_{S0}} . \tag{10.119}$$

It is noted that no restrictions with regard to waveform or frequency of the stator voltages have been imposed so far.

With these definitions the voltage equations (10.111, 10.112) and the expression for the torque Eq. (10.116) assume the following form

$$T_S' \frac{d}{dt}(x_S + j\,y_S) + [1 + j\,\omega_1\,T_S']\,(x_S + j\,y_S)$$

$$- \frac{1}{1 + \sigma_R}\,(x_R + j\,y_R) = \omega_0\,T_S'\,\frac{U_S}{U_{S0}} \,, \qquad (10.120)$$

$$T_R' \frac{d}{dt}(x_R + j\,y_R) + [1 + j\,(\omega_1 - \omega)\,T_R']\,(x_R + j\,y_R)$$

$$- \frac{1}{1 + \sigma_S}\,(x_S + j\,y_S) = 0 \,, \qquad (10.121)$$

$$m_M = 2\,(1 + \sigma_S)\,m_{p0}\,(y_S\,x_R - x_S\,y_R) \,, \qquad (10.122)$$

where m_{p0} is the nominal peak torque for $R_S = 0$, as defined in Eq. (10.86). With the following abbreviations for

Pull–out slip at nominal
frequency and $R_S = 0$

$$S_{pR} = \frac{R_R}{\omega_0\,\sigma\,L_R} = \frac{1}{\omega_0\,T_R'} \,, \qquad (10.123)$$

Pull–out slip of rotor–fed
motor, nominal frequency
and $R_R = 0$

$$S_{pS} = \frac{R_S}{\omega_0\,\sigma\,L_S} = \frac{1}{\omega_0\,T_S'} \,, \qquad (10.124)$$

Mechanical time constant

$$T_m = \frac{J\,\omega_0}{m_{p0}} \,, \qquad (10.125)$$

and the instantaneous normalised frequencies
stator frequency

$$\frac{\omega_1}{\omega_0} \,,$$

mechanical rotational
frequency (speed)

$$\frac{\omega}{\omega_0} \,,$$

rotor frequency

$$\frac{\omega_2}{\omega_0} = \frac{\omega_1 - \omega}{\omega_0} \,, \qquad (10.126)$$

six real differential equations emerge for the direct and quadrature flux components as well as motor speed and position; ε_0 is an arbitrary angle of reference.

$$T_S' \frac{dx_S}{dt} = -x_S + \frac{1}{S_{pS}} \frac{\omega_1}{\omega_0} y_S + \frac{1}{1 + \sigma_R} x_R + \frac{1}{S_{pS}} \frac{U_S(t)}{U_{S0}} \qquad (10.127)$$

$$T_S' \frac{dy_S}{dt} = -\frac{1}{S_{pS}} \frac{\omega_1}{\omega_0} x_S - y_S + \frac{1}{1 + \sigma_R} y_R , \qquad (10.128)$$

$$T_R' \frac{dx_R}{dt} = \frac{1}{1 + \sigma_S} x_S - x_R + \frac{1}{S_{pR}} \frac{\omega_2}{\omega_0} y_R , \qquad (10.129)$$

$$T_R' \frac{dy_R}{dt} = \frac{1}{1 + \sigma_S} y_S - \frac{1}{S_{pR}} \frac{\omega_2}{\omega_0} x_R - y_R , \qquad (10.130)$$

$$T_m \frac{d\left(\frac{\omega}{\omega_0}\right)}{dt} = 2\left(1 + \sigma_S\right)\left(y_S x_R - x_S y_R\right) - \frac{m_L}{m_{p0}}\left(\omega, \varepsilon, t\right), \qquad (10.131)$$

$$\frac{\varepsilon_0}{\omega_0} \frac{d\left(\frac{\varepsilon}{\varepsilon_0}\right)}{dt} = \frac{\omega}{\omega_0} . \qquad (10.132)$$

These equations may be represented in graphical form by the block diagram in Fig. 10.16, where the integrators with feedback are redrawn to form unity gain lag elements. Due to the rotational symmetry of the machine the angles $\varepsilon(t)$, $\lambda(t)$ do not enter the right hand side of the model equations Eqs. (10.127–10.131) with the exception of load torque m_L that may depend on the mechanical angle.

The block diagram describes the dynamic behaviour of the symmetrical motor, fed by impressed stator voltages independent of waveform; the magnitude $u_S(t)$ of the voltage vector is given by Eq. (10.119), its instantaneous angular frequency $\omega_1(t)$ by Eq. (10.96). As special case with $U_S =$ const., $\omega_1 = d\lambda/dt =$ const., the transient condition of the motor operating on the symmetrical three-phase supply results, where U_S is the RMS value of the line- to- neutral voltages.

Because of the transformation into the voltage based coordinate system moving with stator frequency, the steady state values of the flux components and speed, assuming constant load torque, are constant. The steady state values are derived by setting the derivatives in Eqs. (10.127–10.131) equal to zero; the resulting system of algebraic equations is equivalent to the results obtained in Sect. 10.2.

Of course, under dynamic conditions the flux components are no longer constant because there may be transient flux waves which are linked to the stator or rotor windings and, hence, move with the angular velocity ω_1 or $\omega_2 = \omega_1 - \omega$ relative to the coordinate system, where they appear as AC

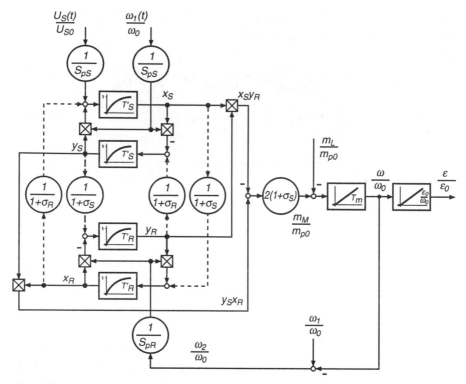

Fig. 10.16. Block diagram of symmetrical induction motor fed by impressed voltages of arbitrary wave form

components of appropriate frequencies; as a result there are torque and speed variations.

When inspecting the general structure of Fig. 10.16 it is recognised that it can be reduced to the form shown in Fig. 10.17 which reminds one of a DC machine, Fig. 5.9. The generation of the electric torque m_M is now much more involved, which is due to the fact that the currents in the rotor winding are magnetically induced rather than impressed through a galvanic contact; also, the angle between the stator- and rotor-ampereturn waves depends on the operating condition, whereas it is essentially fixed by the field- and armature-axes in the case of a compensated DC machine. The presence of AC quantities also tends to blur the view, even though this has been alleviated by the coordinate transformation.

While no analytical solution of the nonlinear differential equations exists, some of the interactions in Fig. 10.16 can be qualitatively interpreted. For example, two feedback loops are recognisable, one belonging to the stator and the other to the rotor, that contain multiplications by ω_1/ω_0 and ω_2/ω_0 respectively. When neglecting for a moment the cross couplings drawn in dashed lines, a homogeneous second order equation may be gained describing

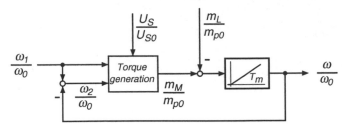

Fig. 10.17. Simplified block diagram of symmetrical induction motor

the flux components, for instance in the stator; by eliminating y_S we find

$$T_S'^2 \frac{d^2 x_S}{dt^2} + 2 T_S' \frac{d x_S}{dt} + \left[1 + \frac{1}{S_{pS}^2} \left(\frac{\omega_1}{\omega_0} \right)^2 \right] x_S = 0 , \qquad (10.133)$$

with the eigenvalues

$$s_{1,2} = \omega_0 \left(-S_{pS} \pm j \frac{\omega_1}{\omega_0} \right) ; \qquad (10.134)$$

hence the pertinent transient is a damped oscillation with the angular frequency ω_1, belonging to a flux wave that is linked to the stator winding and thus remains fixed in space. The damping is achieved by the dynamic short circuit through the voltage source at the stator terminals. Likewise the lower feedback circuit in Fig. 10.16 is found to be described by a differential equation with the eigenvalues

$$s_{3,4} = \omega_0 \left(-S_{pR} \pm j \frac{\omega_2}{\omega_0} \right) . \qquad (10.135)$$

The associated physical effect is a flux wave linked to the rotor winding which moves with slip frequency relative to the coordinate axis.

Considering now the special case $\omega_1 = \omega_2 = 0$, when the mechanically locked motor is fed with direct voltage, and taking the dashed cross-couplings into account, two new non-interacting feedback loops are formed, where each embraces stator and rotor. The differential equations for transients in the x-axis may be combined to

$$T_S' T_R' \frac{d^2 x_S}{dt^2} + (T_S' + T_R') \frac{d x_S}{dt} + \sigma x_S = 0 . \qquad (10.136)$$

Both eigenvalues are negative real, belonging to aperiodic transients; a corresponding equation is valid for the quadrature axis of the machine. Clearly these equations correspond to those of two-winding transformers, describing the decoupled magnetic interactions in the two axes of the machine; under the assumed test conditions and in view of the voltage-based coordinate system,

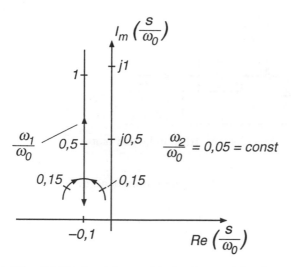

Fig. 10.18. Root locus of induction machine for $\omega_2 = 0.05 = $ const. at different stator frequencies and speeds

the flux in the quadrature axis must be zero, $y_S = y_R = 0$, hence there is no torque. This is in agreement with the practical experience of a symmetrical induction motor with DC excitation at standstill.

When analysing the block diagram in Fig. 10.16 more closely, i.e. for $\omega_1, \omega_2 \neq 0$, the mutual interactions of the two damped oscillators, coupled directly and through the speed, must be taken into account. This leads to a complicated resonant system of 6th order, the response of which is difficult to assess. However, assuming $\omega_1 = $const. and $\omega_2 \approx$ const., i.e. large inertia, there are considerable simplifications because the multiplications with ω_1, ω_2 are reduced to constant factors; assuming complete symmetry between stator and rotor,

$$S_{pS} = S_{pR} = S_p , \quad \sigma_S = \sigma_R = \sigma_0 , \quad T_S' = T_R' = T_0 , \tag{10.137}$$

the characteristic equation pertaining to Eqs. (10.127–10.130) is

$$\left[(T_0\,s + 1)^2 - (1 - \sigma) + \frac{1}{S_p^2}\left(\frac{\omega_1}{\omega_0}\right)^2 \right] \left[(T_0\,s + 1)^2 - (1 - \sigma) + \frac{1}{S_p^2}\left(\frac{\omega_2}{\omega_0}\right)^2 \right]$$

$$+ \frac{1 - \sigma}{S_p^2}\frac{(\omega_1 + \omega_2)^2}{\omega_0^2} = 0 .$$

$$\tag{10.138}$$

The solutions of this equation, i.e. the eigenvalues of the system, have been computed numerically for different values of stator frequency ω_1, assuming the following parameters

$$S_p = 0.10, \quad \sigma = 0.08, \quad \frac{\omega_2}{\omega_0} = 0.05 = \text{const.} \tag{10.139}$$

The result is plotted in Fig. 10.18 in the form of a root locus in the complex s - plane, omitting the conjugate branch of the locus. The shape of the curve indicates that there is little interaction between the two oscillators in Fig. 10.16 for $\omega_1 \gg \omega_2$, approximately leading to the "decoupled" eigenvalues mentioned before; on the other hand, the interactions are strong at low values of the stator frequency, $\omega_1 \approx \omega_2$, resulting in slowly decaying transients. Figure 10.19 depicts the decay of a flux component at different stator frequencies, assuming the same initial conditions; the transient torque is also shown. There is a remarkable effect of stator frequency and speed on the dynamic behaviour of the nonlinear plant. Therefore the induction machine, when operated in a wide speed range, represents a very difficult control plant, which calls for special control procedures; some will be discussed in Chap. 12.

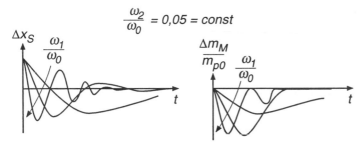

Fig. 10.19. Flux and torque transients for $\omega_2 = 0.05 = $ const at different stator frequencies and speeds

As a qualitative test for the validity of the mathematical model of the induction machine, which was derived with considerable simplifications, a line start at nominal voltage and frequency has been computed by numerical integration of Eqs. (10.127–10.131); the load torque is initially zero but, after steady state speed has been reached, half the peak torque is applied as load. The results are plotted in Fig. 10.20, assuming a low value of inertia so that the start is completed in a few periods of line voltage; the other parameters are the same as before.

Besides the flux components in direct and quadrature direction, the stator currents are of main interest. The associated vector is produced by inverse transformation with the help of Eq. (10.109)

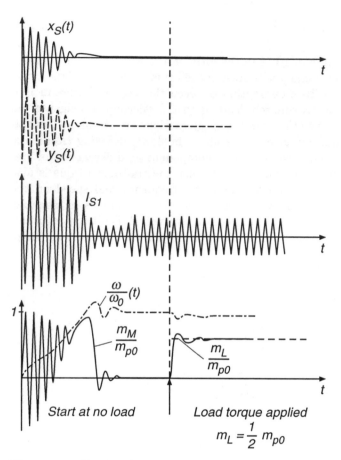

Fig. 10.20. Simulated starting transient of induction motor with low value of inertia

$$i_S(t) = {}^1\underline{i}_S\, e^{j\,\omega_1 t} = \frac{1}{\sigma\,L_S}\left[{}^1\underline{\psi}_S(t) - \frac{1}{1+\sigma_R}\,{}^1\underline{\psi}_R\right]e^{j\,\omega_1 t}$$

$$= \frac{3\sqrt{2}}{2}\,\frac{U_{S0}}{\omega_0\,\sigma\,L_S}\left[\left(x_S - \frac{1}{1+\sigma_R}\,x_R\right) + j\left(y_S - \frac{1}{1+\sigma_R}\,y_R\right)\right]e^{j\,\omega_1 t}\,.$$

$$(10.140)$$

With Eq. (10.55) follows

$$i_{S1}(t) = \tfrac{2}{3}\,\mathrm{Re}\,[\underline{i}_S(t)]$$

$$= \sqrt{2}\,\frac{U_{S0}}{\omega_0\,\sigma\,L_S}\left[\left(x_S - \frac{1}{1+\sigma_R}\,x_R\right)\cos\omega_1 t - \left(y_S - \frac{1}{1+\sigma_R}\,y_R\right)\sin\omega_1 t\right]\,.$$

$$(10.141)$$

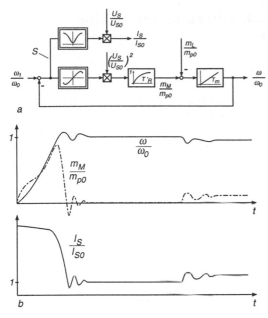

Fig. 10.21. (a) Simplified block diagram of induction machine, **(b)** calculated transients

This is also plotted in Fig. 10.20; it shows the characteristic large inrush current which, as the motor approaches synchronism, settles down to the small sinusoidal no-load current until the load torque is applied. The currents in the other stator phases are similar.

There are strong alternating components of torque which also affect the speed, even though with reduced amplitude because of the mechanical inertia. If the inertia is predominantly on the load side, the shaft and the mechanical coupling may be severely stressed during the start. The speed, as a function of time, follows a similar curve as obtained in Sect. 3.2 when the electrical transients were totally ignored, Fig. 3.12; this explains also the absence of the overshoot in speed which is present in Fig. 10.20.

As an intermediate step between the quasi-steady-state model and the more complete model of the induction motor, the heuristic block diagram in Fig. 10.21 a may be helpful. It differs from the purely mechanical model by the inclusion of a lag T'_R, following the steady-state torque-slip-function. For small values of slip, where the torque function is linear, this clearly resembles the block diagram of a DC machine. A comparison of the starting transient computed with this model shows good qualitative agreement with the transients obtained by integrating the Eqs. (10.127–10.131). Of course, the oscillations superimposed on speed and torque cannot be represented but the mean values are well modelled.

10.4 Induction Motor with Unsymmetrical Line Voltages in Steady State

10.4.1 Symmetrical Components

For intermittent duty drives, such as cranes or hoists, where simplicity of hardware is more important than high efficiency, induction motors with un-symmetrical voltage supply may be used. By way of example, a phasor diagram of sinusoidal but unsymmetrical line-to-neutral voltages is shown in Fig. 10.22 a. Due to the isolated neutral of the symmetrical stator winding, the sum of the instantaneous phase voltages is still zero, Eq. (10.37); hence the phasor diagram forms a closed triangle,

$$\underline{U}_{S1} + \underline{U}_{S2} + \underline{U}_{S3} = 0 \ . \tag{10.142}$$

The same holds naturally if the stator windings are Δ-connected.

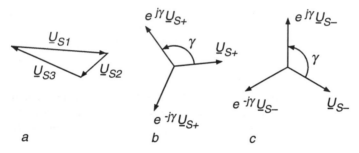

Fig. 10.22. Decomposition of unsymmetrical balanced three-phase system into symmetrical components

It is known, that an unsymmetrical three-phase system may be decomposed into "symmetrical components" defined by

$$\underline{U}_{S1} = \qquad \underline{U}_{S+} + \qquad \underline{U}_{S-} + \underline{U}_0 \ ,$$

$$\underline{U}_{S2} = e^{-j\gamma}\underline{U}_{S+} + \quad e^{j\gamma}\underline{U}_{S-} + \underline{U}_0 \ , \qquad \gamma = \frac{2\pi}{3} \ , \tag{10.143}$$

$$\underline{U}_{S3} = \quad e^{j\gamma}\underline{U}_{S+} + e^{-j\gamma}\underline{U}_{S-} + \underline{U}_0 \ ,$$

where $\underline{U}_{S+}$ is the positive-, $\underline{U}_{S-}$ the negative- and $\underline{U}_0$ the zero-sequence voltage component per phase. The symmetrical components $\underline{U}_{S+}, \underline{U}_{S-}$ form two symmetrical three-phase systems of opposite phase-sequence. Solving Eq. (10.143) yields

$$\underline{U}_{S+} = \tfrac{1}{3} \left[\underline{U}_{S1} + e^{j\gamma} \underline{U}_{S2} + e^{-j\gamma} \underline{U}_{S3} \right] ,$$

$$\underline{U}_{S-} = \tfrac{1}{3} \left[\underline{U}_{S1} + e^{-j\gamma} \underline{U}_{S2} + e^{j\gamma} \underline{U}_{S3} \right] , \tag{10.144}$$

$$\underline{U}_0 = \tfrac{1}{3} \left[\underline{U}_{S1} + \underline{U}_{S2} + \underline{U}_{S3} \right] = 0 .$$

The expression for the positive sequence component is similar to that for the time dependent voltage vector, Eq. (10.34). Since the voltages between the motor terminals and the isolated neutral are not directly available (the motor may be Δ-connected), it is appropriate to introduce the line-to-line voltages instead. By expanding Eq. (10.144) a we find, analogous to Eq. (10.38),

$$\underline{U}_{S+} = \tfrac{1}{3} \big[\underbrace{\underline{U}_{S1} - \underline{U}_{S2}}_{\underline{U}_{12}} - e^{-j\gamma} \underbrace{(\underline{U}_{S2} - \underline{U}_{S3})}_{\underline{U}_{23}} + \underline{U}_{S2} \underbrace{(1 + e^{j\gamma} + e^{-j\gamma})}_{-0} \big]$$

$$= \tfrac{1}{3} \left[\underline{U}_{12} - e^{-j\gamma} \underline{U}_{23} \right] \tag{10.145}$$

and correspondingly

$$\underline{U}_{S-} = \tfrac{1}{3} \left[\underline{U}_{12} - e^{j\gamma} \underline{U}_{23} \right] . \tag{10.146}$$

Two of the line-to-line voltages are sufficient to determine the positive and negative sequence components, because

$$\underline{U}_{12} + \underline{U}_{23} + \underline{U}_{31} = 0 \tag{10.147}$$

is valid by definition. Generally the use of line-to-line voltages does not exclude the appearance of an arbitrary zero sequence component $\underline{U}_0$ but, as seen before, it vanishes with a symmetrical load. If the motor is fed from a line possessing noticeable internal impedance, the symmetrical components of the terminal voltages depend also on the load, i.e. the operating state of the motor.

Should the voltages $\underline{U}_{12}, \underline{U}_{23}, \underline{U}_{31}$ form a symmetrical three-phase system,

$$\underline{U}_{23} = e^{-j\gamma} \underline{U}_{12} , \tag{10.148}$$

the result is

$$\underline{U}_{S+} = \frac{1}{\sqrt{3}} e^{-j\pi/6} \underline{U}_{12} = \underline{U}_{S1} , \quad \underline{U}_{S-} = 0 . \tag{10.149}$$

The case of a symmetrical induction motor fed by unsymmetrical supply voltages is fully contained in the theory developed in Sect. 10.1 but the subsequent discussion of some examples is restricted to steady-state.

When forming the time-varying voltage vector according to Eqs. (10.57–10.59, 10.60) and with the definition (10.34), we find

$$\underline{u}_S(t) = \frac{3\sqrt{2}}{2} [\underline{U}_{S+} e^{j\omega_1 t} + \underline{U}_{S-}^* e^{-j\omega_1 t}], \qquad (10.150)$$

a result, that can be interpreted as the superposition of two constant vectors rotating with equal angular velocity in opposite direction; hence the vectorial sum $\underline{u}_S(t)$ follows a periodic elliptical path, as seen in Fig. 10.23.

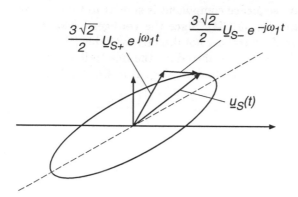

Fig. 10.23. Voltage vector of unsymmetrical balanced supply

In steady state, assuming constant speed, similar definitions hold for the vector of the stator currents,

$$\underline{i}_S(t) = \frac{3\sqrt{2}}{2} [\underline{I}_{S+} e^{j\omega_1 t} + \underline{I}_{S-}^* e^{-j\omega_1 t}], \qquad (10.151)$$

where $\underline{I}_{S+}$, $\underline{I}_{S-}$ are the symmetrical components of the stator currents which are defined in accordance with Eq. (10.143). Likewise the rotor current vector in stator coordinates may be written as

$$\underline{i}_R(t) e^{j\varepsilon} = \frac{3\sqrt{2}}{2} [\underline{I}_{R+} e^{j\omega_1 t} + \underline{I}_{R-}^* e^{-j\omega_1 t}], \qquad (10.152)$$

while the rotor current vector in rotor coordinates is, with $d\varepsilon/dt = \omega = $ const.,

$$\underline{i}_R(t) = \frac{3\sqrt{2}}{2} [\underline{I}_{R+} e^{j(\omega_1-\omega)t} + \underline{I}_{R-}^* e^{-j(\omega_1+\omega)t}]. \qquad (10.153)$$

The speed ω is counted in the direction of the positive sequence field. Hence the frequency of the positive and negative sequence rotor currents is $\omega_1 - \omega$ and $\omega_1 + \omega$ respectively.

The physical interpretation of Eqs. (10.151, 10.152) is that of two separate ampereturns waves rotating with $\pm\omega_1$ in opposite directions and causing corresponding flux density waves in the airgap which in turn induce the voltage components of Eq. (10.150). As long as iron saturation is ignored and the speed may be assumed constant due to a sufficiently large inertia, there is no interaction between the two symmetrical three-phase systems so that linear relations exist between voltage and current components,

$$\underline{I}_{S+} = Y_{S+}\,\underline{U}_{S+}\,,\quad \underline{I}_{S-} = Y_{S-}\,\underline{U}_{S-}\,; \tag{10.154}$$

where Y_{S+}, Y_{S-} are the positive and negative sequence stator admittances per phase. Because of the effective values of slip for positive and negative sequence components, respectively,

$$\frac{\omega_1 - \omega}{\omega_1} = S \quad\text{and}\quad \frac{\omega_1 + \omega}{\omega_1} = 2 - S\,; \tag{10.155}$$

the admittances, derived from Eq. (10.72) with $R_S = 0$, are

$$Y_{S+} = \frac{1}{j\,\omega_1\,L_S}\,\frac{1 + j\,\frac{1}{\sigma}\,\frac{S}{S_p}}{1 + j\,\frac{S}{S_p}}\,,$$

$$Y_{S-} = \frac{1}{j\,\omega_1\,L_S}\,\frac{1 + j\,\frac{1}{\sigma}\,\frac{2-S}{S_p}}{1 + j\,\frac{2-S}{S_p}}\,.$$

When substituting Eqs. (10.151, 10.152) into the torque equation (10.48), four terms arise, two of which are alternating torque components having the frequency $2\,\omega_1$; contributions to the mean torque are only produced by interactions between stator and rotor currents of the same sequence; hence the mean torque, counted in positive sequence direction, is

$$\overline{m}_M = 3\,L_0\,\mathrm{Im}\,\lfloor\underline{I}_{S+}\,\underline{I}_{R+}^{*} + \underline{I}_{S}^{*}\,\underline{I}_{R-}\rfloor = m_{M+} + m_{M-}\,. \tag{10.156}$$

These two components of the mean torque may be derived by the same reasoning as shown in Sect. 10.2, because each is produced by a symmetrical three-phase voltage system.

In accordance with Eq. (10.85) we find

$$\frac{\overline{m}_M}{m_{p0}} = \frac{m_{M+}}{m_{p0}} + \frac{m_{M-}}{m_{p0}}$$

$$= \frac{2}{\frac{S}{S_p} + \frac{S_p}{S}}\left[\frac{U_{S+}}{U_{S0}}\,\frac{\omega_0}{\omega_1}\right]^2 - \frac{2}{\frac{2-S}{S_p} + \frac{S_p}{2-S}}\left[\frac{U_{S-}}{U_{S0}}\,\frac{\omega_0}{\omega_1}\right]^2\,, \tag{10.157}$$

where m_{p0} is again the peak torque, Eq. (10.86), at nominal voltage and S_p the pull-out slip for $R_S = 0$, Eq. (10.71).

Equation (10.157) may be interpreted as the net torque produced by two identical motors coupled to the same shaft which are supplied by symmetrical three-phase voltage systems U_{S+}, U_{S-} of equal frequency but opposite phase sequence, i.e. with two stator terminals of the second motor interchanged, Fig. 10.24.

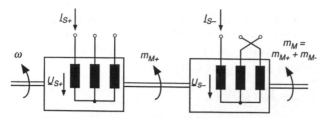

Fig. 10.24. Induction motor with unsymmetrical supply voltages, equivalent scheme

Eq. (10.157) clearly demonstrates the basic disadvantage of all control schemes employing unsymmetrical supply; while the torque is reduced, the power losses are increased by the negative sequence voltage and current components.

By changing the rotor resistance (S_p) and with different circuit connections, resulting in speed-dependent positive and negative sequence voltages, a variety of torque/speed-characteristics can be obtained. This is discussed with the help of three examples.

10.4.2 Single–phase Induction Motor

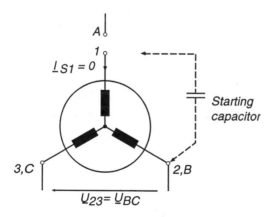

Fig. 10.25. Symmetrical induction motor with single phase supply

By disconnecting one stator terminal of a symmetrical induction motor, the single phase circuit in Fig. 10.25 results, A, B, C are the terminals of the symmetrical line voltages. The constraint imposed by the open circuit is

$$\underline{I}_{S1} = Y_{S+}\underline{U}_{S+} + Y_{S-}\underline{U}_{S-} = 0 ; \tag{10.158}$$

another condition follows from Eqs. (10.145), (10.146),

$$\underline{U}_{S+} - \underline{U}_{S-} = \frac{1}{3}\left(e^{j\gamma} - e^{-j\gamma}\right)\underline{U}_{23} = j\frac{1}{\sqrt{3}}\underline{U}_{23} = \underline{U}_{AM} , \qquad (10.159)$$

where $\underline{U}_{AM}$ is the line-to-neutral voltage of the disconnected line-phase. When combining the two equations, we find

$$\underline{U}_{S+} = \frac{Y_{S-}}{Y_{S+} + Y_{S-}}\underline{U}_{AM} = \frac{Z_{S+}}{Z_{S+} + Z_{S-}}\underline{U}_{AM} ,$$

$$(10.160)$$

$$\underline{U}_{S-} = -\frac{Y_{S+}}{Y_{S+} + Y_{S-}}\underline{U}_{AM} = -\frac{Z_{S-}}{Z_{S+} + Z_{S-}}\underline{U}_{AM} ;$$

$Z_S = 1/Y_S$ is the impedance of a stator phase in symmetrical three–phase operation. The same voltage magnitudes would result if the stator windings of the two coupled motors in Fig. 10.24 were connected in series to the symmetrical three- phase voltages. According to Fig. 10.26 the positive and negative sequence admittances can be taken from the normalised circular diagram (Fig. 10.8), where $\underline{I}_S/\underline{I}_{S0} = Y_S/Y_{S0}$ was plotted for $R_S = 0$ as a function of slip. This indicates that a large difference exists between $Y_{S+} = Y_S(S)$ and $Y_{S-} = Y_S(2-S)$ for small values of pull-out slip, i.e. for a motor with a cage rotor.

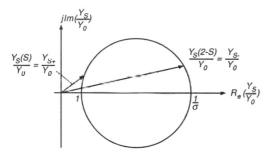

Fig. 10.26. Circular diagram showing positive and negative sequence admittance

If the motor operates near the positive synchronous speed ($S \approx 0$), $|Y_{S+}| \ll |Y_{S-}|$ holds, resulting in $\underline{U}_{S+} \approx \underline{U}_{AM}$, $\underline{U}_{S-} \approx 0$. This means that the negative sequence field is almost completely suppressed by the rotor winding due to the large relative speed $2 - S \approx 2$; hence the motor operates with nearly symmetrical voltages despite the fact that one terminal is not connected. In the vicinity of the negative synchronous speed the conditions are inverted.

The steady-state torque of the single phase motor is obtained by substituting Eqs. (10.160) into Eq. (10.157)

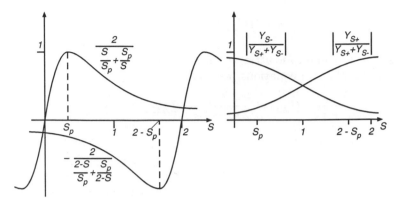

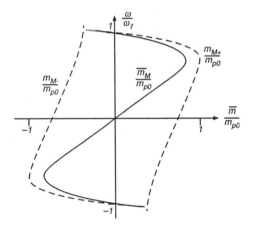

Fig. 10.27. Symmetrical components of single–phase motor

Fig. 10.28. Torque/speed characteristic of single–phase motor

$$\frac{\overline{m}_M}{m_{p0}} = \tag{10.161}$$

$$\left[\frac{2}{\frac{S}{S_p} + \frac{S_p}{S}} \left|\frac{Y_{S-}}{Y_{S+} + Y_{S-}}\right|^2 - \frac{2}{\frac{2-S}{S_p} + \frac{S_p}{2-S}} \left|\frac{Y_{S+}}{Y_{S+} + Y_{S-}}\right|^2\right] \left[\frac{U_{AM}}{U_{S0}} \frac{\omega_0}{\omega_1}\right]^2.$$

Since the torque rises with the square of the voltage, the negative sequence torque is even further reduced than the voltage. Some of the functions contained in Eq. (10.161) are sketched in Fig. 10.27; by combining these curves the torque/speed characteristic of the single-phase motor is obtained, drawn in the usual orientation, Fig. 10.28. The curve is symmetrical with respect to the origin because at $S = 1$, $Y_{S+} = Y_{S-}$ holds and the opposing torque components cancel. Hence the motor develops no net torque at standstill and cannot start without external assistance. If a small motor at standstill is

turned manually in either direction, the corresponding sequence prevails and
the motor continues to accelerate at light load up to the operating speed.

To make the motor self-starting, either some asymmetry has to be built
into it, as is done with shaded-pole motors, where a short circuited turn is
built into the stator winding, or a starting capacitor must be connected to
the third or an auxiliary winding, as is shown in Fig. 10.25; in both cases the
direction of rotation is fixed. The value of the starting torque depends on the
design of the auxiliary winding and the capacitor.

Besides the starting problems, the drawbacks of single-phase operation lie
in the residual pulsating torque as well as reduced utilisation and efficiency
of the motor which are always below that of a symmetrically fed motor. It
is for these reasons that single-phase induction motors are only used at low
power ratings, for example in household appliances.

10.4.3 Single–phase Electric Brake for AC Crane–Drives

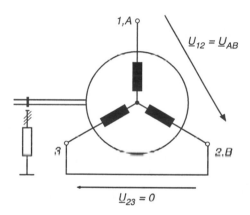

Fig. 10.29. Single phase supply of AC motor for braking duty

A simple intermittent-duty AC crane-drive required to operate as electric
brake in the lowering direction may have the asymmetrical circuit connection shown in Fig. 10.29, where two stator terminals of a symmetrical wound
rotor motor with large rotor resistor are short-circuited while single-phase
voltage is applied between the short-circuit connection and the remaining
stator terminal [G23]. This is one of several possible connections having similar properties. The analysis of the circuit is easy with the help of symmetrical
components.

Introducing the constraints of the circuit

$$\underline{U}_{12} = \underline{U}_{AB} , \quad \underline{U}_{23} = 0 \tag{10.162}$$

into Eqs. (10.145), (10.146) results in

$$U_{S+} = U_{S-} = \tfrac{1}{3} U_{AB} = \frac{1}{\sqrt{3}} U_{AM} , \qquad (10.163)$$

indicating that the symmetrical voltage components are of equal magnitude, independent of speed. Hence, the two machines in the equivalent circuit, Fig. 10.24, would have to be supplied by constant and equal three-phase voltages with two terminals of one motor interchanged; this renders the superposition of the two opposing torques particularly simple.

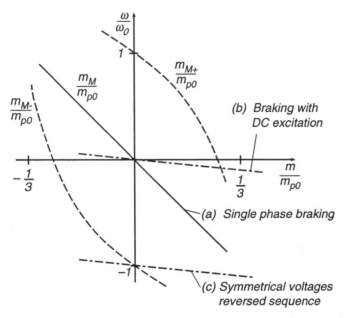

Fig. 10.30. Torque/speed curves of AC motor in unsymmetrical connection

The torque curves are depicted in Fig. 10.30 assuming $S_p = 2.0$ caused by a large external rotor resistor. Due to the reduced voltages the peak torques are only $\tfrac{1}{3}$ of the nominal peak torque, hence the motor is again only partially utilised. Since the resulting torque/speed curves lie in quadrants 2 and 4 the drive can only operate as an electric brake, regeneration is not possible with this scheme.

An advantage of this connection, when used as a lowering brake in crane applications, is the fact that the resulting torque/speed curve (a) passes through the origin; unintentional speed reversal (raising) at light load does not occur as would be possible with symmetrical supply and the rotor resistance still further increased.

Similar braking characteristics are obtained with a smaller rotor resistance by feeding the stator winding with direct current (b). The motor then operates as an eddy-current brake.

Finally, Fig. 10.30 also contains the torque/speed curve with symmetrical voltages but inverted sequence and short-circuited rotor resistance (c). This operation allows regeneration, however at higher than synchronous speed in lowering direction. It could be used as the "Full speed, Lowering" set-point of a contactor-controlled AC crane drive.

10.4.4 Unsymmetrical Starting Circuit for Induction Motor

Smooth acceleration during start-up is important in some applications, such as textile machines, where there is a danger of breaking fibre threads; an asymmetrical circuit like the one shown in Fig. 10.31 can provide a simple solution for this problem. An external impedance Z_1 is inserted between a line phase and one of the stator terminals of a symmetrical cage- or wound-rotor induction motor; the circuit is one of several similar connections. Assuming the impedance Z_1 to be constant, all voltages and currents remain sinusoidal in steady state; the following equations hold:

$$Z_1 \underline{I}_{S1} + \underline{U}_{12} = \underline{U}_{AB} = (1 - e^{-j\gamma}) \underline{U}_{AM} , \tag{10.164}$$

$$\underline{U}_{23} = \underline{U}_{BC} = (e^{-j\gamma} - e^{j\gamma}) \underline{U}_{AM} . \tag{10.165}$$

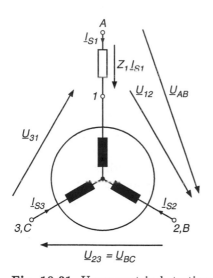

Fig. 10.31. Unsymmetrical starting circuit

The stator current can be expressed with symmetrical components, Eq. (10.158),

$$Z_1 \underline{I}_{S1} = Z_1 [Y_{S+} \underline{U}_{S+} + Y_{S-} \underline{U}_{S-}] ; \tag{10.166}$$

likewise the line-to-line voltages obtained from Eqs. (10.143) are

$$\underline{U}_{12} = \underline{U}_{S1} - \underline{U}_{S2} = (1 - e^{-j\gamma})\,\underline{U}_{S+} + (1 - e^{j\gamma})\,\underline{U}_{S-}\,,$$

$$\underline{U}_{23} = \underline{U}_{S2} - \underline{U}_{S3} = (e^{-j\gamma} - e^{j\gamma})\,\underline{U}_{S+} + (e^{j\gamma} - e^{-j\gamma})\,\underline{U}_{S-}\,.$$

Substituting these equations into Eqs. (10.164, 10.165) and solving for $\underline{U}_{S+}$, $\underline{U}_{S-}$ results in

$$\underline{U}_{S+} = \frac{3 + Z_1\,Y_{S-}}{3 + Z_1\,(Y_{S+} + Y_{S-})}\,\underline{U}_{AM}\,, \tag{10.167}$$

$$\underline{U}_{S-} = \frac{-Z_1\,Y_{S+}}{3 + Z_1\,(Y_{S+} + Y_{S-})}\,\underline{U}_{AM}\,. \tag{10.168}$$

This contains the limiting cases of the symmetrically fed motor ($Z_1 = 0$) and the single phase induction motor ($Z_1 \to \infty$) discussed in the previous paragraph. In general, the positive and negative sequence voltages are complicated functions of the motor admittances which in turn depend on speed and pull-out slip.

The voltage across the series impedance

$$Z_1\,\underline{I}_{S1} = Z_1\,[Y_{S+}\,\underline{U}_{S+} + Y_{S-}\,\underline{U}_{S-}] = -3\,\underline{U}_{S-} \tag{10.169}$$

is a measure of the asymmetry of the stator voltages. With these results the phasor diagram of the motor can be drawn for any chosen operating point.

The torque/speed curves may be computed from Eq. (10.157); their shape depends on the impedance Z_1 and the motor parameters, but they lie inside the area which is bounded by the curves for $Z_1 = 0$ and $Z_1 \to \infty$. This is qualitatively indicated in Fig. 10.32 for a low-power induction motor having fairly large pull-out slip and assuming a resistive impedance, $Z_1 = R_1$. Of course, this and all similar arrangements are associated with additional losses in the external components as well as the motor itself and are only suitable for short duty operation.

The function of the circuit shown in Fig. 10.31 is further illustrated by discussing the phasor diagram of the stator voltages in locked rotor condition. At $S = 1$ we have $Y_{S+} = Y_{S-} = Y_S(1)$. Hence it follows from Eqs. (10.167–10.169) with $Z_1 = R_1$

$$R_1\,\underline{I}_{S1} = \frac{3\,R_1\,Y_S(1)}{3 + 2\,R_1\,Y_S(1)}\,\underline{U}_{AM}\,, \qquad 0 \le R_1 < \infty\,; \tag{10.170}$$

as a limiting case ($R_1 \to \infty$) this includes again the single phase motor

$$\lim_{R_1 \to \infty} (R_1\,\underline{I}_{S1}) = \tfrac{3}{2}\,\underline{U}_{AM}\,, \tag{10.171}$$

with the phasor diagram of the stator voltages at $S = 1$ degenerating into a straight line (Fig. 10.33). Eq. (10.170) represents a linear function which maps the real R_1-axis into a circular arc in the complex plane, extending from $R_1 = 0$ to $R_1 \to \infty$. When the rotor is running, $0 < S < 1$, the phasor diagram changes with the tendency of reducing the voltage $R_1 \underline{I}_{S1}$, thus achieving improved symmetry. This corresponds to the situation found with the single-phase motor.

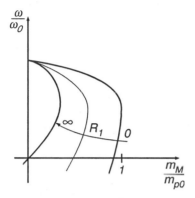

Fig. 10.32. Torque/speed curves of unsymmetrical starting circuit for large pull-out slip

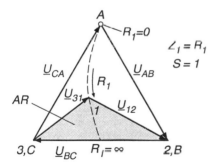

Fig. 10.33. Phasor diagram of asymmetrical circuit at standstill

The shaded area AR of the triangle (Fig. 10.33), formed by the terminal voltages $\underline{U}_{12}$, $\underline{U}_{23}$, $\underline{U}_{31}$ at standstill ($S = 1$), can be expressed as

$$AR = \tfrac{1}{2} \, \mathrm{Im} \, [\underline{U}_{23} \, \underline{U}_{21}^*] \; ; \tag{10.172}$$

which by inserting Eqs. (10.145, 10.146) may be written as

$$AR = k \, [U_{S+}^2 - U_{S-}^2] \, . \tag{10.173}$$

A comparison with Eq. (10.157) shows that AR is a measure of the torque at standstill [28]; it vanishes for $R_1 \to \infty$ and assumes its maximum value at $R_1 = 0$, i.e. with symmetrical supply.

The circuit shown in Fig. 10.31 can be extended by additional controllable impedances to allow continuous speed control, even four-quadrant operation but not regeneration. This has been realised in the past with saturable reactors, the impedance of which could be varied by direct current in specially connected control windings. Because of the high losses these drives were restricted to short duty operation; also the control was rather slow due to the sluggish response of the large reactors, making this type of equipment obsolete today [L29]. A similar control effect, however with greatly reduced size and cost of the components as well as much more rapid response can be achieved by inserting antiparallel thyristors into the stator supply leads of the motor; this causes considerable nonlinear distortion of the motor voltages, resulting in yet increased losses but it may be acceptable in short duty applications.

11. Power Supplies
for Adjustable Speed AC Drives

An important result of the preceding chapter is that the application of AC motors in continuous duty adjustable speed drives calls for static converters of adequate power, generating three-phase voltages of variable amplitude and frequency. This is necessary to maintain at all speeds a low rotor frequency, which is a precondition for acceptable overall efficiency of the drive. Converters of this type are available today employing thyristors, gate turn-off thyristors (GTO), or switched power transistors (MOSFET and IGBT), but the complexity and cost of the converter equipment and control still exceeds that of line-commutated converters of a similar rating. While part of the cost for the converter can be recovered by the reduced cost of the AC motor as compared to a DC machine of comparable rating, the overall cost of the AC drive may still be somewhat higher than that of a standard DC drive. Of course, AC drives exhibit a number of important advantages, most prominent in the case of an induction motor with cage rotor, that will influence the decision in their favour:

- Power rating and speed of an AC motor is not limited by a mechanical commutator,
- Reduced axial length, volume and inertia of an AC motor, as seen in Fig. 10.1,
- Rugged design, reduced maintenance and service requirements,
- Drive is applicable in explosive and contaminated environment.

The last features are of particular interest with traction and servo drives where many driven axes are incorporated into the machine tool or robot. Of course, there are also applications at high power and speed, where DC drives could not be built and only AC drives can provide a solution. Since the cost problems will diminish as the converter- and AC drive- technology matures, it is likely that DC drives will eventually be completely superseded by the more advanced AC drives.

The basic lay-out of an adjustable speed AC drive is shown in Fig. 11.1; it consists of an AC motor, a static converter, which generates a variable voltage, variable frequency AC system and the associated control equipment. The increased complexity of the converter is caused by the fact that inverters with forced commutation may require additional components and a more complicated mode of operation. Similar arguments apply also to the structure

Table 11.1. Adjustable speed AC drives, synopsis

Converters / Machines	DC link converters — Voltage source converters: Transistor-(IGBT) Converter	Thyristor-(GTO) Converter	Current source converters: Thyristor-(GTO) Converter with force comm.	Thyristor-Converter with phase comm.	Cyclo-converter with line-commutation
Synchronous motor with permanent magnets on rotor	Low power (10 kW), very good dyn. performance (servo motors)	Medium power (1 MW), high power density			
Reluctance-motor	low to medium power (100 kW)				
Induction motor with cage - or wound -rotor	Medium power (< 1 MW), high speed, good dyn. perform. (Spindle drives)	Med. to high power (2 MW), Traction	High power (4 MW)		High power (7,5 MW) Low speed good dyn. perform.
Doubly-fed wound rotor motor		Shaft generator on ships (2 MW)		High power (20 MW), subsynchronous (Scherbius)	High power (100 MW), limited speed range
Synchronous motor with field- and damper winding				High power (100 MW), High speed "Converter - motor"	High power (5 MW), Low speed good dyn. perform.

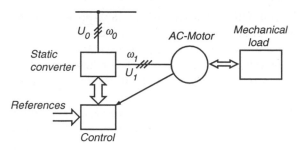

Fig. 11.1. General scheme of AC motor control

of the inner torque control loop which is much more complex than for a DC motor. The reason is that an AC motor with its simple mechanical design represents an involved nonlinear multivariable control plant.The motor must be fed with alternating currents of variable amplitude, frequency and phase; in addition, the rotor currents cannot be measured with ordinary cage motors. This is in contrast to a DC machine, which has a very simple decoupled control structure but a more complex mechanical design. The mechanical part of the control scheme is, apart from the smaller inertia of AC motors, the same for DC- and AC-drives.

Fig. 11.2 indicates that the mechanical commutator of a DC machine acts like an electromechanical converter which feeds the armature conductors with alternating current of a frequency proportional to speed, Fig. 5.2. On an AC drive, the functions of current conversion and torque generation are mechanically separate, thus creating new options for electromechanical interfaces.

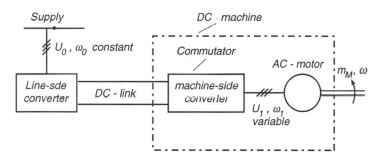

Fig. 11.2. Controlled AC-drive with DC-link converter

Table 11.1 presents an overview of controlled AC-drives, showing five basic forms of power electronic converters for changing from the fixed line- to variable motor-frequency. Four of the circuits contain a DC-link for decoupling the two sides of the converter, whereas the cyclo-converter performs this operation in a single conversion stage. All of the converters are adaptable to

four-quadrant operation, they could also be used with variable speed generators, for instance in wind energy plants. In the vertical column of Table 11.1
five basic types of AC-machines are listed that may be used for controlled
drives; with the exception of the reluctance motor, which is characterised by
salient poles, they can all be treated with the mathematical model derived
in Sect. 10.1.

In principle, most of the 25 possible converter-machine combinations could
be realised but only those, where comments are entered, are presently applied
in practice; since each type exhibits particular advantages, it is unlikely that
the field will eventually narrow to just one or two major circuits as has been
the case with DC-drives. In fact, new combinations may be added as new
switching devices and converter circuits appear on the scene. An example of
the ongoing development are resonant-link converters, presently restricted to
low power applications such as switched mode power supplies; cyclo converters with forced commutation (matrix converters) are also a possible future
option. Only a few typical converter topologies that show promise to stay
will be discussed in this chapter.

11.1 Pulse width modulated (PWM) Voltage Source Transistor Converter (IGBT)

The four-quadrant DC/DC converter shown in Fig. 9.13 a can be extended by
adding two more arms, as seen in Fig. 11.3, to form a three-phase converter
for the supply of AC-motors in the lower and medium power range, from
small high dynamic performance servo drives with speed and position control
(< 10 kW) to most auxiliary drives in industry, ranging up to several hundred
kW. At higher power this two-level circuit is also extended to multi-level
configurations. The converter is suitable for supplying induction as well as
synchronous motors.

With the voltage source converter shown, the terminal voltages of the motor can assume the three possible values U_D, 0 and $-U_D$. When the motor is
regenerating, the mean link current I_D is reversed, calling for a two-quadrant
supply of the DC link. As long as the power level is low or when regeneration occurs only occasionally, for instance during dynamic braking of a servo
drive, the DC link is normally supplied by diode rectifiers and the reverse
power is dissipated in a ballast resistor, Fig. 9.14, as indicated also in Fig.
11.3 with dashed lines. The pulse-width modulated ballast circuit then begins to discharge the link capacitor as soon as the link voltage rises above a
preset value. Of course, any other source of direct voltage could also serve
for feeding the converter, such as a battery on an electrically driven vehicle,
where regeneration is important to save energy.

With bipolar transistors, the switching frequency is limited to a few kHz,
but with low power Insulated Gate Bipolar Transistors (IGBT), as shown in

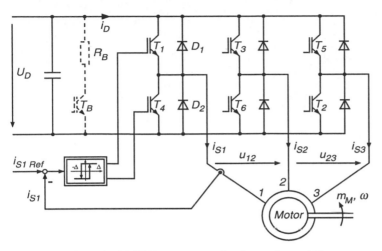

Fig. 11.3. Switched IGBT- converter for three-phase AC motor (simplified circuit)

Fig. 11.3, or Field Effect Transistors (MOSFET), having a switching time in the order of a microsecond, the mean switching frequency may be above 16 kHz, i. e. beyond the audible threshold, so that no objectionable acoustic noise is emitted from the drive. An additional advantage is that the converter then exhibits a large bandwidth for control. If the link voltage U_D is of sufficient magnitude, fast current control loops can be designed which keep the stator currents in line with the alternating reference values; this effectively produces near ideal current sources for the stator windings of the motor, thus eliminating the effects of the stator voltage equation Eq. (10.50) on the dynamics of the drive. As a consequence, considerable simplification of the drive control is achieved because the stator voltage interactions are then dealt with by the current controllers. These could either employ pulse-width modulators operating at constant frequency or simple On/Off comparators having a narrow hysteresis band, also indicated in Fig. 11.3. The resulting current waveform with a single-phase load and On/Off-control (symmetrical modulation) was shown in Fig. 9.16; it has the advantage of fast response but exhibits variable switching frequency which may be undesirable in view of interference problems.

When extending the principle to three-phase loads, as in Fig. 11.3, the hysteresis band is helpful to avoid interactions between the controllers because, in view of the isolated neutral of the machine winding, the stator currents are balanced and one of the controllers is redundant, it is only retained for symmetry.

To produce a preferred switching frequency, a triangular clock signal may be added at the summing point of the current controllers [S29]. Another possibility is the synchronization of the controllers, i.e. allowing a transition to

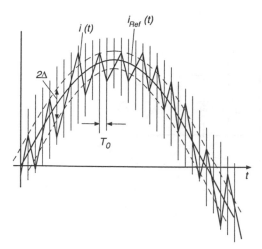

Fig. 11.4. Current waveform produced by a single-phase transistor converter with inductive load impedance and synchronised On-Off controller

occur only at equidistant sampling instants $t = \nu T_0$ defined by a fixed clock frequency $f_0 = 1/T_0$ [B45]. While Fig. 9.16 was derived with a simple free running On/Off-controller, the effect of synchronization is sketched in Fig. 11.4, again for a single-phase converter with inductive load; the amplitude of the ripple superimposed on the load current is now increased, because the switching operations are delayed until the next sampling instants, thus creating waiting intervals ranging between zero and T_0. If synchronised On/Off-controllers are employed, the hysteresis band Δ can be omitted because there is now a minimum time interval T_0 between subsequent switching operations.

When a very low ripple content of the currents is specified, as on machine tool feed drives, the use of a pulse-width modulator in combination with a linear current controller is preferable to On/Off current control. This is demonstrated in Fig. 11.5 where measured current waveforms obtained with the same three-phase transistor converter are compared. Fig. 11.5 a depicts the current produced by a synchronised controller having a sampling period $T_0 = 22\mu s$, whereas the curves in Fig. 11.5 b show the result of linear current control with a pulse-width modulator operating at 5 kHz [H20]. The audible noise is more pronounced with PWM control unless a higher clock frequency is chosen, but the current waveform is much more acceptable. The reason for the increased ripple current is the quantisation of the voltage-time area with the synchronised On/Off - control.

Pulse-width modulators are now available in a variety of integrated circuits, which greatly simplifies the design of PWM converters. There is also the possibility of software-based modulation using fast signal processors which offer unlimited flexibility by combining PWM with sophisticated, such as

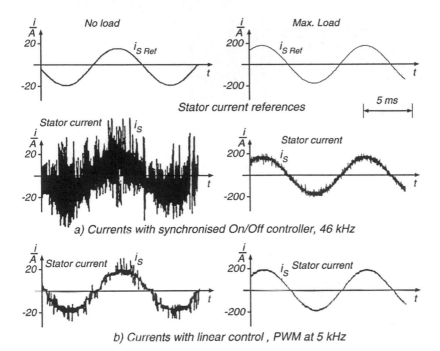

Stator current references

a) Currents with synchronised On/Off controller, 46 kHz

b) Currents with linear control , PWM at 5 kHz

Fig. 11.5. Comparison of current waveforms with different types of controllers;
(a) Current waveform with synchronised On/Off-control, $T_0 = 22\mu s$
(b) Current waveform with linear current controller and PWM,
  operating at 5 kHz

predictive or time-optimal current control, this is of interest with high power converters switching at a lower frequency [H67, H71].

The assumption of virtual current sources for the stator windings is, of course, only valid as long as the ceiling voltage of the converter is not reached. This is demonstrated in Fig. 11.6 for a free running On/Off - controller when the frequency of the current reference is gradually increased. Due to the inductive load impedance, the necessary fundamental component of the voltage rises with frequency according to $\hat{u}_1 = \hat{\imath}_1 \sqrt{R^2 + (\omega_1 L)^2}$. It is seen that, beginning with Fig. 11.6 b, there are signs of saturation as indicated by prolonged intervals where no switching occurs, which is due to insufficient supply voltage. Clearly, this results in a rising control error between the sinusoidal current reference and the piecewise exponential feedback signal and must be taken into account when designing a drive control system.

A very effective method of pulse-width-modulation that is particularly suited for fast switching converters is called vectorial PWM because it represents an attempt to reproduce in a given time interval $\nu T_0 < t \leq (\nu + 1) T_0$ a voltage vector $\underline{u}_{S\,\mathrm{Ref}}(t)$ demanded by the current controllers [P29]. For this it is important to remember that the converter circuit of Fig. 11.3, which in

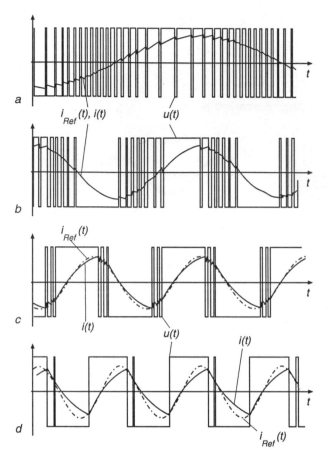

Fig. 11.6. Saturation effects of single-phase switched transistor converter with On/Off- current control supplying a passive R-L load impedance

Fig. 11.7 a is represented by a switching model, must be so operated that none of the legs is short- circuiting the link voltage U_D and that each motor terminal assumes a potential defined by the control. Hence one transistor in each leg must be blocked while the other is conductive, except for the short protective intervals, where both transistors are blocked and the motor current is flowing through one of the shunting diodes (The protective interval resembles the red-light-overlap at a signalled road crossing). To prevent short circuit conditions, each half bridge i is modelled by a reversing switch indicated by a binary variable $S_i = 1, -1$, depending on whether the switch is in the upper or lower position. The switching state (S_1, S_2, S_3) of the complete converter is then described by a three bit binary word having eight different values, including $(1, 1, 1)$ and $(-1, -1, -1)$, called zero vector states, where the motor terminals are short circuited at the upper or lower DC bus. The protective interval which lasts only a few microseconds when transistors are used, can be assigned to a finite switching time in the converter model.

When writing the voltage vector $\underline{u}_S(t)$ in terms of line- to-line voltages (10.38) which can assume the values U_D, 0 and $-U_D$,

$$\underline{u}_S(t) = u_{S1} + u_{S2}\, e^{j\,\gamma} + u_{S3}\, e^{j\,2\,\gamma} = u_{12} - u_{23}\, e^{-j\,\gamma} \tag{11.1}$$

six distinct voltage vectors and two zero vectors result, as seen in Fig. 11.7 b. Clearly, when going from one corner of the hexagon to the next, only one leg of the converter needs to change its state.

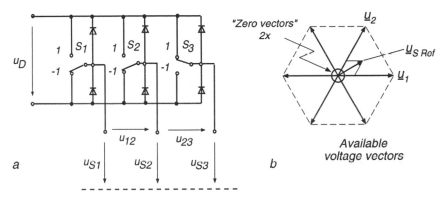

Fig. 11.7. Switching model of converter in Fig. 11.3 and generated output voltage vectors

It is assumed that the converter is switched at a basic clock frequency $f_0 = 1/T_0$, which may either be fixed or correspond to a multiple of the variable fundamental stator frequency, and that the current controllers have determined a command value $\underline{u}_{S\,\mathrm{Ref}}$, valid in the next interval T_0; $\underline{u}_{S\,\mathrm{Ref}}$ is assumed to lie inside the first sector, as shown in Fig. 11.7 b.

Since this commanded voltage vector is normally not coinciding with one of the available voltage vectors, it is to be composed by a switching sequence comprising the adjacent vectors $\underline{u}_1 = \underline{u}_S(S_1, S_2, S_3) = \underline{u}_S(1, -1, -1) = U_D$ and $\underline{u}_2 = \underline{u}_S(1, 1, -1) = U_D\, e^{j\,\pi/3}$, while filling up the rest of the time interval with zero-vectors. The sub-intervals t_1 and t_2 for the two adjacent vectors are to be computed from the following equivalence

$$\underline{u}_{S\,\mathrm{Ref}} = (u_{Sa} + j\,u_{Sb})_{\mathrm{Ref}} = \underline{u}_1\,\frac{t_1}{T_0} + \underline{u}_2\,\frac{t_2}{T_0} = U_D\left(\frac{t_1}{T_0} + e^{j\,\pi/3}\,\frac{t_2}{T_0}\right) ;$$
$$\tag{11.2}$$

in addition $t_1 + t_2 + 2t_0 = T_0$ holds, where t_0 is the zero vector interval.

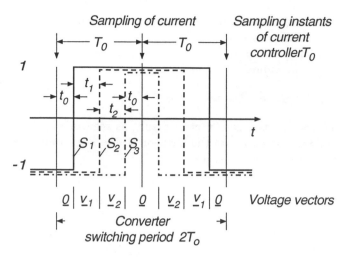

Fig. 11.8. Vectorial PWM with symmetrical switching sequence

Solving for t_1, t_2 results in

$$\frac{t_1}{T_0} = \frac{u_{SaRef}}{U_D} - \frac{1}{\sqrt{3}} \frac{u_{SbRef}}{U_D} ,$$

$$\frac{t_2}{T_0} = \frac{2}{\sqrt{3}} \frac{u_{SbRef}}{U_D} , \tag{11.3}$$

$$2\frac{t_0}{T_0} = 1 - \frac{u_{SaRef}}{U_D} - \frac{1}{\sqrt{3}} \frac{u_{SbRef}}{U_D} \geq 0 .$$

When omitting zero vectors, $t_0 = 0$, the resulting equivalent voltage vector ends on the straight line connecting the two adjacent switching vectors according to

$$\underline{u}_{S\,Ref} = (\underline{u}_1 - \underline{u}_2)\frac{t_1}{T_0} + \underline{u}_2 ; \tag{11.4}$$

the limits for $t_1 = 0$ and $t_1 = T_0$ confirm this result. Hence, by inclusion of t_0 any desired voltage vector inside the hexagon defined by the six switching vectors may be realised. Of course, the adjoining switching vectors must be chosen according to the sectorial position of $\underline{u}_{S\,Ref}(t)$. If the switching frequency f_0 is high enough, fluctuations of the link voltage may be taken into account by substituting a recent measurement of $u_D(t)$ for U_D in the modulating equations (11.3).

To remove the ambiguity with respect to which of the two adjoining switching states should be applied first, it is advantageous to repeat the same switching cycle with reverse sequence so that a symmetrical switching pattern of length $2T_0$ is created. Sampling the motor currents in the center of the short circuit intervals t_0 diminishes the sampling noise, as will be discussed later. The switching sequence pertaining to the example in Fig. 11.7 b is plotted in Fig. 11.8, depicting the potentials of the motor terminals.

Eq. (11.2) describes an idealised situation, where the protective intervals and the inherent delays of the switching devices are neglected. For the actual design of modulators, these effects must be taken into account, particularly the differences between turn - On and turn - Off times, which can cause considerable distortion of the converter characteristics at low output voltage and frequency. Hardware- as well as software- based methods have been proposed for solving this problem [K35, S81, W10].

11.2 Voltage Source PWM Thyristor Converter

The use of transistors (MOSFET, Bipolar) is presently restricted to lower and medium power applications, with IGBT's up to about 1 MW. With thyristors and GTO's, the rating of semiconductor converters extends to even higher power. A proven circuit, resembling the transistor converter in Fig. 11.3 is the thyristor converter shown in Fig. 11.9. Because of the constant link voltage and the low AC impedance of the DC link it is also called a voltage source converter. In view of the higher power and the fact that the motor may have to operate for longer periods of time in the generating region, the line-side converter is usually an active front-end rectifier, able to feed power back to the line by reverse direct current I_D.

When employing gate turn off (GTO) thyristors, which are available in kV, kA- ratings, the converter circuit corresponds directly to the one in Fig. 11.3, with the transistors replaced by GTO's but with normal thyristors that cannot be blocked by gate signals, additional components are needed for implementing forced commutation of the machine-side converter; this is so because the stator windings of an induction motor do not contain current-independent voltage sources as in the case of a line-connected transformer or a synchronous machine with an internal induced voltage. Hence, natural commutation of the machine-side converter feeding an induction motor is not feasible. This is understood when remembering that a converter with delayed firing calls for lagging reactive power which a normal induction machine is unable to supply.

The converter depicted in Fig. 11.9 is an example of a variety of different circuits; it is considerably simplified, showing only the main components. Instead of a pair of transistors as in Fig. 11.3, each phase contains a force-commutated thyristor switch, comprising two main thyristors T, T', with antiparallel diodes D, D', two auxiliary thyristors AT, AT', a commutation capacitor C and an air cored inductor L. The resonant circuits in Fig. 11.9 have the task of creating zero current/negative voltage intervals for the main thyristors T, T' by firing auxiliary thyristors AT, AT'; briefly, a commutating process takes place as follows:

Assume that the main thyristor T initially carries the load current i_{S1} and that the commutating capacitor C is positively charged, $u_c > 0$; then firing of the auxiliary thyristor AT causes T to be blocked so that the load current flows temporarily through AT and C thus recharging the capacitor. At the same time a resonant circuit is formed consisting of AT, C, L and the diode D; when about a half period of the oscillation has elapsed and the charging current tends to reverse its sign, AT and D are blocked, which leaves the capacitor with opposite voltage, ready for the next commutating transient when T' has to be extinguished. The load current is now flowing through D' and the motor terminal 1 is connected to the lower DC bus. If the current pulse through the resonant circuit is of sufficient magnitude as compared to the load current, the commutation is almost independent of the load so that the voltage u_{1N} can be approximated by a rectangular wave $\pm U_D/2$ with short but continuous commutating transients. The load current i_{S1},

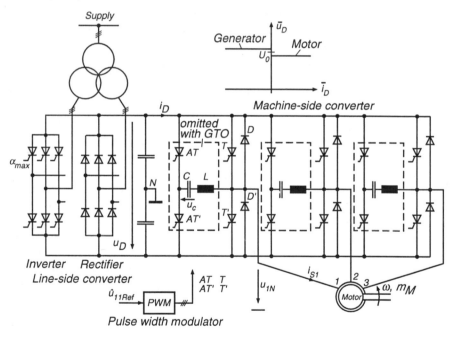

Fig. 11.9. Thyristor converter with constant DC-link voltage (Mc Murray circuit)

being continuous due to the motor leakage reactance, can in principle flow in either direction, even though all four thyristors may be blocked temporarily, because there is always a path open through one of the diodes D, D'. During regeneration, the conducting periods of the diodes are increasing, eventually causing the mean of the link current, $\bar{i}_D$, to change sign which indicates reversed power flow.

It is realised that the initial transient, when i_{S1} is commutated from T to AT, is of course also continuous because small inductances, not shown in Fig. 11.9, limit the rate of change of the thyristor currents. Likewise there are parallel RC snubber circuits connected to each thyristor to limit the rate of change of the voltages. A detailed analysis of converter circuits, taking all these effects into account, is best performed by digital simulation; this is particularly true for transient operation, when the initial conditions, for instance the charge on the capacitor when AT is fired, are changing. As with all converter circuits, the function of the converter is eventually endangered when the required recovery time of the thyristors cannot be maintained, for example due to excessive load current or insufficient initial charge on the commutating capacitor. Hence the accurate analysis of the commutating transients is of major concern to the circuit designer. Special types of thyristors with short recovery time ($< 20\mu s$) are usually specified for pulse-width modulated converters, which limits the power rating to some MW. Also, the occurrence of unacceptable load currents must be prevented by fast control action.

The firing instants for the main- and auxiliary thyristors are determined by voltage and/or current controllers. In view of the limited switching frequency, which is usually below 1 kHz for thyristor converters of higher power, the load currents can no longer be considered close to sinusoidal as was possible with transistor converters operating at a higher switching frequency. Hence PWM–thyristor converters of the type shown in Fig. 11.9 are naturally suited to act as controllable voltage sources, presenting to the load pulse-width modulated rectangular voltages, the fundamental components of which are prescribed by voltage command signals. Because of the large superimposed harmonics, caused by the chosen switching strategy, closed loop control of the fundamental currents at variable frequency is normally not practical; it was different with transistor converters, where the switching frequency is higher and the currents are sufficiently filtered by the load impedances. Therefore, open loop voltage control through a pulse-width modulator is the standard solution, while the current loops are normally closed on the next higher level of drive control, as will be shown in Chap. 12.

The pulse-width modulators may again be of a variety of designs. One frequently used scheme with converters of higher power (Fig. 11.10) employs a triangular sampling signal [J8, M42, S29], the frequency f_0 of which is a multiple n of the stator fundamental frequency f_1 to avoid subharmonics; n is reduced in stages as the stator frequency and voltage rise, to limit the switching frequency and to fully utilise the voltage capability of the converter. A drawback of this otherwise simple scheme is the appearance of large harmonics at the frequencies $f_0 \pm 2f_1$. The load currents are smoothed by the leakage inductance of the motor which should not be chosen too small with this type of converter. All triplen harmonics are eliminated from the currents in steady state due to the isolated neutral of the stator winding. There are still noticeable additional losses as well as sometimes objectionable noise emitted by the converter and the motor. A typical current waveform with a PWM-converter and this type of "synchronous modulation" is shown in Fig. 11.12 b.

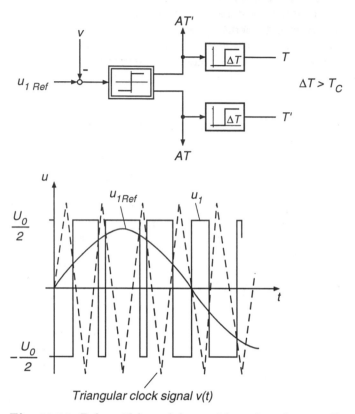

Fig. 11.10. Pulse-width modulator with a triangular sampling signal

This unsatisfactory situation has given rise to the development of a multitude of advanced modulation schemes such as digital pattern generators where binary switching sequences for different amplitudes of the fundamental voltage are kept in a microelectronic memory to be called up in real time for small increments of angle $\tau = \int \omega_1 dt$. This is depicted in Fig. 11.11 [B85, D6, P5, P42, T36]. The patterns are computed off-line with the objective of reducing voltage- and current-harmonics or power losses, torque pulsations or noise under steady state conditions. Clearly, the more often the voltage is reversed per period, the more side conditions can be satisfied, given a prescribed fundamental voltage component.

The restrictions of the converter with regard to the minimum time between two subsequent switching operations must of course be observed to

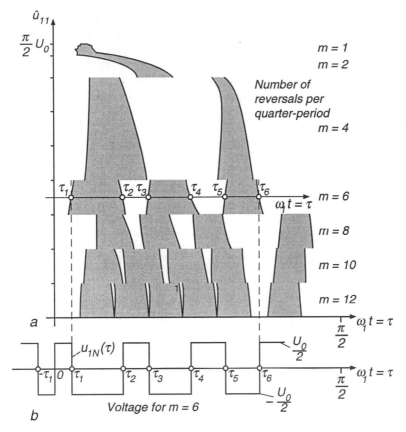

Fig. 11.11. Optimised pulse-width modulation with precalculated switching patterns

allow the commutation to be completed. Also, the losses in the converter caused by each commutation transient should be considered, which means that there is an upper limit for the switching frequency. At the maximum fundamental frequency, which corresponds to full output voltage of the converter when feeding an AC motor, the optimal voltage waveform is a square wave with one reversal per half period; this is included in the switching pattern shown in Fig. 11.11.

When realising a suitable modulator it is of course desirable to keep the required volume of memory small; this can be achieved by storing only one quarter cycle $0 \leq \tau \leq \frac{\pi}{2}$ of the binary pattern for one phase and obtaining the remaining information by transposing and inverting this pattern. On the other hand, considerable angular resolution is needed to satisfy the various conditions with adequate accuracy. Assuming for example a resolution in amplitude of the fundamental voltage of $2^7 = 128$ levels and an angular resolution of $2^{10} = 1024$ steps per quadrant, this results in a memory of 16 kByte which could easily be accommodated by one microelectronic chip of read only memory (ROM). At the very low end of the frequency range some other means of modulation must be employed because even 4 000 steps of angular resolution per period will eventually prove to be insufficient.

In order to avoid continued fluctuations of the voltage amplitude, which would disturb the switching sequences and render the notion of "harmonics" meaningless, the superimposed control should contain a hysteresis band in the amplitude channel; fine and rapid speed control are still possible through the frequency input of the modulator but the voltage would be changed more slowly and temporarily in somewhat coarser steps. Some results of the current waveforms and spectra obtained with different modulators are shown in Fig. 11.12. The curves were recorded on a 22 kw drive supplied by a voltage source thyristor converter [P44].

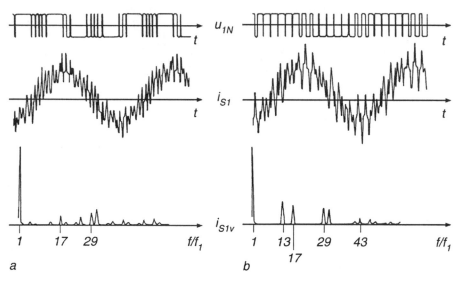

Fig. 11.12. Current waveforms and spectra achieved with different methods of pulse-width modulation. (**a**) Optimised switching pattern; (**b**) Subharmonic modulation with triangular sampling signal

A particular feature of the pulse-width modulator with optimised switching pattern is the fact that the pattern can easily be changed for different applications, placing more emphasis for instance on losses, torque pulsations or noise. With the rapid advances of micro-electronics even on-line optimisation of the individual switching instants has been shown to be feasible which opens the way to optimal modulation under dynamic conditions, so that even on-line current control could become practical with high power converters [H64, H67].

The load-independent commutation, resulting in low impedance motor voltages makes the PWM converter an excellent choice for drives with high dynamic performance at power ratings where high frequency switching is not feasible; an example are high power traction drives with GTO-converters which require good dynamic performance to damp oscillations in the drive train and prevent wheel slippage during starting and braking.

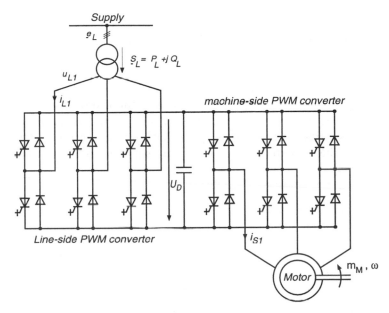

Fig. 11.13. Symmetrical voltage source GTO-converter with forced commutation at machine- and line-side

With increased use of controlled drives the problem of line-side interference by power electronic converters is becoming an important issue for the utilities. A line-commutated converter supplying the DC-link is particularly notorious because it not only produces distorted line currents with low orders of harmonics but draws also substantial reactive currents of line frequency. The first mentioned effect can be alleviated by increasing the pulse number of the converter but the other requires capacitive compensation or, combining both aspects, filters with capacitive fundamental input currents.

Another possibility that is likely to be of strong future interest, is to extend the principle of forced commutation also to the line-side converter. An example is seen in Fig. 11.13 where a PWM voltage source converter with GTO-thyristors at both ends is shown in simplified form. The line-side PWM converter could be controlled such that the line currents i_N are not only approaching sinusoidal waveform but that their fundamental components are also in phase with the line voltages, thus eliminating reactive currents. Single phase converters of this type are common on large AC-traction drives, where the line impedance caused by the single phase catenary is quite large so that voltage fluctuations and distortions would be particularly severe, Fig. 9.17.

The line side converter may also be used as a static compensator by producing controllable, usually capacitive reactive line current, whereas the active component of the line current serves for maintaining a commanded constant or variable link voltage U_D; this is discussed in Sect. 13.2.

11.3 Current Source Thyristor Converters

The converter circuit shown in Fig. 11.14 comprises once again a line-side and a machine-side converter part connected through a DC-link; the two converters operating at different frequencies are now decoupled by a smoothing reactor L_D which, in combination with a current control loop, serves to maintain a direct link current i_D as prescribed by the current reference. Thus the machine-side converter is effectively supplied from a current source [F2, K56, P32].

The circuit offers the following simplifications:

- The unidirectional link current i_D allows the use of a two-quadrant line converter, where reverse power flow is achieved by inverting the mean of the link voltage u_D through delayed firing.
- The link reactor permits the voltage u_D to be temporarily raised or lowered during commutation of the machine-side converter, thus avoiding the auxiliary thyristors; commutation is now performed with capacitors and decoupling diodes.
- Altogether there are 12 thyristors, the same number as would be needed for a reversible DC-drive; also, the thyristors for the machine-side converter need not be of the more expensive fast recovery type necessary for voltage source converters.

The DC-link could also be supplied from a force commutated converter such as a chopper with current control; this is of interest on subway trains fed with constant direct voltage.

The mode of operation of the machine-side converter in Fig. 11.14 differs considerably from that of the voltage source converter, Fig. 11.9. While the motor current supplied from a PWM voltage source converter was at least crudely sinusoidal, it has now a three step waveform, assuming the values $-i_D$, 0, i_D as seen in Fig. 11.15 a; the corners are rounded-off due to finite commutating time. The vector of the stator current follows the trajectory of a six pointed star; this is shown in Fig. 11.15 b which was recorded during a speed reversal of a 22 kW drive [G5]. The commutating transients are more involved now because the motor impedance is part of the commutation circuit which means that the commutating time depends on the loading of the motor and increases at light load.

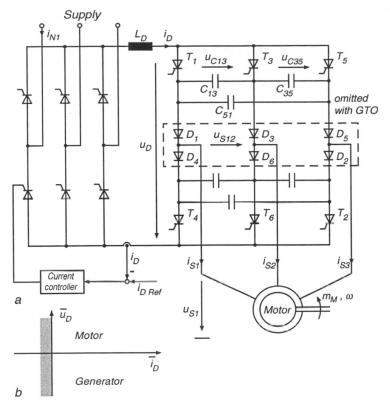

Fig. 11.14. Thyristor converter with direct current link
(a) Circuit (b) Operating regions of DC link

The commutating transient may be briefly described as follows: With no commutation in progress, two thyristors, for example T_1 and T_2 carry the direct current, $i_{S1} = -i_{S3} = i_D$, and the capacitor C_{13} is positively charged as a result of the previous commutation, $u_{C13} > 0$. If thyristor T_3 is now fired, T_1 is extinguished in a rapid transient and T_3 assumes the direct current; this is the starting condition of the commutating transient proper. While the current i_{S1} is now reduced towards zero, i_{S2} is rising towards i_D; during this interval phase 1 of the motor is fed through C_{13} as well as through the series-connected capacitors C_{35}, C_{51}. Eventually diode D_1 is blocked and the commutation is completed with T_2 and T_3 conducting, $u_{C13} < 0$, $u_{C35} > 0$. Clearly the motor transient impedance between terminal 1 and 2 is part of the commutation circuit, as mentioned before. The diodes are required for decoupling to prevent the capacitors from losing their charge, necessary for the next commutation. The commutating interval can be reduced by choosing a motor with low leakage reactance.

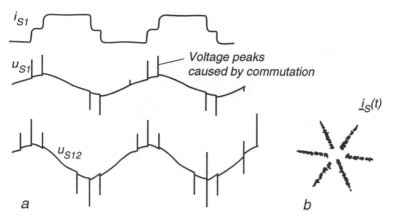

Fig. 11.15. Current and voltage waveforms of a current source converter (a) Wave forms; (b) Stator current vector

Beginning and end of the current steps are manifested by voltage spikes superimposed on the otherwise sinusoidal terminal voltages of the motor; this is an obvious consequence of the changing stator currents. Outside commutation, where

$$\frac{di_{S\nu}}{dt} = 0, \qquad \nu = 1, 2, 3 \tag{11.5}$$

holds, the terminal voltage of the motor is practically sinusoidal since the magnetising current is filtered by the main time constant of the motor.

Depending on the phase of the voltage surges due to commutation, as related to the sinusoidal voltage component, it can be recognised whether the machine is operating as a motor or in the generating region, Fig. 11.15 a; on the DC side this is reflected by the sign of the mean voltage $\overline{u_D}$. The direction of rotation is solely determined by the firing sequence of the machine-side converter. The firing signals may for instance be derived from a ring counter with 6 states, which is stepped forward or backwards as commanded by the speed controller.

With a suitable control structure, this converter is also capable of supplying high dynamic performance drives, operating in four quadrants over a wide speed range [G5, G15]. Besides its simplicity and, considering the link reactor, the ease of protection in case of commutation failure, it has the advantage of little audible noise caused by the absence of "high frequency" modulation. On the other hand, there are substantial additional losses caused by the large harmonic content of the currents. Also the torque pulsations due to the interactions of the stepped stator currents and the sinusoidal flux wave can be objectionable at low speed. In order to alleviate this condition, the machine-side converter may be commutated back and forth by pulse-width modulating the ring counter, i. e. superimposing an alternating component on the slow progression of the state of the counter; this is discussed in Sect. 12.4. Of course, due to the relatively slow commutation at light load, the switching frequency is bound to be much lower than with a voltage source PWM converter [W9]. It is for this reason that this converter is not suited for drives requiring continuous position control such as machine tool feed drives.

Apart from this special restriction, the current-source converter may be controlled nearly as rapidly as a voltage source converter, because the firing instants of the machine-side converter can be quickly shifted; changing the magnitude of the link current involves some delay due to the large smoothing reactor but with adequate ceiling voltage the equivalent lag can be reduced to about 10 ms with a 50 Hz supply, as was discussed in Sect. 8.5.

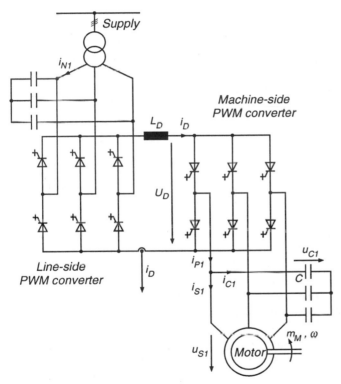

Fig. 11.16. Symmetrical current source PWM converter with GTO-thyristors

As mentioned before, the current vector defined in Eq. (10.5) assumes with this converter a rather unique shape; because of $i_{S\nu} = \pm i_D$ and with one of the motor currents zero (except for commutation intervals), the vector follows a regular six-pointed star with the radius $\sqrt{3}i_D$,

$$\underline{i}_S(t) = i_{S1} + i_{S2}\, e^{j\,\gamma} + i_{S3}\, e^{j\,2\,\gamma} = \sqrt{3}\, i_D\, e^{j\,(\nu\,\gamma/2+\pi/6)}. \tag{11.6}$$

Some measured traces are shown in Sect. 12.4.

When assessing the relative merits of the different converter schemes, it is of course necessary to compare not only the number of components but also their rating and how this is reflected in the total cost.

The circuitry of the current source converter is again considerably simplified when GTO-thyristors are employed; this is indicated in Fig. 11.14 and shown more clearly in Fig. 11.16 for a symmetrical converter, where the load- and the line- currents are pulse-width-modulated and may approach sinusoidal waveforms.

The link current i_D is now switched between the motor terminals and the motor voltages and currents are filtered by capacitors. It is also possible to introduce zero current vectors, bypassing the load through a short circuited leg of the converter which is similar to the zero voltage vector with a voltage source converter; thus, the circuit offers possibilities for high dynamic performance drives. However, in view of the increased complexity of the control plant, caused by the capacitors, sophisticated control algorithms and a powerful microcomputer are needed; in particular, the question of L-C resonances requires detailed study [A18, A19, N17, N18, N19, N21].

11.4 Converter Without DC Link (Cycloconverter)

The converters discussed so far had in common that line-side and machine-side converters operated at different frequencies and were decoupled by a DC-link containing energy storage devices in the form of low-pass filter components. This twofold conversion provides a degree of independence in controlling the converter ends, but at the same time causes additional power losses. The double conversion may be avoided with direct AC/AC-converters which perform the conversion from the constant frequency, constant line voltage to variable frequency, variable motor voltage in one stage and without resorting to intermediate energy storage [49, 58].

The reversible line-commutated AC/DC converter without circulating current shown in Fig. 9.4 is capable of operating in all 4 quadrants of the u_a/i_a–plane supplying an inductive load (to assure continuous current flow). Hence, by providing it with closed loop control of the output current and applying a low frequency current reference, the converter will act as a low frequency current source. Clearly, 4-quadrant capability is necessary when supplying a motor because, when a sinusoidal current i_a is fed to a R-L-load impedance, the operating point in the voltage/current-plane describes an elliptical path around the origin, passing through all quadrants. When reversing the sequence of the current references, the sequence of the three-phase output system is inverted. To avoid undesirable interactions between the three current controllers (one of which is redundant because of the zero sum of the currents), they may be chosen as P-instead of PI- controllers thus allowing some flexibility.

By supplying each phase of the motor winding from a reversible converter, a low frequency AC drive system is formed, as seen in Fig. 11.17. Again, the neutral of the stator winding remains isolated to exclude zero sequence, such as triplen harmonic, currents. The main feature of this circuit is that only standard line-commutated thyristor-converters are required which have been in use for many years with DC drives and HVDC systems up to the highest power ratings. Also, the cost of the thyristors is reasonable since no particular specifications with regard to turn-off time are necessary with line-commutated converters.

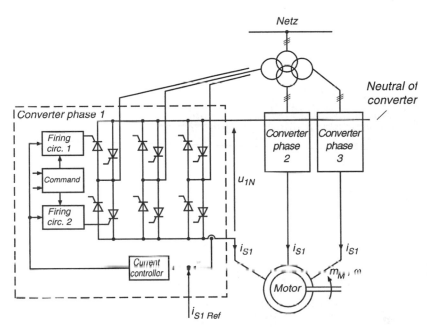

Fig. 11.17. Six-pulse cycloconverter for a three-phase motor load

On the other hand, the large number of thyristors seems at first sight staggering; for a six-pulse converter with three-phase output a minimum of 36 thyristor branches is required as well as a transformer with three complete sets of three-phase secondary windings. This indicates that cycloconverters are mainly of interest for large drives, where parallel thyristor branches would be necessary with other converter circuits. An important restriction is the limitation of output frequency which is caused by the discrete nature of the control process and the presence of a carrier frequency, since the output voltages are assembled from sections of the line voltages. As the output frequency rises, the output currents are tracking the sinusoidal references with increasing errors and consequent distortion. The frequency range

$$0 \le f_2 \le f_{2\max} \approx \frac{p\,f_1}{15} \tag{11.7}$$

is usually considered for operation of a cycloconverter, where f_1 is the line frequency and p is the pulse number. With a 50 Hz grid and a three-phase bridge circuit ($p = 6$) this results in $f_{2\,max} \approx 20$ Hz. At this frequency, a period of the output voltage would on the average consists of 15 sections of line voltages. Of course, if a three-phase supply of higher frequency is available, the range of output frequency is extended accordingly; this may be the case on vehicles, aircraft or ships, when a diesel- or turbine-driven generator provides on-board power for variable frequency AC drives. Also, a somewhat wider frequency range could be achieved by cycloconverters with circulating current, as discussed in Sect. 9.1 [H28].

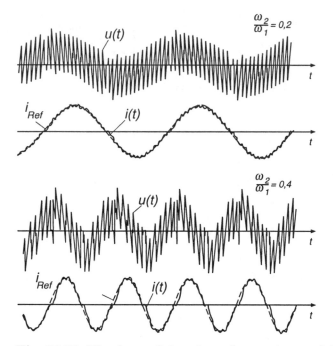

Fig. 11.18. Waveforms of six-pulse cycloconverter supplying a single phase-load at different output frequencies (simulation)

It is of interest to note that the nature of the load supplied by the cycloconverter is unimportant as long as it contains a sufficiently large inductive impedance to assure continuous current flow. In particular, there is no difference whether the load is active or passive, i.e. whether the machine operates as a generator or motor because operation in all four quadrants is possible.

When considering the effects of a cycloconverter on the line-side currents it is helpful to remember that a symmetrical three-phase system of sinusoidal

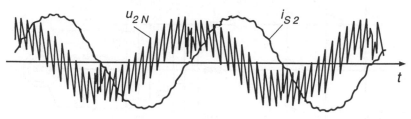

Fig. 11.19. Measured output current and voltage of a six-pulse cycloconverter with three-phase inductive load at 10 Hz

voltages and currents results in constant total power. Since the cycloconverter contains only switches but no storage devices (apart from the unavoidable leakage inductances, snubber circuits, etc.), the total three-phase input power corresponds to the output power. Hence it is mainly the effects of harmonics that have to be considered. In addition there will be reactive power at the line side which is inherent in the control of line-commutated converters by delayed firing.

In Fig. 11.18 the results of a computer simulation are shown, where a reversible six-pulse converter with current control, operating on the 50 Hz grid, supplies a predominantly inductive single-phase impedance with current of 10 Hz and 20 Hz. At these frequencies there is still a reasonable agreement between reference current and feedback signal, even though some phase shift becomes noticeable in the 20 Hz-trace, but it might be corrected by a lead-term in the reference channel. The 2 ms-zero current interval, which is necessary for safe operation when activating the opposite converter bridge, is clearly seen. With a three-phase output and symmetrical load, all triplen harmonics would be removed from the load current. Due to the inductive load the voltage rises with increasing frequency. This will eventually lead to additional distortion of the voltages and currents caused by the ceiling voltage of the converters.

Finally in Fig. 11.19 oscillograms of the measured output current and line-to-neutral voltage of a cycloconverter are shown; the converter is supplied from the 50 Hz three-phase line and feeds a symmetrical three-phase inductive load at 10 Hz [A13].

For large drives, synchronous motors are often preferred over induction motors since there the reactive power does not have to be supplied through the converter and because a larger airgap is desirable for mechanical reasons. Cycloconverters are very suitable for this duty, provided the stator frequency is sufficiently low, which is often the case with large low speed motors such as required for rolling mill and mine hoist drives; synchronous drives will be discussed in Chap. 14.

As an alternative to DC link PWM converters, force commutated cycloconverters, called matrix converters [E8, S47], may become attractive at lower power rating, because the energy storage components in the DC-link

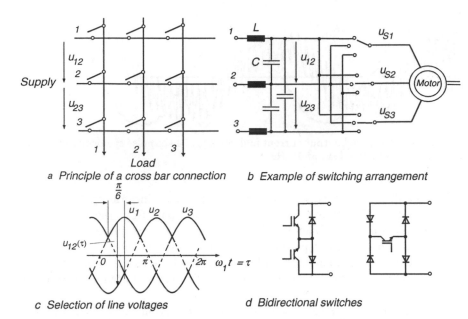

a *Principle of a cross bar connection*   b *Example of switching arrangement*

c *Selection of line voltages*   d *Bidirectional switches*

Fig. 11.20. Principle of a PWM matrix converter for generating variable frequency voltages directly from the line voltages

are avoided, thus facilitating large scale integration with semiconductors. The principle of a matrix converter is shown in Fig.11.20 a in the form of a cross bar arrangement, where the three load terminals are alternatively connected to the three supply terminals. This requires nine bidirectional switches, Fig.11.20 d, to be pulse-width-modulated at some kHz; a possible arrangement is seen in Fig. 11.20 b, where in each time interval $\omega_1 t = \pi/6$ the highest line-to-line voltage assumes the role of the former DC link voltage, as shown in Fig. 11.20 c. This switching strategy permits the highest possible output voltage; at the same time the reactive line-side current is reduced since the current flows only in the centre region of the line voltage periods. The higher frequency components of the line currents can be removed by capacitive filters. As usual, it must be prevented that the line voltages are temporarily short circuited but this is excluded with the arrangement of switches in Fig. 11.20 b. The equivalence for vectorial modulation which corresponds to Eq. (11.2), then reads

$$\underline{u}_{S\,\mathrm{Ref}} = (u_{Sa} + j\,u_{Sb})_{\mathrm{Ref}} = u_{ik}(\tau)\left(\frac{t_1}{T_0} + e^{j\,\pi/3}\,\frac{t_2}{T_0}\right),\qquad(11.8)$$

where $u_{ik}(\tau)$ is the highest line-to-line voltage in each $\pi/6$-interval. In addition $t_1 + t_2 + 2t_0 = T_0$ holds, where t_0 is the zero vector interval, when the motor is short circuited at one of the line terminals, Fig. 11.8.

12. Control of Induction Motor Drives

When comparing the dynamic model of a separately excited DC machine, Eqs. (5.1–5.4), Fig. 5.4, with that of an AC induction machine, Eqs. (10.50–10.53), Fig. 10.16, it is obvious that the latter represents a much more complex control plant. This is caused by the fact that the main flux and the armature current distribution of a DC machine are fixed in space and can be directly and independently controlled while with an AC machine these quantities are strongly interacting and move with respect to the stator, the rotor as well as each other; they are determined by the instantaneous values of the stator currents, two of which represent independent control variables. An additional complication stems from the fact that the rotor currents cannot be measured with ordinary cage rotors. Hence the AC motor is a nonlinear multi-variable control plant that kept control engineers puzzling for a long time.

The differences in control dynamics of a DC and an AC motor are best explained by the mechanical models shown in Fig. 12.1. In Fig. 12.1 a, corresponding to a DC motor with a mechanical load, a disk is driven by the tangential force f_T acting on a pin P which can be moved in a radial slot by a radial force f_R acting against a spring; caused by velocity-dependent friction, the radial motion of the pin is assumed to be relatively slow.

Between this mechanical arrangement and the electric model of a DC machine the following analogies hold:

$$f_T \,\hat{=}\, \text{armature current}, \quad R\,f_T \,\hat{=}\, \text{electrical torque},$$
$$R \,\hat{=}\, \text{main flux}, \quad R\,\omega \,\hat{=}\, \text{induced voltage (e.m.f.)},$$
$$f_R \,\hat{=}\, \text{field voltage}, \quad R\,\omega f_T \,\hat{=}\, \text{electrical power}.$$

Controlling torque, speed and angular position of the disk is straight-forward if f_T and f_R can be independently chosen; the analogy applies to operation below base speed as well as in the field-weakening range with the limitation being either saturation (length of the radial slot) or maximum induced voltage (circumferential velocity of pin).

A similar arrangement applies to the AC motor, Fig.12.1 b; however, the pin is now driven by three connecting rods that apply forces f_1, f_2, f_3 in three fixed directions spaced by 120°. To produce a smooth circular motion while at the same time keeping the pin at a given radius, a well coordinated set of alternating forces is required for producing constant radial

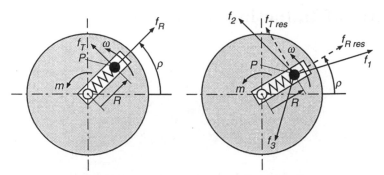

Fig. 12.1. Mechanical model of (a) DC and (b) AC motor

and tangential resultant forces. With the analogy that the orientation of the pin (R, ϱ) corresponds to the position of the fundamental flux wave and the forces correspond to the stator currents, it is easy to see why an AC motor is so much more difficult to control than a DC motor.

This is one reason why it has taken a long time before generally accepted solutions for controlling AC motors have emerged, which had long been available for DC motors; instead, a large number of contending proposals existed, most of which were only useful for special cases or particular applications. However, since the beginning of the 70's, new methods adapted to vectorial machine models have been proposed that had the potential of forming the basis for a unified theory of controlling AC machines. In the following chapters, some of these methods will be discussed.

12.1 Control of Induction Motor
Based on Steady State Machine Model

For many mechanical loads, such as pumps or fans, there is no need for high dynamic performance, as long as the speed can be varied with good efficiency over the desired speed range; this permits the use of steady state machine models for the design of the drive control. Two of the schemes that have been proposed for this purpose and are in frequent use, will be discussed first. They are based on the assumption that the induction motor is supplied by a PWM-voltage source converter, producing symmetrical three- phase voltages and currents of a fundamental frequency related to the desired speed.

In view of the simple and very effective control schemes for DC motors it was appropriate to search for similar methods that might be applicable to the control of AC motors. A salient feature with DC motor control was to maintain constant flux below base speed to fully utilise the given motor size with regard to torque, which was controlled through the armature current; a similar approach may be followed here. When for the moment neglecting the stator resistance and assuming symmetrical sinusoidal stator voltages U_1 of frequency ω_1, the electric torque of the induction motor is given by Eq.(10.85) where the normalised slip derived from Eqs.(10.67), (10.71) is

$$\frac{S}{S_p} = \frac{(\omega_1 - \omega)\,\sigma\,L_R}{R_R} = \omega_2\,\sigma\,T_R\,. \tag{12.1}$$

By varying the stator voltages in proportion with frequency,

$$\frac{U_1}{U_{S0}} = \frac{\omega_1}{\omega_0}\,, \tag{12.2}$$

the stator flux is kept constant, hence the steady state torque is

$$m_M = m_{p0}\,\frac{2}{\dfrac{S}{S_p} + \dfrac{S_p}{S}} = 2\,m_{p0}\,\frac{\omega_2\,\sigma\,T_R}{1 + (\omega_2\,\sigma\,T_R)^2}\,. \tag{12.3}$$

The torque/speed curves are obtained by shifting this reference curve along the frequency axis according to $\omega = \omega_1 - \omega_2$, while maintaining its shape; this is seen in Fig. 12.2.

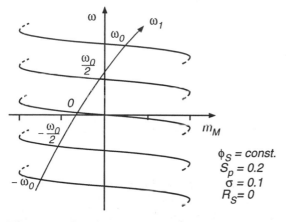

Fig. 12.2. Steady state torque/speed curves at different stator frequencies, assuming constant stator flux and neglecting stator resistance

If only the centre portion of the curves having negative slope is used, i. e. by limiting the rotor frequency to values below pull-out slip, the curves exhibit a marked resemblance to the steady state torque/speed curves of a DC- machine with constant flux, Fig. 5.5 a); hence, the rotor frequency may be regarded as representing torque, serving as a substitute for the armature current [J17, B31].

As long as the voltage drop across the stator resistor can be neglected, Eq. (12.2) is adequate for maintaining constant stator flux; however, large errors occur at low speed, where the resistive voltage component will eventually exceed the induced voltage; hence, a correction of the stator voltage according to Fig. 12.3 a is required,

$$U_S = \left| j \frac{\omega_1}{\omega_0} U_{S0} + R_S \underline{I}_S \right| ; \tag{12.4}$$

if the operating range is to be extended to low speed. With Eq. (10.72) this results in

$$U_S = \left| \frac{\omega_1}{\omega_0} + \frac{R_S}{j \, \omega_0 \, L_S} \frac{1 + j \, \omega_2 \, T_R}{1 + j \, \omega_2 \, \sigma \, T_R} \right| U_{S0} . \tag{12.5}$$

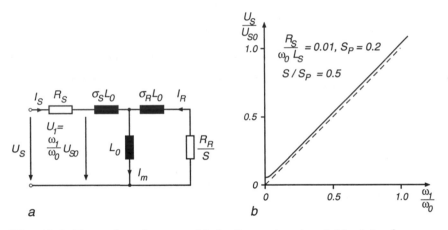

a   b

Fig. 12.3. Torque/speed curves of induction motor at variable stator frequency; (a) equivalent circuit, (b) stator voltage as a function of stator frequency for constant stator flux at $S = S_p/2$

The curve in Fig. 12.3 b is calculated for $R_S/\omega_0 \, L_S = 0.01$, $S = 0.1$. Deviations from the idealised Eq. (12.2) are noticeable mainly at low frequency and at $\omega_1 = 0$, the necessary voltage is caused only by the stator resistance, which depends on the winding temperature. In Fig. 12.4 it is seen how this function can be incorporated into a simple speed control scheme of an induction motor. The speed controller is producing a reference for the rotor frequency which is taken to represent torque; it must be limited to the range where the torque/speed relationship is approximately linear, to avoid pull- out. By adding the signal for measured speed, a reference for the stator frequency ω_1 is obtained. Naturally, such a simple control scheme is only suitable for speed adjustment in steady state, not for high performance drives.

Open loop compensating control schemes like the one described are sensitive to side effects such as temperature changes that are not accounted for. An improvement would be to convert from impressed voltages to currents to

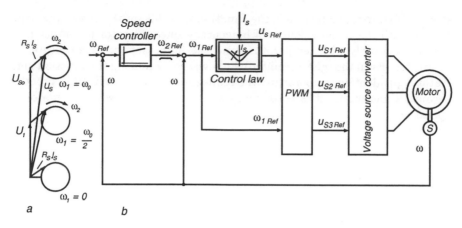

Fig. 12.4. Speed control of an induction motor with impressed stator voltages and open loop flux control, based on a steady state model of the machine

eliminate the effects of stator resistance and the primary leakage reactance; this would also simplify the dynamics of the drive by removing Eq. (10.50) from the model equations. In addition, the assumption of a sinusoidal supply is much more realistic with stator currents than with voltages produced by a switched power converter. As explained in Chap. 11, either a PWM voltage source converter or a cycloconverter with current control could be used, covering a wide range of power ratings.

When supplying the motor with impressed sinusoidal currents, it is advantageous to aim for constant air-gap- or, better yet, rotor-flux, because this eliminates not only the effects of stator resistance but also of leakage inductance; in the equivalent circuit in Fig. 10.7 b, this corresponds to the condition of maintaining a constant rotor- based magnetising current I_{mR} or rotor flux, independent of speed and load. The pertinent stator currents are obtained from the equivalent circuit, resulting in

$$\underline{I}_S = \left[1 + \frac{j\,\omega_1\,(1 - \sigma)\,L_S\,S\,(1 + \sigma_R)^2}{R_R} \right] \underline{I}_{mR} = (1 + j\,\omega_2\,T_R)\,\underline{I}_{mR} \,.$$

(12.6)

To maintain a given rotor flux, the magnitude (RMS- value) of the stator currents must follow a function that depends on the rotor frequency ω_2 and contains only rotor parameters,

$$I_S = \sqrt{1 + (\omega_2\,T_R)^2}\, I_{mR} \,.$$

(12.7)

This is achieved with the help of another function generator which produces a reference according to Eq. (12.7) for the amplitude of the stator currents. The function is plotted in Fig. 12.5 for two values of the rotor time constant; it is symmetrical to $\omega_2 = 0$, i.e. for operation below and above synchronous speed. Of course, the fact that the function depends on the rotor time constant constitutes again a source of inaccuracy since the rotor resistance may vary considerably with temperature and the inductance with the degree of saturation. Even though this might be alleviated by making the function generator dependent on temperature, the realisation would be difficult in practice because a measurement of rotor temperature is not easy with the motor in operation.

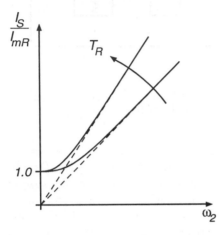

Fig. 12.5. Amplitude of stator currents required for constant rotor flux

With the definition of the rotor- based magnetising current in Fig. 10.7 b

$$\underline{I}_{mR} = \underline{I}_S + (1 + \sigma_R)\, \underline{I}_R \tag{12.8}$$

the steady state torque according to Eq. (10.81) may be written in the form

$$m_M = 3\, L_0\, \text{Im}\,[\underline{I}_S\, \underline{I}_R^*] = 3\,(1 - \sigma)\, L_S\, \text{Im}\,[\underline{I}_S\, \underline{I}_{mR}^*]\,. \tag{12.9}$$

Assuming a correctly tuned function generator, the magnetising current will be maintained at nominal value, corresponding to the ideal no-load current, hence

$$\underline{I}_{mR}\, \underline{I}_{mR}^* = I_{S0}^2\,, \tag{12.10}$$

which on insertion into Eq. (12.9) yields

$$m_M = 3\,(1 - \sigma)\, L_S\, I_{S0}^2\, \text{Im}[\underline{I}_S/\underline{I}_{mR}]\,. \tag{12.11}$$

From this follows with Eq. (12.6)

$$m_M = 3\,(1 - \sigma)\, L_S\, I_{S0}^2\, \omega_2\, T_R\,. \tag{12.12}$$

This confirms the earlier observation that the torque/speed curves may be shifted along the speed axis without changing their shape, Fig. 12.2. Since it is now the rotor flux instead of the stator flux that is maintained at constant value, the effect of stator leakage is eliminated, resulting in a reduced slope of the curves.

The next task in the design of the AC motor control consists in converting the current amplitude reference $I_{S\,Ref}$ produced by the function generator to a system of three-phase variable frequency AC signals to be used as current references for the power converter and motor. This is a modulation process for which various realisations exist; two possibilities will be discussed here, one analogue and the other digital.

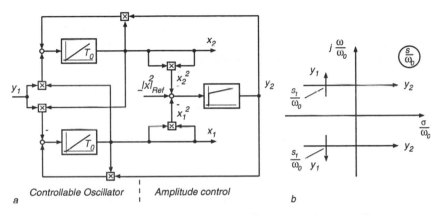

Controllable Oscillator   Amplitude control

a

b

Fig. 12.6. Variable frequency, variable amplitude two-phase oscillator.
(a) Block diagram, (b) Eigenvalues in s-plane

The block diagram of an amplitude-controlled, variable frequency two-phase oscillator is shown in Fig. 12.6 a which is suitable for generating the required AC reference signals [H26]; similar structures were contained in Fig. 10.16. The input signals y_1, y_2 and the output signals x_1, x_2 are connected by the following nonlinear differential equations

$$T_0 \frac{dx_1}{dt} = -y_1\,x_2 + y_2\,x_1 ,$$
$$T_0 \frac{dx_2}{dt} = y_1\,x_1 + y_2\,x_2 . \tag{12.13}$$

Elimination of x_2 results in a second order equation

$$T_0^2 \frac{d^2 x_1}{dt^2} - 2\,y_2\,T_0 \frac{dx_1}{dt} + (y_1^2 + y_2^2)\,x_1 = -T_0 \frac{dy_1}{dt}\,x_2 + T_0 \frac{dy_2}{dt}\,x_1 . \tag{12.14}$$

Assuming y_1, y_2 =const., a linear homogeneous differential equation is obtained, the characteristic equation of which is

$$T_0^2\, s^2 - 2\, y_2\, T_0\, s + y_1^2 + y_2^2 = 0\,. \tag{12.15}$$

With $\omega_0 = 1/T_0$ the two complex roots are

$$\frac{s_{1,2}}{\omega_0} = y_2 \pm j\, y_1\,; \tag{12.16}$$

the real and imaginary parts are subject to decoupled adjustment by y_1, y_2, as seen in Fig. 12.6 b.

$y_1 = $ const. $\neq 0$ and $y_2 \equiv 0$ give rise to an undamped oscillation of frequency $\omega_1 = y_1 \omega_0$; changing the sign of y_1 results in an inversion of the phase sequence. Correspondingly, the amplitude of the oscillation decays for $y_2 < 0$ and rises for $y_2 > 0$. Since a constant amplitude $|\underline{x}| = \sqrt{x_1^2 + x_2^2}$ is desired in steady state, this calls for an amplitude control loop as is indicated in Fig. 12.6 a, where y_2 serves as actuating input. Even though the control loop is nonlinear, its stability is uncritical so that a PI-controller proves adequate; when entering the square of the amplitude reference, $|\underline{x}|_{Ref}^2$, the root function can be omitted.

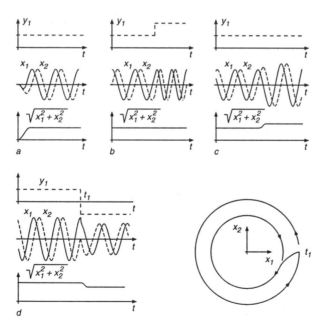

Fig. 12.7. Transients of controlled two-phase oscillator

Some computed transients of the controlled oscillator are plotted in Fig. 12.7, showing time functions $x_1(t)$, $x_2(t)$ as well as the path of the state vector $\underline{x}(t) = x_1(t) + j x_2(t)$ in the state plane. The curves indicate that the oscillator responds rapidly and in a well damped transient to changes of y_1

and $|\underline{x}|^2_{Ref}$. Following a step change of y_1, there are no transients at all as the oscillator simply continues from the existing initial conditions with the new frequency. The state vector $\underline{x}(t)$ is a continuous function that is well suited for use as current reference, as it enables the current control loops to track the references with small dynamic errors.

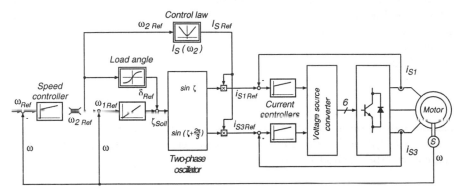

Fig. 12.8. Speed control scheme with impressed stator currents and open loop control of rotor flux

The two-phase orthogonal signals x_1, x_2 are now combined to form an equivalent three-phase system w_1, w_2, w_3 to be used as current references

$$\underline{x}(t) = x_1(t) + j\, x_2(t) = \tfrac{2}{3}\left[w_1(t) + e^{j\,\gamma}\, w_2(t) + e^{-j\,\gamma}\, w_3(t)\right] \,, \quad \gamma = \frac{2\,\pi}{3}. \tag{12.17}$$

To render the transformation unique and exclude a zero sequence component,

$$w_1(t) + w_2(t) + w_3(t) = 0 \tag{12.18}$$

is added as a side condition. A linear combination of x_1, x_2 results in a set of symmetrical three-phase reference signals for the stator currents,

$$w_1(t) = x_1(t) = \frac{i_{S1\,Ref}}{I_{S0}}\,,$$

$$w_2(t) = -\tfrac{1}{2}\, x_1(t) + \frac{\sqrt{3}}{2}\, x_2(t) = \frac{i_{S2\,Ref}}{I_{S0}}\,, \tag{12.19}$$

$$w_3(t) = -\tfrac{1}{2}\, x_1(t) - \frac{\sqrt{3}}{2}\, x_2(t) = \frac{i_{S3\,Ref}}{I_{S0}}\,.$$

A block diagram of the complete speed control system is drawn in Fig. 12.8, where the reference signals for the stator currents are supplied by the oscillator; the amplitude is derived from the function- generator which in turn is controlled by the desired rotor frequency $w_2 = w_1 - w$. By making the output

of the function generator dependent on stator frequency, field weakening can
also be implemented with this scheme.

The other actuating input of the controlled oscillator, $y_1 = \omega_1 T_0$, determining the frequency of the stator currents is obtained by combining the
measured speed signal ω with the rotor frequency reference ω_{2Ref} as produced by the speed controller. To restrict the operation of the motor to the
"linear" range, the rotor frequency reference is again limited as shown in Fig.
12.8. Thus the stator frequency is determined on the basis of the measured
speed and the required torque.

For larger motors having low pull-out slip ($S_p < 0.1$), the signal for the
stator frequency is derived as the sum of two greatly differing quantities,
$\omega_1 = \omega + \omega_{2Ref} \approx \omega$, which makes the realisation of an accurate torque limit
difficult; to improve this, the ω_1-signal may be formed digitally. In fact, with
the microelectronic components available today the complete control scheme
can be realised on a digital basis. This is of particular interest with the field
orientated control described in the next chapter; only a digital equivalent of
the controllable oscillator is shown here.

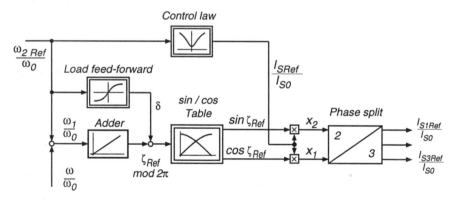

Fig. 12.9. Digital signal processing for obtaining variable frequency, variable amplitude current references

Part of the digital control scheme is seen in Fig. 12.9, where sections of
the trigonometric functions are stored as a function table in a memory chip,
the address of which is the angle $\zeta = \int \omega_1 dt$ in suitable increments. The
two variables $\cos\zeta$, $\sin\zeta$ moving on a unit circle are then multiplied by the
output of the function generator for the magnitude of the stator currents,
before the signals are split into three phases. The dynamic response may be
greatly improved by a feed forward signal based on the load angle δ; this is
similar to replacing an I- by a PI-controller. When choosing

$$\delta = \arctan \omega_2 T_R \tag{12.20}$$

according to Eq. (12.6), the angular position of the stator ampereturns- wave is correctly advanced under load, while the flux wave remains at its former position. This simple example shows how digital techniques can be employed to great advantage.

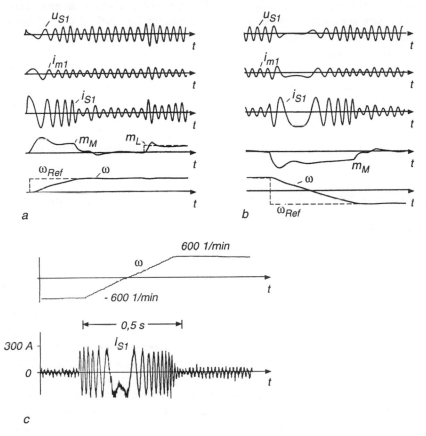

Fig. 12.10. Transients of induction motor drives
(a), (b) Simulated results for starting, loading and speed reversal with the control scheme in Fig. 12.8.
(c) Measured transient of experimental drive with the digital speed control according to Fig. 12.9

The control scheme shown in Fig. 12.8 was simulated on a digital computer with Eqs. (10.50)–(10.53) representing the motor, while the control loops for the stator currents were simplified by unity gain, first order lag elements [T14]. The results are plotted in Fig. 12.10, showing a starting transient with subsequent loading and a speed reversal when inverting the speed reference. Due to the rotor frequency limit, the speed varies approximately linearly with time. It is interesting to note that the magnetising current is fairly well maintained during the transients, even though the design of the control scheme is based on a steady state model of the motor.

A digitally controlled experimental drive using a 5 kW induction motor supplied by a voltage source transistor converter with fast current control was tested with the control scheme in Fig 12.9; a recorded speed transient with a high inertia load is shown in Fig 12.10 c [H20]. Clearly, as the stator frequency is tracking the speed, the motor cannot be "pulled out of step" but, when overloaded, operates at the torque limit.

Even though acceptable results can be obtained with a well tuned control that is based on a steady state model of the machine, these restrictions are undesirable for high dynamic performance drives, they may be avoided with the control strategies described in the following sections. Basing the control on the dynamical models of the machine, AC drives can be designed whose dynamic performance is equal or superior to that of DC drives, considering the larger bandwidth of the converter and the reduced inertia of the motor.

12.2 Rotor Flux Orientated Control of Current-fed Induction Motor

12.2.1 Principle of Field Orientation

The heuristic control scheme described in the preceding section is, in spite of the acceptable results obtained, not best suited for high dynamic performance drives. Because of the open loop flux control it would be difficult or impossible to operate the motor with full torque at low speed or even standstill which excludes, for example, its application as a servo-drive. o remove these restrictions it is necessary to return to the dynamic model equations of the induction motor, based on instantaneous currents and voltages, as derived in Sect. 10.1. Of particular interest must be the interaction of stator and rotor currents resulting in flux and torque. When initially assuming that the stator currents are impressed by fast current control loops considerable simplifications result since the stator voltage equation (10.50) becomes a concern of the current controllers and can be omitted from the model equations of the drive. The restrictions of steady state are now avoided which is a precondition for achieving rapid transients. The power supply could be a transistor converter with high switching frequency and adequate ceiling voltages or a cycloconverter, resulting in very fast response of the stator currents.

The expression for the electric torque (10.48)

$$m_M = \frac{2}{3} L_0 \, \text{Im} \left[\underline{i}_S \, (\underline{i}_R \, e^{j\,\varepsilon})^* \right] \tag{12.21}$$

was obtained by describing the interaction between the rotor currents and the flux wave resulting from stator currents. Since the rotor currents cannot be measured with cage motors, it is appropriate to substitute for the rotor current vector $i_R e^{j\varepsilon}$ an equivalent quantity that could be measured with stator based sensing equipment. A good choice, as will be shown later, is the rotor flux, Eq. (10.32), or the equivalent magnetising current vector, defined in stator coordinates,

$$\underline{\psi}_R(t)\, e^{j\,\varepsilon} = L_0\, \underline{i}_{mR}(t) = L_0\, i_{mR}(t)\, e^{j\,\varrho(t)} = L_0\,[\underline{i}_S(t) + (1+\sigma_R)\,\underline{i}_R\, e^{j\,\varepsilon}]\,. \tag{12.22}$$

Eliminating the rotor current from Eq. (12.21) results in

$$m_M(t) = \frac{2}{3}\frac{L_0}{1+\sigma_R}\, \text{Im}\,[\underline{i}_S\,(\underline{i}_{mR} - \underline{i}_S)^*]\,, \tag{12.23}$$

which is simplified to

$$m_M(t) = \frac{2}{3}\,(1-\sigma)\,L_S\, \text{Im}\,[\underline{i}_S\, i_{mR}^*]\,. \tag{12.24}$$

Inserting the magnetising current vector according to Eq. (12.22) yields

$$m_M(t) = \frac{2}{3}\,(1-\sigma)\,L_S\, i_{mR}\, \text{Im}\,[\underline{i}_S\, e^{-j\,\varrho}]\,, \tag{12.25}$$

where

$$\underline{i}_S\, e^{-j\,\varrho} = i_S\, e^{j\,\delta} = i_{Sd} + j\, i_{Sq} \tag{12.26}$$

represents the stator current vector in a moving frame of reference defined by the rotor flux or the magnetising current vector $\underline{i}_{mR}$.

The angular relationships of the current vectors are depicted in Fig. 12.11; the stator current vector in field coordinates or the "field oriented" stator current vector is seen to consist of two orthogonal components,

$$i_{Sd} = \text{Re}\,[\underline{i}_S\, e^{-j\,\varrho}] = i_S\, \cos\delta$$
$$i_{Sq} = \text{Im}\,[\underline{i}_S\, e^{\,j\,\varrho}] = i_S\, \sin\delta$$

one in the direction of the magnetising current vector and the other perpendicular to it. In steady state condition i_{Sd} and i_{Sq} are constant apart from converter induced ripple, i.e. the stator current vector $\underline{i}_S$ and the magnetising current vector $\underline{i}_{mR}$ rotate in synchronism. The steady state current trajectories in Fig. 12.11 were taken from actual measurements with the drive shown in Fig. 12.28.

Introducing the quadrature current component into Eq. (12.25) leads to a simple expression

$$m_M(t) = k\, i_{mR}\, i_{Sq}\,, \qquad k = \frac{2}{3}\,(1-\sigma)L_S\,, \tag{12.27}$$

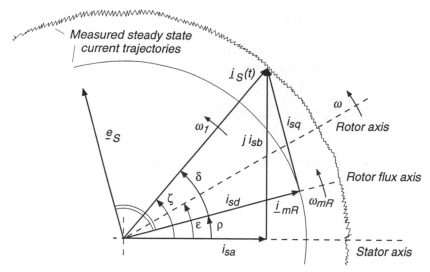

Fig. 12.11. Angular relations of current vectors in steady state

which gives a hint, why the transformation of the stator current vector into field coordinates, also called field orientation, is the key to rapid control of AC machines. This principle has been proposed by Blaschke [B33–B38], extending and generalising earlier work by Hasse [H26].

Clearly Eq. (12.27) reminds one of the expression for the electric torque of a DC machine, Eq. (5.3), where i_{mR} corresponds to the main flux Φ_e and i_{Sq} to the armature current i_a. As seen in Fig. 12.11, the magnetising current i_{mR} is controlled by the direct component i_{Sd} of the stator current vector, which may be compared to the field voltage of the DC machine because, as will be seen, there is also considerable magnetic lag between i_{Sd} and i_{mR}. Hence, i_{Sd}, i_{Sq} are the two independent input quantities controlling the torque of the motor; δ represents a load angle that vanishes at no load.

The $d - q$-axes established by $\underline{i}_{mR}$ differ from those commonly used with synchronous machines where the Park transformation is based on the rotor position ε; here the frame of reference is the rotor flux vector which moves across the stator at stator frequency and across the rotor with slip frequency.

To obtain the complete model of the AC motor in field coordinates, $\underline{i}_{mR}$ is now used for elimination of the rotor current from the rotor voltage equation (10.51) with $u_R = 0$,

$$R_R \, \underline{i}_R + L_0 \frac{d}{dt} \underbrace{[(1 + \sigma_R) \, \underline{i}_R + \underline{i}_S \, e^{-j\varepsilon}]}_{\underline{i}_{mR} \, e^{-j\varepsilon}} = 0 \, ; \qquad (12.28)$$

with $T_R = L_R/R_R$, this leads to

$$T_R \frac{di_{mR}}{dt} + (1 - j\omega T_R) \underline{i}_{mR} = \underline{i}_S ,$$ (12.29)

where $d\varepsilon/dt = \omega$.

The definitions (10.7) and (12.22) yield also the instantaneous angular velocities of the magnetising and stator current vectors

$$\frac{d\varrho}{dt} = \omega_{mR}(t) ,$$ (12.30)

$$\frac{d\zeta}{dt} = \omega_1(t) = \omega_{mR} + \frac{d\delta}{dt} .$$ (12.31)

Multiplying Eq. (12.29) by $e^{-j\varrho}$ and expanding the left hand expression results in

$$T_R \frac{di_{mR}}{dt} + j\,\omega_{mR} T_R\, i_{mR} + (1 - j\,\omega\, T_R)\, i_{mR} = \underline{i}_S\, e^{-j\varrho} ,$$ (12.32)

which may be split into real and imaginary parts,

$$T_R \frac{di_{mR}}{dt} + i_{mR} = i_{Sd} ,$$ (12.33)

$$\frac{d\varrho}{dt} = \omega_{mR} = \omega + \frac{i_{Sq}}{T_R\, i_{mR}} = \omega + \omega_2 .$$ (12.34)

Eqs. (12.27, 12.33, 12.34), together with the mechanical equations (10.52, 10.53) constitute a model of the induction motor in field coordinates, as described by the block diagram in Fig. 12.12. The two field-orientated input currents are produced by transforming the stator currents into a coordinate system defined by the flux angle ϱ. This is performed in two steps, first converting the three stator currents to an orthogonal two-phase AC system,

$$\underline{i}_S(t) = i_{S1} + i_{S2}\, e^{j\gamma} + i_{S3}\, e^{j\,2\gamma} = i_{Sa} + j\, i_{Sb} .$$ (12.35)

With the condition for balanced three phase currents, $i_{S1} + i_{S2} + i_{S3} = 0$, this results in

$$i_{Sa}(t) = \tfrac{3}{2} i_{S1}(t) ,$$
$$i_{Sb}(t) = \tfrac{\sqrt{3}}{2} [i_{S2}(t) - i_{S3}(t)] .$$

These AC currents are then, by transformation into field coordinates, converted to DC quantities in steady state

$$\underline{i}_S(t)\, e^{-j\varrho} = (i_{Sa} + j\, i_{Sb}) (\cos\varrho - j\,\sin\varrho) = i_{Sd} + j\, i_{Sq} ;$$ (12.36)

hence we have

$$i_{Sd}(t) = i_{Sa}\, \cos\varrho + i_{Sb}\, \sin\varrho ,$$
$$i_{Sq}(t) = i_{Sb}\, \cos\varrho - i_{Sa}\, \sin\varrho .$$

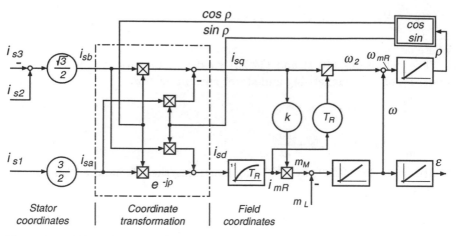

Fig. 12.12. Block diagram of induction motor in field coordinates, assuming impressed stator currents

The flux angle ϱ, obtained by integration of Eq. (12.34), is an internal variable; the coordinate transformation essentially constitutes a demodulation.

Once the transformation has been completed, the dynamic structure of the AC motor is quite straight-forward, not unlike that of a DC machine. In particular there is a lag time constant T_R comparable to the field lag of a DC machine; therefore the magnitude of the rotor flux or the equivalent magnetising current vector, i_{mR}, is not suitable for fast control action influencing torque, this task should be assigned to the quadrature current i_{Sq}. The leakage time constant in the quadrature axis, which would be the equivalent of the DC armature time constant, is not effective in Fig. 12.12 considering the assumed impressed stator currents.

For achieving high dynamic performance control of the AC drive, the same strategy as with the DC machine is appropriate: The magnetising current i_{mR} should be maintained at maximum level, limited below base speed by saturation of the iron core and above base speed by the ceiling voltage of the converter, while the torque should be controlled through the quadrature current. When highest efficiency of a drive is of utmost importance, as in battery supplied vehicles, the flux level could be reduced when operating the motor at part load for extended periods, in effect balancing reduced iron losses against increased conduction losses in the motor and converter. This was mentioned in Sect. 5.3.3 on combined armature and field control of DC motors. Of course, there is an increased delay when the flux level first needs to be raised in case of sudden torque demand.

For alleviating the complexities of the AC control plant, it can now be attempted to cancel the coordinate transformation by an inverse transformation at the control side of the converter; this idea, which is part of the principle of control in field coordinates as suggested by Blaschke, is illustrated in simplified form in Fig. 12.13. By modulating the signals at the control side with the reconstructed flux angle $(e^{j\varrho})$ and subsequent 2/3 phase split the corresponding interactions in the control plant are cancelled as long as the flux angle is correct and the delay of the converter and the residual lag of the current loops are negligible, i.e. the converter acts as a current source.

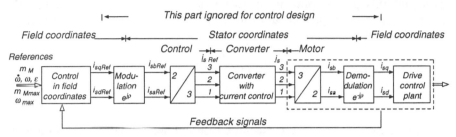

Fig. 12.13. Principle of control in field coordinates

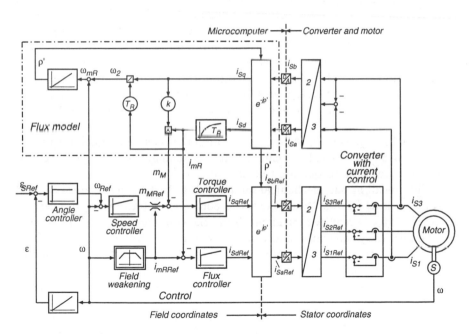

Fig. 12.14. Speed/position control in field coordinates of AC motor

With these assumptions the design of the remaining control system becomes very similar to that of a DC motor control scheme; this is shown in Fig. 12.14 for a position control configuration. The three current references for the controlled converter are obtained from the two-phase current references $(i_{Sa}, i_{Sb})_{Ref}$ with the help of Eq. (12.35),

$$i_{S1\,Ref} = \tfrac{2}{3}\, i_{Sa\,Ref}$$
$$i_{S2\,Ref} = -\tfrac{1}{3}\, i_{Sa\,Ref} + \tfrac{1}{\sqrt{3}}\, i_{Sb\,Ref}\,, \qquad\qquad\qquad (12.37)$$
$$i_{S3\,Ref} = -\tfrac{1}{3}\, i_{Sa\,Ref} - \tfrac{1}{\sqrt{3}}\, i_{Sb\,Ref}\,.$$

Preceding this phase splitting operation is the modulation of the signals $(i_{Sd}, i_{Sq})_{Ref}$, calling for four multiplications with trigonometric functions of the unbounded flux angle ϱ' mod 2π,

$$
\begin{aligned}
(i_{Sa} + j\, i_{Sb})_{\text{Ref}} &= (i_{Sd} + j\, i_{Sq})_{\text{Ref}}\, e^{j\,\varrho'} \\
&= i_{Sd\,Ref}\,\cos\varrho' - i_{Sq\,Ref}\,\sin\varrho' + j\,[i_{Sq\,Ref}\,\cos\varrho' + i_{Sd\,Ref}\,\sin\varrho']\,.
\end{aligned}
$$
$$(12.38)$$

To achieve cancellation of the machine-internal demodulation, the available flux signal ϱ' must be in close agreement with the actual position ϱ of the fundamental flux wave; this aspect of flux acquisition still has to be discussed.

Proceeding in Fig. 12.14 further to the left, it is seen that the remaining control structure is the same as that of a DC machine. Instead of the armature current controller, a torque controller, whose synthetic feedback signal is derived on the basis of Eq. (12.27) from flux and stator currents, generates the quadrature axis current reference. Speed and position control loops may be superimposed. With a digital realisation, the position controller can be a simple proportional controller, as will be discussed in Sec. 15.2. It would also be possible to omit the torque controller and generate i_{SqRef} directly with the speed controller.

In the direct axis there is a flux controller, the reference signal of which is reduced above base speed to achieve field weakening; the flux reference also serves to reduce the torque limit at high speed.

When recalling the mechanical model in Fig. 12.1 b, the strategy of field orientation is as follows: When the speed controller determines that a certain value of torque is required for maintaining the commanded speed, references for the necessary tangential and radial forces are computed which are then, by coordinate transformation, converted to reference quantities for the forces f_1, f_2, f_3. The transformation must be based on the instantaneous position of the pin; hence the need for flux acquisition in a field orientated control scheme.

Clearly, the principle of control in field coordinates looks like an effective way of decoupling the complex multivariable control structure of the induction machine. Its main advantages are:

- Direct access to flux and torque permitting controlled field weakening and torque limit.
- Provided the angle ϱ' used for modulation is sufficiently accurate, the motor is self-controlled and cannot be pulled out of step; this corresponds to the

function of the commutator on a DC-machine with current limit. When overloaded, the motor is stalled with maximum torque.

- In contrast to the schemes discussed in Sect. 12.1 the decoupling is effective in dynamic as well as steady state conditions.
- In steady state the drive controllers process DC quantities which makes the control less sensitive to unavoidable phase shifts.

The speed control system contained in Fig. 12.14 has been simulated on a computer, using the same set of parameters as for Fig. 12.10a,b [T14]. The current control loops were again approximated by unity gain first order lag elements. When comparing the starting, loading and reversing transients in Fig. 12.15 with those of Fig. 12.10 a,b, the superior performance of the field orientated control scheme is obvious. The flux is maintained constant and the electrical torque shows a rapid rise, limited only by the lag of the converter; when the speed controller is clamped, the torque is precisely fixed at its maximum value.

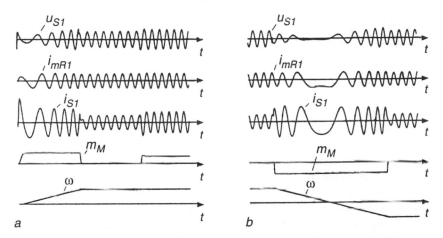

Fig. 12.15. Computed transients of AC motor controlled in field coordinates

Of course, these are only simulated results which have been computed with a highly idealised model of the control plant; to realise a field orientated control scheme in practice, several serious problems first have to be solved:

- Acquisition of frequency independent flux signals, representing amplitude and position of the fundamental flux wave.
- Effects of residual lag of the current controlled converter.
- Implementation of the complex signal processing shown in Fig. 12.14.
- Estimating the effects of a detuned flux model.

These problems will now be discussed separately.

12.2.2 Acquisition of Flux Signals

Clearly, having up-to-date information on the magnitude and phase of the fundamental flux wave is of paramount importance since this is the basis of coordinate transformation, leading to decoupled control of the currents [B37, L7].

For obtaining a frequency-independent flux measurement one could attempt to measure the flux density in the airgap of the machine directly by placing suitably spaced magneto-galvanic or magneto-resistive devices such as Hall- sensors on the face of stator teeth; by interpolating the local samples of flux density, an estimate of the magnitude and position of the airgap flux wave could be obtained. Then by adding a voltage component proportional to the stator current vector, a signal representing the rotor flux $\underline{\psi}_R \, e^{j\,\epsilon}$ or the magnetising current $\underline{i}_{mR}$ would result, Eq. (12.22). However, apart from the fact that the tiny Hall sensors are mechanically fragile devices that would not stand up very well under severe vibrations and thermal stress, there are large harmonics caused by the rotor slots, the frequencies of which change with speed. This calls for adjustable filters that would be difficult to design, particularly under the condition of zero phase shift. Also, the torque signal computed from this information is likely to be unreliable since torque is a surface-related integral quantity which is difficult to estimate on the basis of a few local field measurements. Another disadvantage of this scheme is that the motor would have to be fitted with these sensors, so that it would no longer be a standard motor that could easily be exchanged for another motor taken from stock.

The active semiconductor elements in the motor could be avoided while at the same time suppressing the undesirable slot harmonics, if sensing coils having a width equal to full pole pitch would be installed in the stator, for example enclosed in the wedges covering the stator slots; this geometrically filters the disturbances by the rotor slots and produces signals proportional to flux change which, after integration, could serve as a measure of the main flux. Again by using several sensing coils, displaced around the circumference of the stator, and adding voltages proportional to stator currents, signals representing rotor flux could be obtained.

This scheme was tried in the laboratory [G1, G2] and found to work well down to very low frequency (> 0.5 Hz), where the drift of the integrators eventually became too large; this limitation precludes the application of this scheme to drives requiring position control. Again, a specially prepared motor would have to be used.

The last mentioned disadvantage is avoided by measuring terminal voltages, i.e. using the stator windings as sensing coils. This, however, complicates the situation even further because of the resistive voltage drop, that dominates at low frequency and must be compensated prior to integration; as the stator resistance changes with temperature this measuring scheme can become quite involved. Tests without temperature compensation have indi-

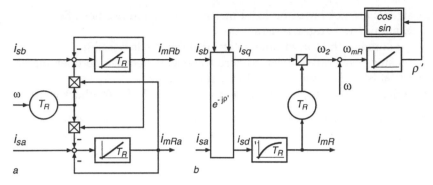

Fig. 12.16. Flux model for obtaining the magnetising current vector
(a) in stator coordinates and **(b)** in field coordinates

cated a practical lower frequency limit of about 3 Hz with a 50 Hz motor.
The situation may be improved when performing the integration in field coordinates, as discussed in Sect. 12.5.

Another quite different approach is based on the model equations of the
motor. The equation representing rotor voltages, Eq.(12.29), a first order
vectorial differential equation, may be used for computing $i_{mR}(t)$ in stator
coordinates on- line. The measured stator currents i_S and speed ω could serve
as input functions, with the motor being described by a single parameter, the
rotor time constant T_R.

By again introducing orthogonal two phase current components

$$\underline{i}_S(t) = i_S(t)\, e^{i\,\zeta(t)} = i_{Su}(t) + j\, i_{Sb}(t)\,, \tag{12.39}$$
$$\underline{i}_{mR}(t) = i_{mR}(t)\, e^{j\,\varrho(t)} = i_{mRa}(t) + j\, i_{mRb}(t)\,, \tag{12.40}$$

the block diagram of Fig. 12.16 a results which illustrates in graphical form
the interactions of Eq. (12.29). The structure is similar to the controllable
oscillator shown in Fig. 12.6 for $y_1 = \omega\, T_R$, $y_2 = -1$.

A more direct method of computing the magnitude and position of the
fundamental flux wave is to immediately convert the flux model into field
coordinates, as described by Eqs. (12.33, 12.34) [S41,S42]. Again, stator currents and speed serve as input signals while the transformed stator currents
i_{Sd}, i_{Sq} and the magnetising current i_{mR} are now directly available as output
signals for controlling flux and torque. The flux model in field coordinates,
Fig. 12.16 b, is preferable because of its simplicity and potential accuracy;
also the output signals are constant in steady state. Of course, the influence
of the rotor time constant T_R is still present; this parameter may have to be
updated during operation. Fig. 12.16 b corresponds directly to the model of
the motor shown in Fig. 12.12 and is included as a "flux model" in Fig. 12.14.

The use of a dynamic flux model such as shown in Fig. 12.16 b offers
several advantages:

- Flux signals are not based on local measurements and are not affected by slot harmonics and stator resistances.
- Use of a standard motor without additional sensors is possible.
- Flux sensing is operative down to zero frequency, because no open ended integration is needed which would be subject to drift.
- The model immediately generates the variables needed for feedback control.

However, there are stringent accuracy requirements, for example with respect to the modulo 2π integration of ϱ; this can best be solved by digital computation in a microprocessor as will be shown later. The dependence of the rotor model on the time constant T_R constitutes a source of error because an incorrect orientation ϱ' of the flux wave leads to undesirable coupling between the d- and q-axes and could endanger the control in field coordinates, eventually even resulting in instability.

Another possibility for obtaining a flux signal is the evaluation of the stator voltage equation, Eq. (10.50), when sensing the terminal voltages and currents. However, after eliminating the unaccessible rotor currents, Eq. (12.45) results which contains the rotor flux in derivative form so that an open ended integration would again be necessary; because of the necessary $i_S R_S$- compensation and integrator drift this is likely to produce unreliable results at low speed. This option is not pursued here, it will be further discussed in Sect. 12.5.

12.2.3 Effects of Residual Lag of the Current Control Loops

When explaining the principle of control in field coordinates, Fig. 12.13, it has been assumed that the stator windings are supplied from perfect current sources, i.e. that the lag of the current control loops can be ignored. This calls for a cycloconverter or a voltage source converter with high frequency pulse-width modulation, allowing rapid access to the currents through the reference signals, as well as adequate ceiling voltage throughout the operating range. Clearly the second condition is in conflict with an economical design of the converter because it would preclude a full utilisation of the available voltage at high speed. On the other hand, if the lag of the current controlled converter is not negligible, the decoupling by the inverse model becomes inaccurate and undesirable coupling terms may arise which will eventually render the field orientated control inoperative; to avoid this, two options are available:

- Inclusion of a lead- lag network with the transfer function

$$F(s) = \frac{T_1 s + 1}{T_2 s + 1}, \qquad T_2 < T_1, \tag{12.41}$$

in the reference channel of the stator current control loops.
- Addition of decoupling terms in field coordinates, i.e. prior to the transformation into stator coordinates.

For the following it is assumed that the first approach has been taken but that its potential is exhausted, leaving a residual lag T_e that cannot be reduced any further; hence the second option must be pursued.

The situation is depicted in Fig. 12.17 showing the two opposite transformations which are supposed to cancel and the unavoidable lag effects placed in between. With the vectorial notation for the currents in stator- and field

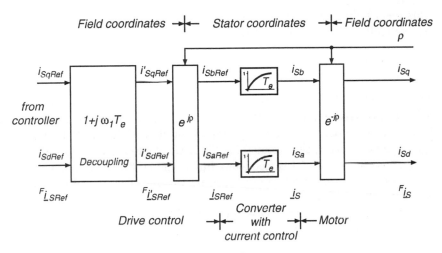

Fig. 12.17. Removal of undesired coupling effects caused by the residual lag of current control loops

coordinates

$$\underline{i}_S(t) = i_S(t)\, e^{j\varsigma} = i_{Sa}(t) + j\, i_{Sb}(t)$$

$${}^F\underline{i}_S(t) = \underline{i}_S\, e^{-j\varrho} = i_{Sd}(t) + j\, i_{Sq}(t)$$

and, correspondingly, for the reference signals,

$$\underline{i}_{S\,\mathrm{Ref}} = {}^F\underline{i}'_{S\,\mathrm{Ref}}\, e^{j\varrho}\,, \tag{12.42}$$

the lag effect of the converter may be described by

$$T_e\,\frac{d\underline{i}_S}{dt} + \underline{i}_S = \underline{i}_{S\,\mathrm{Ref}}\,. \tag{12.43}$$

With the assumption that the lag time constant T_e has already been reduced to a relatively small value, the undesirable coupling is mainly due to phase shift rather than amplitude. In steady state condition, with $i_S = $ const., $\omega_{mR} = \omega_1$ and inserting Eqs. (12.42, 12.43), this yields

$$^F\underline{i}_S = \frac{1}{1 + j\,\omega_1\, T_e}\, {}^F\underline{i}'_{S\,\mathrm{Ref}}\,; \tag{12.44}$$

when specifying a decoupled transmission,

$$^F\underline{i}_S \overset{!}{=} {}^F\underline{i}_{S\,\mathrm{Ref}} \, ,$$

the result is

$$^F\underline{i'}_{S\,\mathrm{Ref}} = (1 + j\,\omega_1\,T_e)\,{}^F\underline{i}_{S\,\mathrm{Ref}} \, .$$

This means that corrective terms should be applied to the output signals of the controllers. Written in orthogonal components, this calls for

$$i'_{S\,d\,\mathrm{Ref}} = i_{S\,d\,\mathrm{Ref}} - \omega_1\,T_e\,i_{S\,q\,\mathrm{Ref}} \, ,$$
$$i'_{S\,q\,\mathrm{Ref}} = i_{S\,q\,\mathrm{Ref}} + \omega_1\,T_e\,i_{S\,d\,\mathrm{Ref}} \, .$$

The practical realisation of this additive correction is quite easy since DC signals are involved in steady state. For simplification, the speed signal ω may be used instead of ω_1 with low slip motors. In view of the limited ceiling voltage of the converter, the residual lag T_e of the current control loops is likely to increase with the stator frequency but if the control is implemented in a microcomputer, this effect can be offset by varying the decoupling parameter T_e as the speed changes.

The rating of the converter is determined by the specified values of stator voltages, currents and frequency. After eliminating the rotor current from Eqs. (10.50, 12.22), the stator voltage reads

$$\underline{u}_S(t) = R_S\,\underline{i}_S + \sigma\,L_S\,\frac{d\underline{i}_S}{dt} + (1 - \sigma)\,L_S\,\frac{d\underline{i}_{mR}}{dt} \, . \tag{12.45}$$

In steady state condition, i.e. with constant speed and sinusoidal currents,

$$\omega_{mR} = \omega_1 \, , \quad i_S = \mathrm{const} \, , \quad i_{mR} = \mathrm{const.},$$

hence

$$\underline{u}_S(t) = (R_S + j\,\omega_1\,\sigma\,L_S)\,\underline{i}_S + j\,\omega_1\,(1 - \sigma)\,L_S\,\underline{i}_{mR} \, . \tag{12.46}$$

This confirms that the fundamental component of the stator voltage rises with frequency, Fig. 12.3 b. When the available voltage range is nearly exhausted at the frequency ω_{10} – some margin must be left for control – a further increase of the speed is only possible by reducing the magnetising current, i.e. by field weakening. This can be achieved with

- an open loop scheme employing a speed dependent reference $i_{mR\mathrm{Ref}}(\omega)$ as shown in Fig. 12.14 or
- an auxiliary control loop which limits the magnitude of the stator voltage by reducing the flux reference i_{mRRef}; a similar scheme has earlier been described for DC machines, Fig. 7.13.

When the field is reduced, the ceiling voltage of the converter should normally be fully utilised, resulting in gradual cessation of pulse-width modulation so that each leg of the converter eventually produces square wave output voltages, Fig. 11.6. The stator currents would then be no longer sinusoidal but the field orientated control still remains functioning as the computation of the flux and the magnetising current is based on the measured stator currents. Because of the limited voltage and stator currents, the maximum torque is, of course, reduced in the field weakening region, Eq. (10.85).

12.2.4 Digital Signal Processing

The signal processing required for the field-orientated control, Fig. 12.14, as well as the flux acquisition using a dynamic model, Fig. 12.16 b, is of considerable complexity; in particular, the various multipliers and function generators needed for the coordinate transformation are difficult to adjust accurately when realised with analogue components. On the other hand, on-line digital control was economically impractical as long as it had to be performed by large and costly process computers. These obstacles have delayed widespread application of the sophisticated control schemes needed for AC drives [L44].

However, this situation has changed dramatically since digital processors have now shrunk to minute size thanks to the progress of microelectronics. The reduction in volume and cost and the advances in processing power have in the meanwhile reached a point where all the signal processing required for a high dynamic performance AC drive can be executed by a single microprocessor which, together with the associated peripheral components, finds room on a postcard sized printed circuit board. The main advantage of digital control with microprocessors is that the same standard hardware can be used for many different applications because the function of the processor is determined by a flexible and individually adaptable program, i.e. by software; once the program has been developed and verified it can be duplicated at negligible cost and transferred to a read-only memory chip connected to the processor.

A precondition for the digitalisation of signal processing is, of course, an adequate resolution of the signals with respect to amplitude and time, i.e. sufficient word length and computing speed. The current 16 bit-processors are quite satisfactory from these points of view; a 16-bit representation of variables corresponds to an internal resolution of one part in more than 65,000; if 10- or 12- bit analogue-to-digital converters are employed for converting the analogue signals from the plant, such as stator currents or motor speed, the effects of quantisation are usually unnoticeable even though there are still exceptions, when higher internal resolution, such as 32 bit, proves to be necessary with some variables. Most microprocessors still use integer valued arithmectic but processors with floating point arithmetic are becoming more common now, particularly if the control programs are to be written in higher level language such as in C, to make them independent of the processor hardware used. Reduced word length, such as 10 bit, is usually adequate for the D/A-converters, i.e. at the output of the microprocessor because the D/A-converter is part of the plant and thus under the surveillance of the digital controller, whereas errors in the A/D-converter produce false signals.

The choice of the required resolution in time is more complex as it depends not only on the type of drive but also on the particular function within the drive control. Clearly, when sampling and processing non-sinusoidal currents having a fundamental frequency of 100 Hz, a higher sampling frequency is needed than for the digital realisation of speed or position control, where the rate of change is restricted by the inertia of the drive. As long as computational speed of the processors was a major consideration which is no longer the case with todays microprocessors, an advantage could be gained by employing different sampling periods for the different tasks. For example, for an experimental 1.5 kW induction motor servo drive having a PWM-transistor power supply and an 8086 control unit, the following sampling periods proved commensurable with fast control transients [S41].

a) Sampling period of 1 ms for
   A/D–conversion of stator currents
   Updating of flux model
   Conversion to field–coordinates
   Current and torque control
   Conversion of d/q–currents to stator coordinates
   Output of current references
b) Sampling period of 5 ms for
   Flux control including field weakening
   Speed control
   Position control

A sampling time of 1 ms for the stator currents is adequate up to a stator frequency of about 100 Hz, when the currents are sampled and processed ten times per period; at this frequency the flux vector is computed at 36° intervals which can make the use of a simple extrapolation scheme desirable. At still higher stator frequency, faster sampling of the variables is required. For todays high dynamic performance servo drives, a sampling period of $100\mu s$, corresponding to a 10 kHz sampling frequency may be specified; this can be achieved with signal processors. Another possibility are application specific integrated circuit (ASIC's), where all the control software is incorporated as custom- hardware on special chips, [K34 - K38], [T11].

An incremental position sensor as described in Chap. 15.2 can serve for sensing angular position as well as speed; when the forward/reverse pulses are accumulated in a reversible counter, which is initially set by a reference pulse, the angular position of the shaft can be detected at the sampling instants by the microprocessor. Forming the difference between two subsequent samples provides a measure of the average speed in the last interval. Naturally, an encoder, delivering at any instant the absolute angular position of the shaft would be preferable for speed and position sensing as it removes the ambiguity with regard to initial position, but this advantage has to be balanced against increased cost and complexity.

With a high resolution incremental speed/position sensor producing 8000 increments per revolution, the speed signal has a frequency of 200 kHz at the nominal speed of 1500 1/min for a 4-pole, 50 Hz motor. Hence a 5 ms sampling period of the speed controller results in a speed measurement of 1000 counts at nominal speed but correspondingly less at lower speed. Since this is usually inadequate with regard to resolution and accuracy, a quasi-analogue sensing scheme providing finer resolution at low speed could be employed [S45, K11]; another option is to convert the sensing scheme from frequency- to time-measurements which allows better resolution at low speed; this is discussed in Sect. 15.2.

The design of linear sampled data control systems is part of the general control theory that has been exhaustively covered in the literature; there is no lack of proven design methods, e.g. [43]. A look at Figs. 12.13 and 12.14 reveals that even though the control plant is highly nonlinear, a considerable degree of decoupling and linearisation is achieved through the method of field orientation. In addition, the principle of cascade control, identical to that applied with DC drives, serves for consolidation by allowing a step-by-step design of the control structure. All that has been said about the design of controlled DC drives, including the combination of feed-forward and feed-back control, remains applicable here, because the difference between a DC and an AC drive is, besides the changed parameters, confined to the block "Torque control loop". Further simplification is due to the fact that relatively high sampling frequencies can be chosen with a suitable microprocessor; as a result, the control functions in the outer loops are practically continuous so that the added complexity of sampled data design procedures is avoided.

As an example, the difference equation of a PID-controller relating the sampled control error $e(\nu)$ to the actuating signal $y(\nu)$ may be written in straight parallel form

$$y(\nu) = G \left[e(\nu) + \frac{T}{T_i} \sum_{-\infty}^{\nu} e(\mu) + \frac{T_d}{T} \left[e(\nu) - e(\nu - 1) \right], \right] \tag{12.47}$$

where T is the sampling interval, G the controller gain and T_i, T_d the time constants of the integrating and derivative channels respectively. The equation is to be solved in real time by the microcomputer. The integrating term would normally be computed recursively in the form

$$y_i(\nu) = \sum_{-\infty}^{\nu} e(\mu) = y_i(\nu - 1) + e(\nu). \tag{12.48}$$

It is important that all the variables are bounded to prevent numerical over-flow with integer arithmetic, or undesirable signal overshoot if floating point arithmetic is employed.

The discrete transfer function of this controller is, letting $z = e^{Ts}$,

$$F(z) = \frac{Y(z)}{E(z)} = G\left[1 + \frac{T}{T_i}\frac{z}{z-1} + \frac{T_d}{T}\frac{z-1}{z}\right]. \tag{12.49}$$

12.2.5 Experimental Results

Figures 12.18 and 12.19 show speed and position control transients of a 1.5 kW induction motor with a switched transistor power supply and micropro-cessor control; the motor was not loaded during the tests. The 8086-based control algorithm comprised altogether 8 k bytes of read-only-memory [S44]. The response of the drive is illustrated by the fact that after a fraction of one revolution the motor reaches field weakening speed. The smooth and rapid transients, resembling the simulated results in Fig. 12.15, indicate that field orientation is indeed a powerful concept of controlling AC drives with high dynamic performance. During the transients the waveforms of the stator currents are far from sinusoidal.

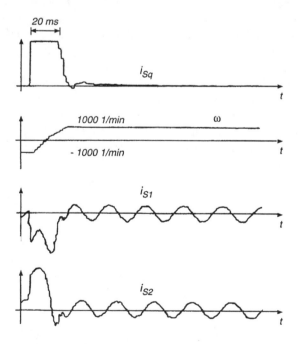

Fig. 12.18. No–load speed reversal of 1.5 kW induction motor with transistor converter and microprocessor control

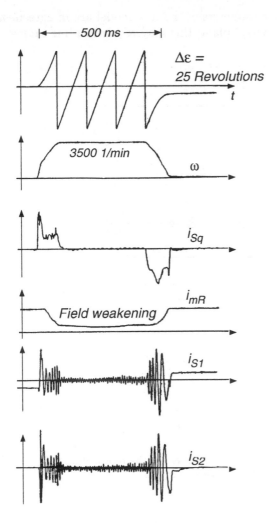

Fig. 12.19. Step–response of a position controlled 1.5 kW induction motor drive

12.2.6 Effects of a Detuned Flux Model

The parameter T_R can be expected to change slowly due to changes of rotor temperature, but instantaneously when operating into the field weakening range, when the motor core becomes desaturated. The interaction is described by Fig. 12.20, where it is seen that the flux model of Fig. 12.16 b, controlled by the measured stator currents and speed, functions as an open loop observer in parallel to the motor dynamics. It produces estimates of the flux angle ρ' and the field orientated current components i'_{Sd}, i'_{Sq} and i'_{mR} to be used as feedback variables for the control. The estimated feedback signals can of course only be correct if the parameters of the flux model are in agreement with the parameters of the control plant; this may require on-line tuning of the model.

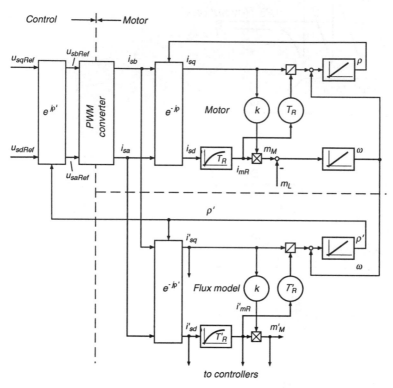

Fig. 12.20. Flux model as an open loop observer for the motor flux, the field orientated currents and the electrical torque

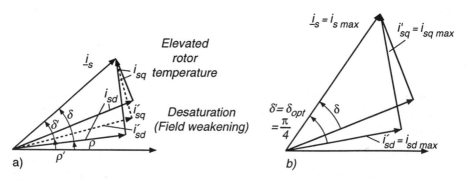

Fig. 12.21. Vector diagram of detuned flux model

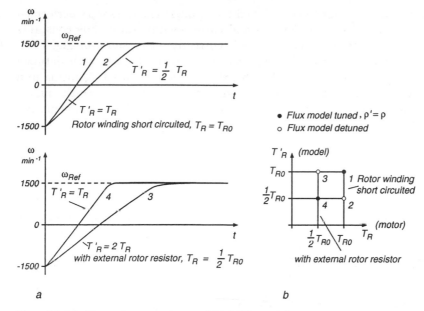

Fig. 12.22. Reversing transients of 22 kW wound rotor motor
with correctly set and with detuned flux model.
(a) Reversing transients with current limit;
(b) operating points in T_R, T_R'- plane

The vector diagram in Fig. 12.21 indicates the deviation of the load angle
$\delta' = arctan\, \omega_2 T_R'$ produced by a flux model, that may have been tuned to
the motor at nominal temperature and flux level, and the actual load angle
of the motor $\delta = arctan\, \omega_2 T_R$. The deviation is

$$\Delta\delta = \delta' - \delta = \arctan \frac{\omega_2(T_R' - T_R)}{1 + \omega_2{}^2 T_R' T_R},\qquad(12.50)$$

$\Delta\delta > 0$ für $T_R < T_R'$ caused by overheated motor
$\Delta\delta < 0$ für $T_R > T_R'$ caused by cold or desaturated motor
(field weakening)

If the flux model is tuned at rated load of the motor the maximum deviation may cover a range of $0.75 T_{R0} < T_R < 1.5 T_{R0}$ with a converter-fed
motor where full line voltage starts do not occur.

The effect of a detuned flux model is concealed as long as the motor is only
partially loaded, because the speed controller would generate the quadrature
current reference i'_{SqRef} needed for producing the torque, even though the
flux may be somewhat increased or reduced. However, when the motor is
fully loaded and i'_{SqRef} is clamped, there is an actual loss of torque, i.e. the
motor cannot be fully loaded. This is demonstrated in Fig. 12.22, where a
22 kW wound-rotor motor with additional inertia was controlled through a

speed reversal with tuned and detuned flux model. The tests were performed with two values of rotor resistance and corresponding rotor time constants T_R, [H31, H32]. When the flux model is correctly set (1 and 4 in Fig. 12.22 a,b), full torque is produced, whatever the rotor resistance; with detuned flux model (2 and 3 in Fig. 12.22 a,b), the torque is reduced, resulting in a lower acceleration. Of course, with an increased rotor resistance the motor operates at higher rotor frequency and lower efficiency when producing the same torque, as was shown in Fig. 10.11.

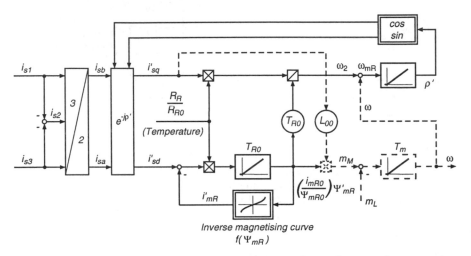

Fig. 12.23. Nonlinear flux model, containing a nonlinear function for saturation effects and a control input for on- line tuning the rotor resistance

The flux model in Fig. 12.16 b may be further refined by including a nonlinear function representing saturation [S44] and an additional input for a temperature- dependent rotor resistance, as shown in Fig. 12.23 [H32]. The saturation curve could be identified for a given motor in the factory or prior to commissioning in the field, whereas the slow variations of the rotor resistance can be tuned on- line, based on temperature measurements or some identification procedure. An experimental result is seen in Fig. 12.24, where the same motor used before is accelerated at current limit through a wide speed range with a linear flux model according to Fig. 12.16 b and with the nonlinear model of Fig. 12.23; the variation of the magnetising current i_{mR} is also shown in both cases. Clearly, taking the magnetic nonlinearity into account results in a better tracking of the flux and again increases the maximum torque at the given current level.

A variety of schemes has been suggested for identifying the varying motor parameter T_R, most of which call for additional measurements by making use of the stator voltage equation e.g. [G8].

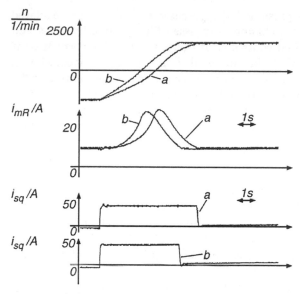

Fig. 12.24. Reversing transients of 22 kW induction motor between field weakening speeds, **(a)** with constant and **(b)** with saturation-dependent flux model

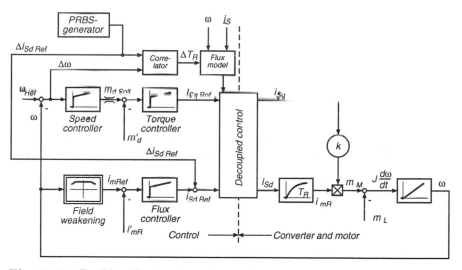

Fig. 12.25. T_R- Identification through correlation

An identification method without the need of a voltage measurement [G5, G6] is based on the fact that the output signals i'_{SdRef}, i'_{SqRef} of the controllers in Fig. 12.25 allow decoupled control of the plant in the d-, q-axes only if the angle ϱ' used for the modulation is in agreement with the actual position ϱ of the flux wave. Hence a low level characteristic test disturbance injected into the presumed i'_{SdRef} channel must be uncorrelated with the noise in the real i_{Sq}-channel, expressed for example by the speed error at the input of the speed controller. If there is noticeable correlation it can serve as an indication that the supposed d'-axis is not orthogonal with the actual q-axis and, disregarding other sources of error, the parameter T'_R in the flux model needs to be corrected to track the plant. The correlation process is manageable when it can be performed in the microprocessor used for controlling the drive. For example, by employing a pseudo-random binary sequence (PRBS) as a test disturbance, the generation of the signal, its correlation with the speed error and the adjustment of the parameter in the flux model are straight–forward, requiring very little additional computing time. The sign of the cross correlation function points to the direction in which T'_R should be adjusted. The sampling time of the PRBS should be chosen sufficiently small to allow the field lag to attenuate the signal, i.e. to avoid an immediate effect on the torque through variation of i_{mR}. On the other hand, building up a meaningful correlation function requires sufficient averaging time for suppressing irrelevant noise components.

Another R_R- adaptation method that is somewhat easier to implement was described in [H31,H32,S42]; it is based on a comparison of voltage signals obtained by measurements and derived from the flux model. Eq. (12.45) may be written in the form

$$\underline{e}_S(t) = (1 - \sigma) L_S \frac{di_{mR}}{dt} = \underline{u}_S(t) - R_S \underline{i}_S - \sigma L_S \frac{di_S}{dt}, \tag{12.51}$$

where $\underline{e}_S(t)$ is the "voltage behind the transient stator impedance". Assuming R_S to be known, for instance from a temperature sensor in the stator winding, and σL_S to be relatively unaffected by saturation, $\underline{e}_S(t)$ may be derived from a measurement of terminal voltages and currents. When converting this equation to field coordinates defined by the rotor flux model in Fig. 12.20 or 12.23, we find for the direct component of the voltage

$$e_{Sd} = (1 - \sigma) L_S \frac{di_{mR}}{dt} = u_{Sd} - R_S i_{Sd} - \sigma L_S \frac{di_{Sd}}{dt} + \omega_{mR} \sigma L_S i_{Sq}, \tag{12.52}$$

which reduces in steady state to

$$e_{Sd} \approx u_{Sd} - R_S i_{Sd} + \omega_{mR} \sigma L_S i_{Sq} \approx 0. \tag{12.53}$$

As the angle of transformation ρ', being derived from an initially detuned model, may be incorrect this condition is not automatically satisfied but may be used for self- tuning the model according to

$$T_{adapt} \frac{d(R_R/R_{Ro})}{dt} = \frac{e_{Sd}}{U_{So}}. \tag{12.54}$$

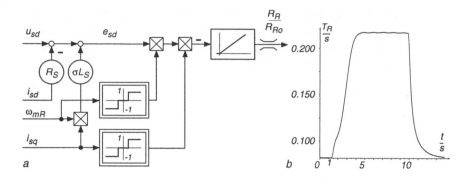

Fig. 12.26. R_R- self- tuning employing terminal voltages. (a) Self-tuning scheme; (b) experimental results with 22 kW wound-rotor motor

This scheme is shown in Fig. 12.26 a; to obtain convergence, the signs of ω_{mR} and i_{Sq} have to be correctly chosen. Also, there are lower practical limits of speed and load, for which the adaptation yields reliable results. When the rotor current is close to zero at low load, a correct setting of T'_R is not important, provided the parameter is quickly adjusted when load is applied. A test result is seen in Fig. 12.26 b taken on a 22 kW wound rotor motor, with external rotor resistors being switched in and out. The lag of a few seconds would normally be acceptable for correcting a temperature induced variation of the rotor resistance.

12.3 Rotor Flux Orientated Control of Voltage-fed Induction Motor

The field orientated control scheme in Sect. 12.2 was introduced with the simplifying assumption of current sources in the stator circuits; this was achieved by fast control loops for the stator currents, making use of MOSFET- or IGBT- converters with ample ceiling voltage and short access time due to high switching frequency. Of course, these conditions cannot be met at maximum speed, where rectangular voltage waveforms are desired for fully utilising the converter; hence the concept of impressing currents becomes impractical in the field weakening speed range. Another area of application, where the approach with current sources and, in steady state condition, nearly sinusoidal stator currents has to be abandoned is that of pulse-width modulated voltage-source converters for larger drives employing thyristors and GTO-thyristors. The switching frequency of such converters is usually below 1 kHz, as compared to 10 kHz for IGBT- or 20 kHz for MOSFET- converters. As a consequence, the harmonics of the stator currents may be of appreciable magnitude and the assumption of nearly "instantaneous" current control becomes questionable.

As a result, the stator voltage equation (10.50) now has to be taken into account as well; by inserting the magnetising current vector, Eq. (12.22), representing the rotor flux, Eq. (12.45) results. Converting the voltage vector to field coordinates as in Eq. (12.26),

$$\underline{u}_S(t)\,e^{-j\varrho} = u_{Sd} + j\,u_{Sq}\,,\tag{12.55}$$

and splitting in real and imaginary components yields

$$\sigma\,T_S\,\frac{di_{Sd}}{dt} + i_{Sd} = \frac{u_{Sd}}{R_S} - (1-\sigma)\,T_S\,\frac{di_{mR}}{dt} + \sigma\,T_S\,\omega_{mR}\,i_{Sq}\,,\tag{12.56}$$

$$\sigma\,T_S\,\frac{di_{Sq}}{dt} + i_{Sq} = \frac{u_{Sq}}{R_S} - (1-\sigma)\,T_S\,\omega_{mR}\,i_{mR} - \sigma\,T_S\,\omega_{mR}\,i_{Sd}\,.\tag{12.57}$$

These equations describe the interactions between the field-oriented stator voltages and currents; together with Eqs. (12.27, 12.33, 12.34) and the equations for the mechanical part of the plant (10.52, 10.53) they form the mathematical model of the voltage-fed induction motor in field coordinates. This model is contained in graphical form in the upper part of Fig. 12.27. The additional right-hand side terms in Eqs. (12.56, 12.57) constitute undesirable coupling terms which could be compensated by suitable feed-forward signals at the output side of the current controllers [G3, G4].

The principle of field orientated control, applied to this scheme, is shown in the lower part of Fig. 12.27. With the assumption that the control dynamics of the voltage source converter can be approximated by a small delay that is related to the clock frequency of the modulator, the transformations within the plant are again cancelled by inverse operations performed on the reference side calling for a transformation from the field-orientated into the stator coordinate system,

$$\underline{u}_{S\,\mathrm{Ref}} = (u_{Sd} + j\,u_{Sq})_{\mathrm{Ref}}\,e^{j\varrho'}\,;\tag{12.58}$$

ϱ' is again the angle of the fundamental flux wave as determined by a flux acquisition scheme, for instance the flux model shown in Fig. 12.16 b. The main difference between the control schemes in Fig. 12.14 and Fig. 12.27 is that the current control is now performed in field coordinates, where the controllers are processing DC signals in steady state, thus avoiding the problem of phase shift of AC current loops. Also, the digital part of the control may now include the pulse-width-modulator and can be extended down to the generation of the switching signals for the converter.

The remainder of the control system is similar to the one shown in Fig. 12.14. The structure in Fig. 12.27 is somewhat redundant since the task of i_{Sd}- and i_{Sq}-control could also be assumed by the flux and torque controllers, omitting the current controllers. Field weakening is achieved with a function generator defining a speed-dependent reference for the magnetising current; limiting the stator voltages with a superimposed voltage control loop would be equally possible.

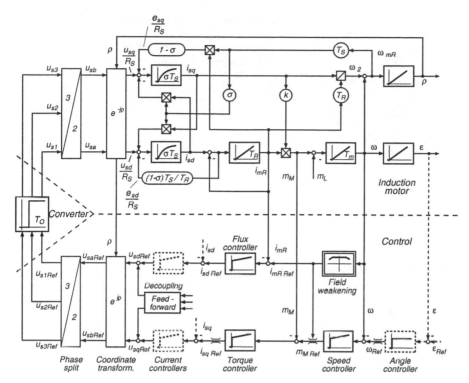

Fig. 12.27. Block diagram of induction motor control in field coordinates, assuming voltage source converter (- - - optional)

A 22 kW induction motor fed by a voltage source transistor converter with vectorial PWM and controlled by a signal processor programmed in C language has been designed and tested in the laboratory; the somewhat modified control structure is seen in Fig. 12.28 [K52]. Field-weakening is now performed by a voltage limiter, which corresponds to the solution shown in Fig. 7.13 for a DC motor. The converter was a 30 kVA experimental unit with 540 V DC link voltage and a clock frequency of 1.37 kHz; the sampling frequency for the inner loops is twice the clock frequency, Fig. 11.8; an integer fraction of this frequency was used for the outer control loops.

A reversing transient of the speed controlled drive is depicted in Fig. 12.29, with the inertia of the drive increased by a DC dynamometer; it confirms the expectations by showing nearly perfect decoupling of the $d - q$- axes; this is manifested by the constant magnetising current and the rapid rise to maximum value of the quadrature current representing torque. While the stator currents exhibit large transient terms, the magnetising current remains practically sinusoidal; this is due to the lag T_R, acting as a low pass filter.

The rise time of the torque of about $2ms$ corresponds to the limits imposed by the control dynamics of the PWM-converter; this rapid response is supe-

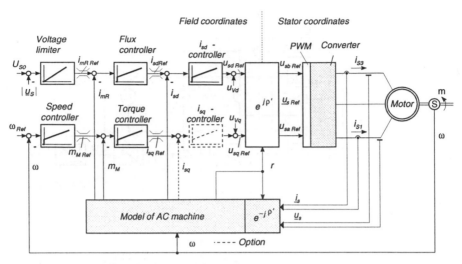

Fig. 12.28. Microcomputer control of induction motor
with a voltage source PWM–converter

rior to that of a high-performance DC drive with a 6-pulse line-commutated
converter at 50 Hz. The pertinent loci of the stator- and the magnetising
current vectors $\underline{i}_S(t)$, $\underline{i}_{mR}(t)$ are seen in Fig. 12.30. Again, the magnetising current vector reverses on a near perfect circular path, while the stator
current vector is temporarily boosted to produce the desired torque.

The block diagram shown in Fig. 12.27 is simplified in so far as the internal
feedback signals seem to be derived directly from the motor, while in reality
they are generated by the flux model computed in parallel to the motor,
Fig. 12.20. This observer-like structure is shown in more detail in Fig. 12.31.
Hence one should distinguish between the real but not accessible variables,
for instance ρ, and the model-based approximate signals such as ρ'. Since the
coordinate transformations at the input of the converter and in the flux model
are performed by the same angle ρ', the actual stator frequency ω_1 is in exact
agreement with the frequency ω_{mR} generated in the flux model. In steady
state it corresponds (assuming a two pole machine) to the extrapolated no-
load speed of the motor; if the measured speed signal ω is correct, then the
estimated rotor frequency is also in agreement with the actual rotor frequency,
$\omega_2' = \omega_2$. Of course, should the model parameter T_R be detuned, the flux could
deviate from the desired value; at reduced flux, the rotor frequency would be
too high and vice versa, but the torque would still be correct, unless the
currents are limited, otherwise the speed controller (assuming it to be free of
drift) initiates the necessary changes.

In the lower part of Fig. 12.31 the on-line R_R-tuning scheme is seen that
was described in Fig. 12.26; also an algorithm is indicated for estimating the
angle $\varphi' < 0$ between the current and voltage vectors in steady state; as

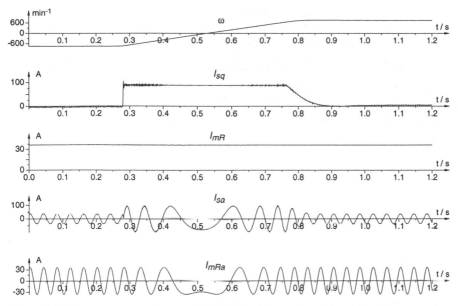

Fig. 12.29. Reversing transient of 22 kW induction motor drive with voltage source converter and signal processor control at no load

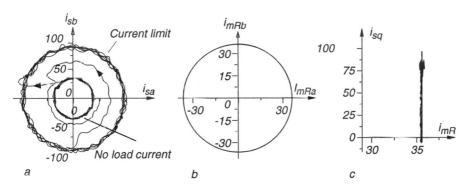

Fig. 12.30. Stator current and magnetising current vectors of PWM–converter drive during the speed reversal shown in Fig. 12.29
(a), (b) Currents in stator coordinates, **(c)** in field coordinates

shown in Fig. 12.32 the needed signals are available in the control computer, where they are generated with the same transformation angle ρ'. With the help of Eqs. (10.72, 10.76) this results in

$$\tan\left(\varphi' + \frac{\pi}{2}\right) = \cot\left(-\varphi'\right) = \frac{(1-\sigma)\,\omega_2\,T_R}{1+\sigma\,(\omega_2\,T_R)^2} \; ; \tag{12.59}$$

from the resulting quadratic equation for $(\omega_2\,T_R)'$ and the known value of ω_2 another estimate of the model parameter T_R' could be derived.

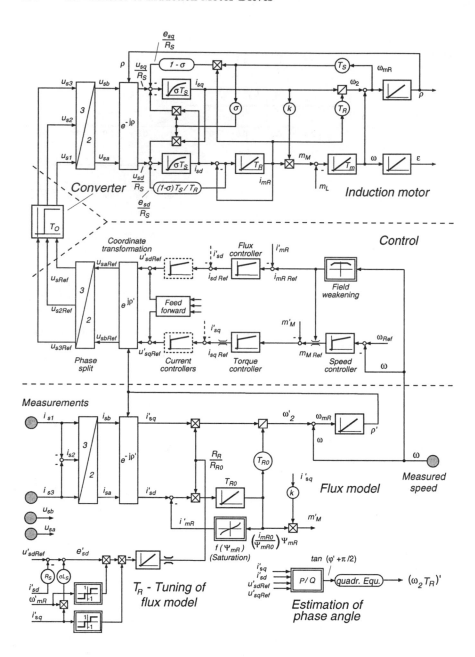

Fig. 12.31. Signal flow diagram of induction motor drive
with saturation-dependent flux model and on-line tuning of R_R

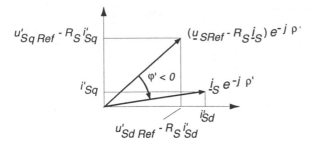

Fig. 12.32. Computing the steady state phase angle φ' from measured voltages and currents in field coordinates

12.4 Field Orientated Control of Induction Motor with a Current Source Converter

The current source thyristor converter described in Sect. 11.3 is of interest for many AC drive problems because of the simplicity of the power circuit. As shown in Fig. 11.14 the control functions are naturally performed in polar coordinates, where the magnitude of the stator current vector, being in noncommutation state directly proportional to the link current, is controlled through the line side converter, while the angle $\zeta(t)$ of the stator current vector,

$$\underline{i}_S(t) = i_S(t)\, e^{j\,\zeta(t)} = \sqrt{3}\, i_D(t)\, e^{j\,\zeta(t)} , \qquad (12.60)$$

is determined by the switching state of the machine side converter; in view of the link reactor and the reactances of the motor, $i_S(t)$ and $\zeta(t)$ are both continuous functions. Outside the commutation intervals $\zeta(t)$ is identical with the switching state $\xi(t)$ of the machine side converter, where $\xi(t)$ advances with increments of $60°$ in forward or reverse direction. Control of $\xi(t)$ mod 2π could be accomplished by switching a hardware or software ring counter with six states.

To this basic control structure which is motivated by the power converter circuit, the principle of field orientation can also be applied; one possible solution is drawn in Fig. 12.33. In the upper right hand corner it contains the model of the induction motor in field coordinates assuming impressed stator currents. The direct and quadrature current components i_{Sd}, i_{Sq} are obtained from i_S and the load angle $\delta = \zeta - \varrho$ by a polar/rectangular conversion (P/R); the inverse operation (R/P) is seen at the output of the flux and torque controllers. The transformation into stator coordinates is achieved by simply adding again to the load angle reference δ_{Ref} the angle ϱ mod 2π of the flux wave, as obtained by some method of flux acquisition. The output signals of the drive control are the reference values for the magnitude and angle of the stator current vector, i_{SRef} and ζ_{Ref} respectively. The former is supplied to the control loop for the direct link current which is represented

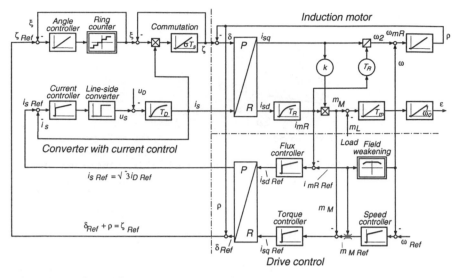

Fig. 12.33. Simplified control scheme of induction motor supplied by current source converter

in Fig. 12.33 in simplified form. T_D is the filter time constant of the DC link circuit, U_D the effective supply voltage of the machine side converter which acts as a disturbance to the current control loop; U_D depends on speed, flux and stator current of the motor.

A problem arises in the second control channel where a finely discretised reference angle ζ_{Ref} is to be tracked by a feedback signal having only six discrete angular states. An approximate solution is again to employ pulse-width modulation, i.e. switching the ring counter back and forth between two adjacent states [G5, G15, W9]; this may be achieved with the help of an angle control loop. For this purpose the state ξ of the ring counter which is readily available in digital form can be used as feedback signal. The current angle ζ follows the lead from the ring counter with a commutation lag that is related to the leakage reactance of the motor; the time required for commutation depends on the link current because at reduced current it takes longer to recharge the capacitors; this is qualitatively modelled in Fig. 12.33 by a current dependent gain factor. The reversible switching of the ring counter takes place only at low speed, it ceases as the stator frequency rises and the angle controller reverts to a unidirectional progression.

The pulse-width modulation achieved by the angle control loop has the added benefit of reducing the effect of torque pulsations at low speed which are caused by the interactions of the stepped stator ampereturns–wave and the continuously moving flux wave. However, as was mentioned before, the frequency of the pulse-width modulation is much lower than in the case of a voltage source converter, where the commutation is practically independent of the motor reactances. Clearly the current source converter is not best suited for continuous position control, where extended operation at very low speed or even standstill with controlled torque may be required. The situation is different on discontinuously operated drives with position control such as needed for boring machines or screw-down drives on reversing rolling mills; in these applications the positioning drive is made inoperative and the brakes are set, once the desired angular position has been reached.

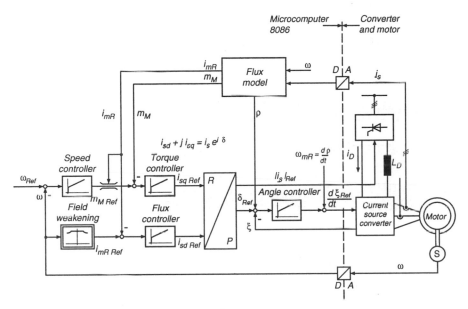

Fig. 12.34. Microcomputer control of induction motor with a current source converter

The speed control scheme of an experimental 22 kW drive, employing a commercial current source converter and an 8086 single board computer is seen in Fig. 12.34 [G5, G6]. The main portion of the block diagram is similar to the one shown in Fig. 12.14; differences are evident in the output section, where an analogue reference for the link current and a pulse sequence for controlling the ring counter are now generated; another addition is the feed-forward signal to the angle control loop to reduce its velocity error. It is noted that the converter could be controlled at the lower level by a hardware sequencer which maintains minimum separation of the firing pulses to assure complete commutation; these protective functions are not shown in the diagram. At an advanced stage of development they can also be realised by software or special integrated hardware.

A reversing transient of the 22 kW drive at half rated speed is recorded in Fig. 12.35 which again shows the good decoupling of the $d - q$-axes achieved

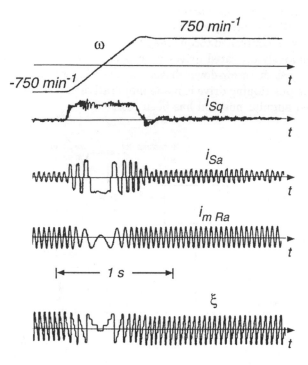

Fig. 12.35. Reversing transient at half nominal speed of a 22 kW current source converter drive

by field orientated control. The rise time of the torque is about 10 ms, the magnitude of the flux is maintained nearly constant. As seen in the lower traces the magnetising current is practically sinusoidal while the stator currents have a highly distorted waveform. The motor was again connected to a DC dynamometer at no load.

The behaviour of the drive at low speed is demonstrated by the transient in Fig. 12.36 where a slow speed ramp causes the pulse-width modulation to become effective around zero speed. This is clearly seen from the trace of the switching state $\xi(t)$ of the ring counter; as soon as the speed leaves the narrow band around standstill, the modulation ceases and the angle controller continues with a unidirectional switching sequence.

Characteristic loci of the current vectors are seen in Fig. 12.37 during a speed reversal at very low speed. Initially the stator current vector is stepping around the six-pointed star; when a reversal of the speed is commanded, the magnitude increases sharply and the direction of the rotation is inverted. After a few steps the speed reversal is completed and the magnitude of the current vector is reduced to its former no-load value.

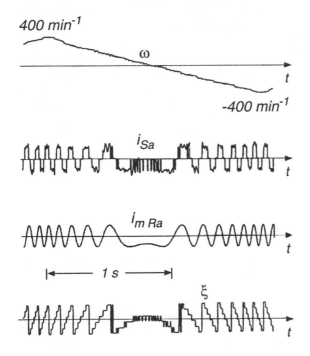

Fig. 12.36. Slow speed reversal of current source converter drive

The locus of the magnetising current vector, Fig. 12.37 b, is again hardly affected by the switching operation, even though the radius of the locus is not quite as perfectly constant as in the case of the voltage source converter drive, Fig. 12.30 b, indicating a somewhat reduced dynamic performance; this is of course due to the lower switching frequency of the current source converter.

In general it can be stated that both types of thyristor-fed induction motor drives show similar behaviour in steady state operation and for large amplitude transients. Differences do exist, however, at very low speed, where the current source converter causes a slightly cogging motion while the voltage source PWM converter drive operates smoothly even at standstill. A noteworthy feature of the current source converter is that it performs with very little audible noise which is in contrast to some voltage source PWM converters unless a higher switching frequency beyond 16 kHz or special modulation techniques are employed.

Further progress is achieved with the circuit in Fig. 11.16, where a current source PWM–converter with GTO–thyristors is supplying an AC motor having filter capacitors across its terminals [A18, A19, N17, N18, N19, N21]. With a suitable choice of the capacitors and the switching frequency this results in greatly improved waveforms of the motor voltages and currents than is

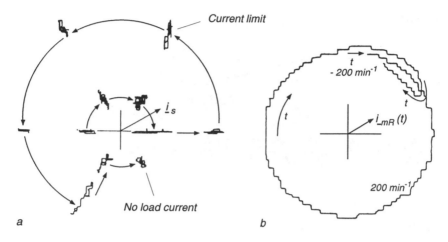

Fig. 12.37. Stator and magnetising current vectors of current source converter drive during a speed reversal at 200 1/min

possible with either voltage source or current source converters discussed so far. As a consequence, the additional copper- and iron- losses in the motor are reduced; there are also lower voltage stresses on the insulation of the windings.

Since the converter is now feeding a low impedance load, the commutation again becomes independent of the motor, as with a voltage source converter, and can operate at a comparable switching frequency, producing a smooth motor torque. Instead of zero voltage intervals there is now the possibility of employing zero current intervals where the link current bypasses the motor, thus creating additional degrees of freedom for designing PWM strategies. At the same time, the advantages of regeneration by inverting the link voltage instead of the current and the ease of protection, with the link reactor limiting the rise of the current in case of malfunction of the converter, are the same as in the original current source converter circuit in Fig. 11.14. The circuit can also be combined with a symmetrical PWM line-side converter as shown in Fig. 11.16; since the diodes present in Fig. 11.14 are no longer needed, a very simple circuit with only 12 GTO–thyristor branches results. However, controlling the drive in Fig. 11.16 presents additional problems in view of the capacitors and the extended machine dynamics.

The stator voltage differential equation (10.50), after substituting the rotor based magnetising current i_{mR} for the rotor current $i_R\,e^{j\,\varepsilon}$, Eq. (12.51), was

$$\sigma\,T_S\,\frac{di_S}{dt} + i_S = \frac{u_S}{R_S} - (1-\sigma)\,T_S\,\frac{di_{mR}}{dt} = \frac{u_S}{R_S} - \frac{e_S}{R_S}\,; \qquad (12.61)$$

the voltage e_S is interpreted as the induced voltage "behind the transient impedance" of the motor,

$$e_S(t) = (1 - \sigma) L_S \frac{d\underline{i}_{mR}}{dt} = (1 - \sigma) L_S \left[\frac{d i_{mR}}{dt} + j\,\omega_{mR}\,i_{mR} \right] e^{j\varrho} .$$

$$(12.62)$$

This is now supplemented in Fig. 11.16 by the stator current equation

$$C \frac{d\underline{u}_S}{dt} = \underline{i}_P - \underline{i}_S \,, \qquad\qquad (12.63)$$

where $\underline{i}_P$ is the current vector supplied by the GTO–converter,

$$\underline{i}_P = \sqrt{3}\,i_D(t)\,e^{j\,\xi(t)} \,, \qquad\qquad (12.64)$$

and $\underline{i}_S = i_S(t)\,e^{j\,\zeta(t)}$ is the vector of the stator currents. As with a voltage source converter, Fig. 11.7, the angle $\xi(t)$ of the converter current can assume six discrete equidistant values; in addition there are zero vectors $\underline{i}_P = 0$ caused by the short circuits of the DC-link, when the link current bypasses the load.

The block diagram in Fig. 12.38 describes the extended dynamic system, using a hybrid representation of the motor, where the stator side quantities are left in stator coordinates, while the remaining motor dynamics are in field coordinates. The block diagram of the GTO–converter, in accordance with its mode of operation, could also be drawn in polar form.

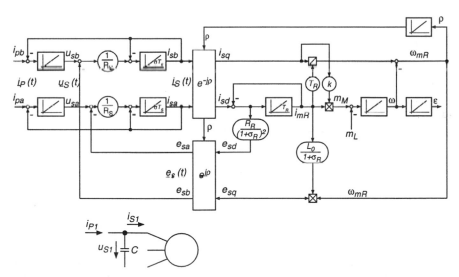

Fig. 12.38. Block diagram of current source PWM converter with GTO–thyristors and capacitive filter

As mentioned before, the control of this type of drive is quite complex, it requires a high performance signal processor to achieve a stable and well

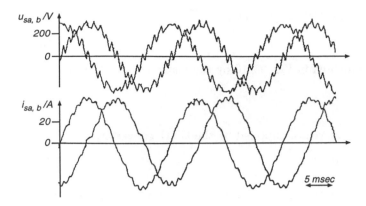

Fig. 12.39. Waveforms of terminal voltages and currents of a 55 kW induction motor with current source GTO converter operating at a stator frequency of 50 Hz. Reactive power of the capacitive filter is 24 kVA, switching frequency of converter is 500 Hz

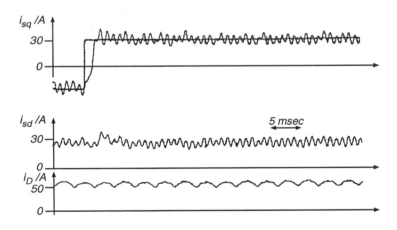

Fig. 12.40. Step response of torque control loop with the 55 kW current source GTO converter drive used for Fig. 12.38

damped response. Some steady state and dynamic results, that have been obtained with a 55 kW motor using a model based predictive control strategy [A18, A19], are seen in Figs. 12.39 and 12.40. Clearly, this could be a very promising approach to future AC motor drives, combining smooth waveforms of voltages and currents with excellent dynamic performance. These advantages are also beneficial for the design of the motor, because the electrical stresses of the winding insulation are reduced when the steeply rising voltages caused by the "hard-switching" converter are avoided and the motor is relieved from the task of serving as a filter choke.

At higher power ratings such as for low speed piston compressor-, mine hoist- or rolling mill drives, cycloconverters may also offer a solution as long as their frequency limit is not exceeded. The control is similar to the one described in Sect. 12.2 where control loops for the stator currents provide nearly impressed and close to sinusoidal currents [H28].

If the range of stator frequency rules out the use of a cycloconverter, a synchronous motor with a load commutated current source converter could be used; this is discussed in Sect. 14.3.

12.5 Control of an Induction Motor Without a Mechanical Sensor

12.5.1 Machine Model in Stator Flux Coordinates

After the basic problems of controlling induction motors with high dynamic performance had been solved by applying the principle of field orientation, where information on the magnitude and orientation of the flux wave is derived from a "flux model", i.e. a mathematical model computed on a microcomputer in real time, the attention of researchers has turned towards simplification as well as refinement of these quite sophisticated control methods. One point where further progress was felt to be desirable is the omission of the mechanical speed/position sensor needed with many of these control schemes; for instance, the rotor-flux model introduced in Fig. 12.16 b calls for a measurement of speed, even in cases where the specification of only a moderate accuracy would not make a speed sensor mandatory. While electrical measurements are usually acceptable since the sensors can be placed anywhere, preferably inside the converter cabinet, a mechanical sensor is often considered a nuisance because of space restrictions or the added cost and complexity, particularly with smaller motors. Of course, a certain loss of accuracy and dynamic response must be accepted when an estimate of speed serves as a feedback signal, hence such schemes may not meet high performance demands, for instance as required by machine tool feed drives, where sensors will be inevitable.

One reason, why the omission of a mechanical sensor has become such a prominent topic, is surely the fact that the powerful microcomputers now used for the control are applicable to the estimation of not measurable internal quantities; hence it makes sense to employ their potential also for substituting the mechanical sensor. Had microcomputers been available for controlling DC machines, the same demands would have emerged there.

The various proposals for the design of controlled AC drives without a mechanical sensor have in common that only terminal quantities, i.e. stator voltages and currents are measured from which the information on flux and speed of the motor must be derived, based on a nominal knowledge of the important motor parameters. An ideal estimation algorithm should produce the necessary signals on the flux wave as well as mechanical speed and, in addition, provide information on changing motor parameters, so that the flux- and speed- models remain tuned to the motor. Of course, it is unlikely that all these expectations can be fulfilled at the same time, still it is a goal worth pursuing and, when applying powerful numerical procedures, there may indeed be effective ways for solving some of these problems, e.g. [78b, H46, H69, H72, H73, O5, O10, T4, 62].

The three tasks mentioned may be summarised under the headings "Observation" and "Identification"

• Acquisition of flux wave with regard to magnitude and angular position,
• Reconstruction of speed,
• Estimation of machine and load parameters;

they are posing different demands with regard to response time as well as accuracy. The estimation can only be based on the available mathematical model of the motor, a set of six nonlinear real- valued differential equations with uncertain parameters, Eqs.((10.50–10.53).

Since flux acquisition is the most time- critical, being a basis for the torque control, it is reasonable to select a coordinate frame of reference that can be directly employed for the subsequent control; the choice is either a stator flux- or a rotor flux- based coordinate system. Both have their advantages and disadvantages as discussed, for example in Sect. 12.2 and 13.1. Since rotor currents cannot be measured with cage motors and all measurements have to be stator based, a stator flux coordinate system related to the equivalent circuit in Fig. 10.7 c appears interesting [D26, K74, S63]. However, the question of choosing the coordinate system is sometimes overrated because the difference between stator- and rotor-flux is leakage flux that cannot be accurately modelled anyway, Eqs. (10.25, 10.32). Still, it seems interesting in view of the stator based measurements to try also a stator-flux orientated approach.

With the definition of a stator flux vector

$$\underline{\psi}_S(t) = L_0\, \underline{i}_{mS},\tag{12.65}$$

where

$$\underline{i}_{mS} = i_{mS}(t)\, e^{j\,\mu(t)} = (1 + \sigma_S)\, \underline{i}_S + \underline{i}_R\, e^{j\,\epsilon}\tag{12.66}$$

represents a stator based magnetising current, the unaccessible rotor current may be eliminated from Eqs.(10.50, 10.51), resulting in two equations for $\underline{i}_{mS}$ and $\underline{i}_S$,

$$\frac{d\underline{\psi}_S}{dt} = L_0\frac{d\underline{i}_{mS}}{dt} = \underline{u}_S - R_S\,\underline{i}_S\tag{12.67}$$

and

$$\sigma\, L_S \frac{d\underline{i}_S}{dt} + L_S\,(1/T_R - j\omega\sigma)\,\underline{i}_S = L_0\,\frac{d\underline{i}_{mS}}{dt} + L_0\,(1/T_R - j\omega)\,\underline{i}_{mS}\,. \tag{12.68}$$

The "induced voltage behind the stator resistance" is integrated in Eq.(12.67) resulting in i_{mS} and μ. This is the moving frame of reference defining "stator flux coordinates"; the stator current vector in field coordinates is then

$$\underline{i}_S\,e^{-j\mu} = i_{Sd} + ji_{Sq}\,. \tag{12.69}$$

When separating from Eqs. (12.67, 12.68) an inner voltage vector which is defined by measurable quantities,

$$\underline{v}_S = \underline{u}_S - (R_S + \frac{1+\sigma_S}{1+\sigma_R}R_R)\,\underline{i}_S - \sigma\,L_S\frac{d\underline{i}_S}{dt}\,, \tag{12.70}$$

the motor speed may be estimated from

$$j\omega L_S\,(\frac{i_{mS}}{1+\sigma_S} - \sigma\underline{i}_S) = \underline{v}_S + \frac{R_R}{1+\sigma_R}\,i_{mS}\,. \tag{12.71}$$

Substituting the rotor current in Eq.(10.52) by $\underline{i}_{mS}$, yields the motor torque

$$m_M(t) = \frac{2}{3}\,L_0\mathrm{Im}[\underline{i}_S\,\underline{i}_{mS}^*] = \frac{2}{3}\,L_0\,i_{mS}\,i_{Sq}\,. \tag{12.72}$$

These equations still correspond to the original model of the induction motor, Eqs. (10.50–10.53), only now written in terms of stator- flux coordinates; they are graphically represented at the right hand side of Fig. 12.41 with the measurable terminal quantities $\underline{u}_S$ and $\underline{i}_S$ in stator- and the motor dynamics in stator flux- coordinates.

12.5.2 Example of an "Encoderless Control"

Proceeding in Fig. 12.41 from the motor terminals to the left, a voltage source converter is assumed, feeding the motor with two orthogonal AC source voltages u_{Sa} and u_{Sb}, which in turn are controlled by command signals u_{SaRef} and u_{SbRef}; the converter is represented by its effective delay. If the reference voltages, available as digital signals in the control computer, can be substituted for the terminal voltages, a difficult measuring problem is avoided, because the voltages produced by a voltage source converter have a wide bandwidth, calling for costly A/D converters; another option worth trying is the reconstruction of the terminal voltages from the DC link voltage and switching signals, taking the protective intervals into account.

The part further to the left of the diagram represents an example of an "encoderless control"; its inputs are the measured electrical signals indicated

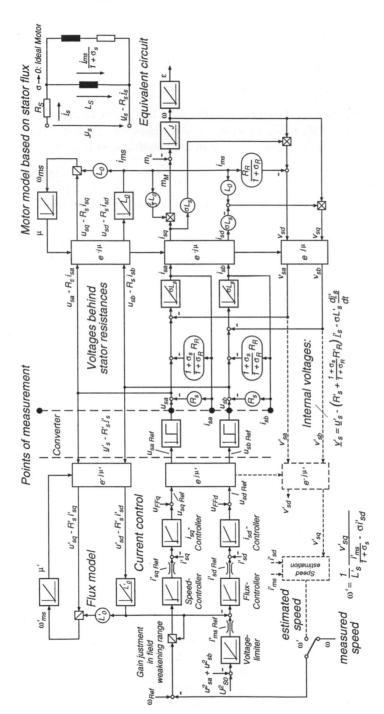

Fig. 12.41. Dynamic model of an AC motor in stator- flux coordinates and example of a control scheme without a mechanical sensor

by the orthogonal set $u_{Sa}, u_{Sb}, i_{Sa}, i_{Sb}$ of AC quantities. Integration of the "voltages behind the stator resistances", $u_{Sa} - R_S i_{Sa}$ and $u_{Sb} - R_S i_{Sb}$ produces the stator flux; transformation of Eq. (12.67) into field coordinates results in

$$L_0\left(\frac{di_{mS}}{dt} + ji_{mS}\frac{d\mu}{dt}\right) = (\underline{u}_S - R_S\underline{i}_S)\,e^{-j\,\mu}$$
$$= (u_{Sd} - R_S i_{Sd}) + j(u_{Sq} - R_S i_{Sq})\,, \tag{12.73}$$

where

$$\frac{d\mu}{dt} = \omega_{mS} \tag{12.74}$$

is the instantaneous angular velocity of the stator flux vector. Hence, the equations of the stator flux model are

$$L_0\frac{di_{mS}}{dt} = u_{Sd} - R_S i_{Sd}\,, \tag{12.75}$$

$$L_0\frac{d\mu}{dt} = \frac{u_{Sq} - R_S i_{Sq}}{i_{mS}}\,. \tag{12.76}$$

Deriving the stator flux from terminal voltages and currents is difficult because the open integrations at low stator frequency are subject to sensing errors and integrator drift; on the other hand, if the operation is performed inside a feedback loop, as is the case when being carried out in field coordinates, the effects of integrator drift resemble normal offsets, common in analogue signal processing. The result of the flux modelling are estimates of the magnetising current vector expressed by i'_{mS}, μ', and ω'_{mS}. Clearly when measuring stator voltages and currents, the flux model does not require a speed signal; this makes the scheme fit the needs of "encoderless" control. The flux model shown in the upper left hand part of Fig. 12.41 may be interpreted as the inverse of the corresponding part of the motor dynamics, creating a copy of the stator flux; its function resembles that of a phase locked loop (PLL). The signal processing may be done with an analogue/digital sensing circuit, where the trigonometric functions are represented digitally in a ROM, to be multiplied in D/A converters with analogue voltage signals [J15, J16].

Once estimates of the magnitude and orientation of stator flux are established, it is easy to compute with a microcomputer the remaining quantities needed for a two- channel field orientated control structure. This is seen in the lower left hand part of Fig. 12.41, where a voltage limiter generates the flux reference i'_{mSRef} as a command value for a flux controller producing i'_{SdRef} ; similarly, the output of the speed controller is the reference for the quadrature current, i'_{SqRef}. The stator current control in field coordinates corresponds to that shown in Fig. 12.27. Clearly, if the voltages $\underline{v}'_S(t)$ containing relevant information on the rotor can be derived from the terminal quantities, there is a good chance of obtaining estimates of the motor speed ω' and the rotor resistance R'_R. Reconstruction of $\underline{v}'_S$ with the help of Eq. (12.71) involves a differentiation of the stator currents which can again be implemented in analogue form. When transforming Eq. (12.70) into field coordinates

$$\underline{v}_S\, e^{-j\mu} = v_{Sd} + jv_{Sq} = \frac{-R_R\, i_{mS}}{1+\sigma_R} + \omega\sigma L_S i_{Sq} + j\omega L_S\left(\frac{i_{mS}}{1+\sigma_S} - \sigma i_{Sd}\right),$$
$$(12.77)$$

it is seen that, assuming valid estimates of the field orientated currents, the speed can be estimated on the basis of v'_{Sq}, whereas a change of the rotor resistance might be detected from v'_{Sd}. This is indicated at the lower end of the control section in Fig. 12.41.

On the other hand, should a mechanical sensor be available for speed control, the structure in Fig. 12.41 becomes a normal field orientated control scheme, based on stator- instead of rotor-flux, where ω' is replaced by the measured speed ω. The flux acquisition is unaffected, still being based on measured voltages and currents. Also, the outer layers of control are the same, whether the control is rotor flux- or stator flux orientated; they could again be extended to include a position control loop.

The advantages of stator flux coordinates are that the rotor resistance enters the flux model only indirectly, the stator temperature causing changes of the stator resistance can be measured and that the effect of saturation is already contained in the voltages. On the other hand, the stator flux model is more sensitive at low frequency to detuned parameters such as R_S compared with the rotor flux model shown in Fig. 12.16 b; also, measurements of the terminal voltages, the accuracy of which is quite critical at low speed, are needed as input and the magnitude of the magnetising current is produced by an integration with nonlinear feedback instead of a linear lag term.

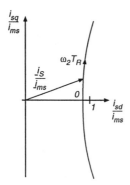

Fig. 12.42. Stator flux orientated stator current vector as a function of rotor frequency

It is of interest to note that the current components i_{Sd}, i_{Sq} defined in stator flux coordinates are not completely decoupled, as was the case with rotor flux coordinates; there is a slight coupling effect similar to armature reaction with a DC motor. This becomes apparent when transforming Eq. (12.68) into field coordinates; in steady state, i.e. with constant values of $i_{mS}, i_{Sd}, i_{Sq}, \omega, \omega_{mS} = \omega_1$, and $\omega_2 = \omega_{mS} - \omega$ we find

$$i_{Sd} = \frac{i_{mS}}{1 + \sigma_S} + \omega_2\, \sigma\, T_R\, i_{Sq}\,, \tag{12.78}$$

$$i_{Sq} = \omega_2\, T_R\, (\frac{i_{mS}}{1 + \sigma_S} - \sigma\, i_{Sd}). \tag{12.79}$$

Solving for i_{Sd}, i_{Sq} results in

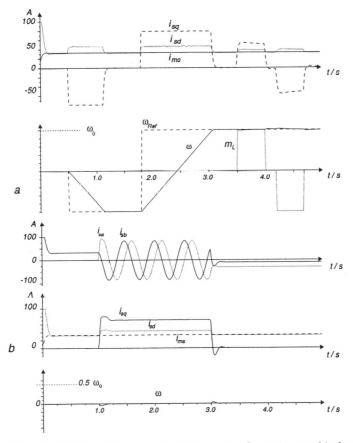

Fig. 12.43. Simulation results of the stator flux orientated induction motor control scheme shown in Fig. 12.41, using **measured** speed feedback
(a) Starting, reversing and loading transients; **(b)** loading transients at zero speed

$$i_{Sd} = \frac{1 + \sigma\,(\omega_2\,T_R)^2}{1 + (\omega_2\,\sigma\,T_R)^2}\,\frac{i_{mS}}{1 + \sigma_S}\,, \tag{12.80}$$

$$i_{Sq} = \frac{(1 - \sigma)\,\omega_2 T_R}{1 + (\omega_2\,\sigma\,T_R)^2}\,\frac{i_{mS}}{1 + \sigma_S}\,, \tag{12.81}$$

where $\omega_2\,T_R$ is a measure of torque, as found in Eq.(12.3). The two functions are combined in Fig. 12.42; if the stator flux is to be maintained at constant level, i_{Sd} must increase with the load. This could be achieved with a flux controller.

12.5.3 Simulation and Experimental Results

Simulation results with the scheme in Fig. 12.41 and based on parameters of a 22 kW motor are seen in Figs. 12.43 and 12.44; they show starting and

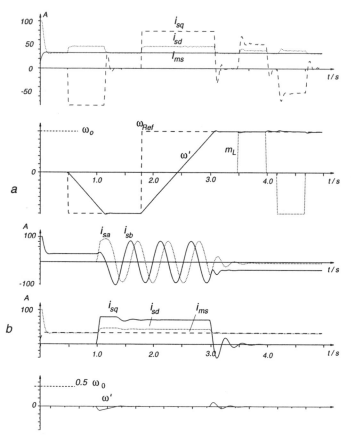

Fig. 12.44. Simulation results of the stator flux orientated induction motor control scheme shown in Fig. 12.41, using **estimated** speed feedback
(a) Starting, reversing and loading transients; **(b)** loading transients at zero speed

reversing as well as loading transients at zero speed reference. In Fig. 12.43 the "measured" speed ω and in Fig. 12.44 the "estimated" speed ω' serves as feedback signal. Naturally, there is little difference as long as the motor parameters are accurately known but, as mentioned before, the flux model is sensitive to detuned parameters at low speed. The only minor difference in the simulation results should not lead to overoptimistic conclusions.

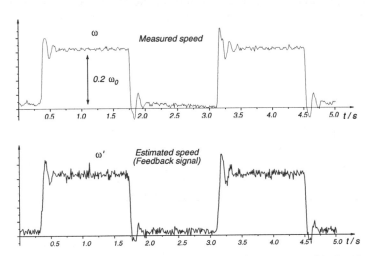

Fig. 12.45. Experimental results of a stator flux orientated induction motor control scheme similar to Fig. 12.41, using **estimated** speed feedback, transients following step changes of speed reference

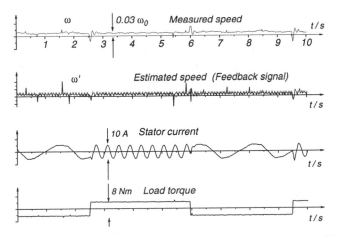

Fig. 12.46. Experimental results of a stator flux orientated induction motor control scheme similar to Fig. 12.41, using **estimated** speed feedback, transients following step changes of load torque at very low speed

Some experimental results with an "encoderless" induction motor drive have been obtained with a 1.5 kW standard motor, fed by a commercial IGBT inverter switched at 8.8 kHz and a control scheme similar to the one shown in Fig. 12.41 [J16]. Transients following step changes of the speed reference are seen in Fig. 12.45, with the measured speed also plotted for comparison. Fig. 12.46 shows the transients following step changes of the load torque corresponding to ± 0.4 of nominal torque at the very low speed of 0.03 ω_o, indicating a remarkably good performance under these difficult test conditions.

12.6 Control of an Induction Motor Using a Combined Flux Model

From the preceding section it became apparent that employing stator voltages as inputs to a "voltage model" based on the stator voltage equation, Eq.(10.50), has weaknesses in the low speed region because the terminal voltages decrease with speed until they are eventually dominated by the voltage drops across the temperature dependent stator resistors and have little bearing on the magnetic fields in the motor. Also, use of stator flux coordinates creates additional coupling terms between the stator current components in field coordinates. With the "current model" based on the rotor voltage equation, Eq.(10.51) and with rotor flux orientation corresponding to Fig. 12.16 b, these coupling terms are not present and zero speed is within the validity range of the model; however, a speed signal is needed, which is undesirable and should be made superfluous. Hence one could consider a combined solution, where the two flux models in rotor flux coordinates are computed simultaneously and a speed- dependent selection of their results is made, such that the signals from the current model are preferred at low speed and those from the voltage model at higher speed.

The "voltage model" is obtained by converting Eq.(12.51) into rotor field coordinates and splitting it in real and imaginary parts (for the real part this was already done in Eq.(12.52). This results in

$$e_{Sd} = (1 - \sigma) L_S \frac{di_{mR}}{dt} = u_{Sd} - R_S\, i_{Sd} - \sigma L_S \frac{di_{Sd}}{dt} + \omega_{mR}\, \sigma\, L_S\, i_{Sq},$$

$$(12.82)$$

and

$$e_{Sq} = (1 - \sigma) L_S\, i_{mR}\, \omega_{mR} = u_{Sq} - R_S\, i_{Sq} - \sigma L_S \frac{di_{Sq}}{dt} - \omega_{mR}\, \sigma\, L_S\, i_{Sd},$$

$$(12.83)$$

where

$$\underline{e}_S(t)\, e^{-j\varrho} = e_{Sd} + j\, e_{Sq} \qquad\qquad (12.84)$$

are again the "voltages behind the transient stator impedance". Eqs. (12.82, 12.83) constitute the "voltage model" in rotor-flux coordinates.

The "current model" was defined by Eqs.(12.33, 12.34) as

$$T_R \frac{di_{mR}}{dt} + i_{mR} = i_{Sd}, \tag{12.85}$$

$$\frac{d\varrho}{dt} = \omega_{mR} = \omega + \frac{i_{Sq}}{T_R i_{mR}} = \omega + \omega_2, \tag{12.86}$$

where the measured speed ω should now be substituted by a suitable estimate ω' obtained from the combined evaluation algorithm. Of course, when both models are simultaneously processing the measured currents and voltages, they are bound to produce somewhat different results for the field orientated quantities, indicated in the following by an additional subscript S and R. The control will employ a weighted mean of these values, depending on the estimated speed. There are countless possibilities to realise this by software; for instance, a speed dependent selection could be realised with an observer in the frequency-domain [J5, J6] processing AC signals in stator coordinates or, as is alternatively suggested here, an algebraic weighting function may be employed.

An example of how the combined flux estimation could be implemented is shown in Fig. 12.47. In order to avoid a differentiation of the current signals in the voltage model, an observer-like structure is chosen, where the signals e'_{SdS}, e'_{SqS} are obtained implicitly by matching the predicted and measured current signals i'_{SS} and i_S. This results in a lag effect that should be duplicated in the current model for obtaining consistent results.

The combined flux estimation vector containing i'_{mR}, ω'_{mR}, i'_{Sd} and i'_{Sq} which serves for the control of the drive may be defined as the weighted mean of the voltage and current model estimates. For the example of the angular velocity of the flux vector this results in

$$\omega'_{mR} = f(\omega')\,\omega'_{mRR} + (1 - f(\omega'))\,\omega'_{mRS}, \tag{12.87}$$

where $0 \le f(\omega') \le 1$ is a speed dependent weighting function; corresponding definitions are used for all other estimates derived by the flux models. With a suitably shaped function $f(\omega')$ as shown in Fig. 12.47 the control is purely "current model based" at zero speed and converges to a "voltage model based" operation at higher speeds, when the terminal voltages are of sufficient magnitude and the uncertainty of the resistive voltage drop can be neglected. As indicated in Fig. 12.47 the estimated speed signal ω' to be used in the weighting function $f(\omega')$ as well as for the encoderless speed control may be obtained by combining Eqs.(12.83, 12.86),

$$\omega' = \frac{1}{(1 - \sigma)\,L_S\,i_{mR}}\left(e_{Sq} - \frac{R_R\,i_{Sq}}{(1 + \sigma_R)^2}\right). \tag{12.88}$$

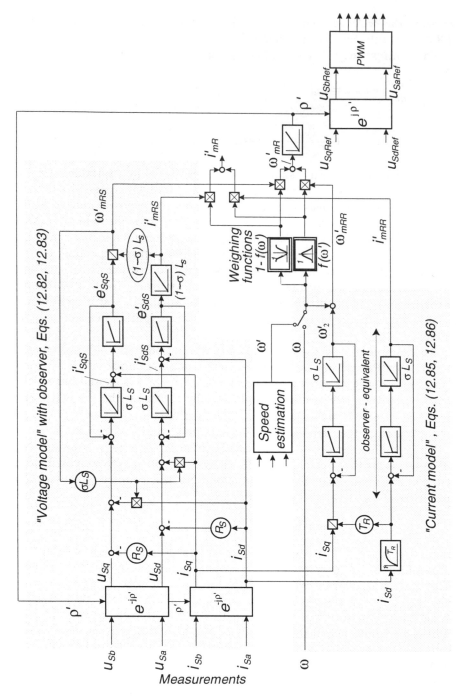

Fig. 12.47. Combined voltage/current model for rotor flux estimation

If the flux model is implemented in stator coordinates, processing variable frequency AC quantities, the speed dependent transition of the estimates from the current-based to the voltage-based section could be done dynamically by using the frequency response of the observer; again, at low speed the results of the current model would dominate whereas the signals obtained from the voltages would prevail at elevated speed.

13. Induction Motor Drive
with Reduced Speed Range

Some mechanical loads, such as fans or centrifugal pumps, exhibit a strong dependence of the load torque on speed, so that a limited speed control range suffices for achieving the desired effect. The same is true for rotating converters employing a flywheel, Sect. 7.4; since the kinetic energy of the rotating masses varies with the square of the speed, there is little incentive in varying the speed by more than, say 20 or 25%. Similar situations exist if an electrical island grid with a fluctuating frequency, for instance a railway grid, is to be supplied from the public constant frequency line through a rotating converter, or conversely, when a variable speed generator is feeding power into the constant frequency grid, as may be the case with wind energy plants or pumped storage hydro sets; using variable speed there would improve the aerodynamic efficiency of the wind rotor with changing wind velocity or the hydraulic efficiency of a pump/turbine with varying head.

For applications of this kind a wound-rotor induction motor drive presents an interesting solution; if the stator is connected to the constant frequency grid while the rotor is fed with slip frequency by a static converter, its rating is mainly determined by the desired speed range $\Delta\omega$ and can be kept relatively small. Still, the slip power may be substantial with large machines, calling for an efficient control scheme.

Beginning at about the turn of the century, a variety of drive circuits has been developed that made use of auxiliary machines such as rotary frequency changers [L23], they are of course obsolete today; with the progress of power electronics new solutions have become possible here too, one of which will be discussed in Sect. 13.1. The simplified mathematical model of the symmetrical induction machine derived in Chap. 10 can be used for this purpose with minor modifications.

13.1 Doubly-fed Induction Machine with Constant Stator Frequency and Field-orientated Rotor Current

When direct current is supplied to the sliprings of a wound-rotor induction motor, the stator of which is connected to the constant frequency line, the motor assumes the properties of a synchronous machine. Constant torque can only be produced if the rotor is in synchronism with the stator field; the machine can also deliver reactive power, but at the same time the problems of the line-fed synchronous machine become apparent, which includes starting and synchronisation as well as oscillatory transients and pull-out torque. This is still true, if instead of the direct current, a constant frequency alternating current excitation is applied to the rotor winding; only the speed of the motor must be different for the stator and rotor ampereturn waves to move again in synchronism. For these reasons doubly-fed machines operated at constant rotor frequency are not particularly attractive.

The situation is different, however, if the AC excitation of the rotor is made dependent on the line voltage vector and the angular position of the rotor. The machine then looses its synchronous characteristics entirely and can operate at variable speed and reactive line current; the speed transients can be well damped [A13, A14, A31, D32, H36, K17, L13, L23, L27, L40, W11, W20, W21, W22]. The basic circuit in Fig. 13.1 a shows a wound-rotor motor, the stator of which is connected to the symmetrical three-phase supply with nominal line voltage U_{S0} and frequency ω_0 while the rotor is fed with impressed three-phase currents of variable amplitude, frequency and phase. The power supply in the rotor-circuit could either be a current controlled cycloconverter or a DC- link converter, Chap. 11. Thus the shaded operating range of the torque/speed plane in Fig. 13.1 b becomes accessible. Voltage rating and frequency of the converter are directly dependent on the width of the desired speed range.

The impressed rotor currents are derived on the basis of the rotor angular position and the vector of the line voltages; speed and reactive power control with current limit are superimposed. It would also be possible to choose a motor with a two phase rotor winding instead of the usual three phases; the operation would be the same, except for the harmonics on the line side caused by the switching of the converter [B27]. For the following considerations a three-phase rotor with $N_R = N_S$ is assumed, as in Sect. 10.1.

The mathematical model of the symmetrical AC machine, Eqs. (10.50–10.53), that was derived with various simplifying assumptions can be employed here also, it only needs to be adapted to the new constraints. The rotor winding is now fed by a converter,

$$R_R \underline{i}_R + L_R \frac{d\underline{i}_R}{dt} + L_0 \frac{d}{dt}\left(\underline{i}_S\, e^{-j\,\varepsilon}\right) = \underline{u}_R(t)\,, \tag{13.1}$$

where

$$\underline{u}_R(t) = u_{R1} + u_{R2}\, e^{j\,\gamma} + u_{R3}\, e^{j\,2\,\gamma} = u_{R12} - u_{R23}\, e^{-j\,\gamma} \tag{13.2}$$

is the vector of the rotor voltages supplied by the converter to the sliprings of the machine. When again assuming that the converter is equipped with fast responding current control loops, the rotor currents $\underline{i}_R(t)$ are in effect produced by current sources; hence, as long as the converter is able to impress the rotor currents, i.e. possesses adequate ceiling voltage and control

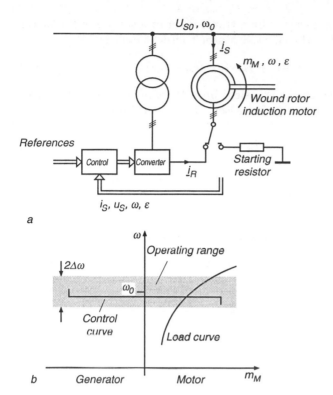

Fig. 13.1. Wound-rotor induction motor with variable frequency rotor supply (a) Circuit; (b) Operating range

bandwidth, the rotor voltage equation Eq. (13.1) has no significance with regard to the dynamics of the drive; this results in considerable simplification of the drive control. The situation is in fact quite similar to the case of the stator-fed induction motor; an added simplification is that now both, stator and rotor currents, can be readily measured. It is thus appropriate to follow again the field orientated approach.

In analogy to Sec. 12.5 the stator voltage equation (10.50) is written in the form

$$R_S \underline{i}_S + L_0 \frac{d}{dt} \left[(1 + \sigma_S) \, \underline{i}_S + \underline{i}_R \, e^{j\,\varepsilon} \right] = \underline{u}_S \,, \tag{13.3}$$

where according to Eq. (12.66)

$$\underline{i}_{mS}(t) = (1 + \sigma_S) \, \underline{i}_S + \underline{i}_R \, e^{j\,\varepsilon} = i_{mS}(t) \, e^{j\,\mu(t)} \tag{13.4}$$

is an extended magnetising current vector responsible for the stator flux including leakage, Fig. 10.7 c. Should the symmetrical three-phase supply to which the stator is connected, exhibit a noticeable internal impedance,

$R_N + j\omega_0 L_N$, for instance caused by a transformer or a section of the power line, it can be included in the effective stator transient impedance

$$R_S + j\,\omega_0\,\sigma_S\,L_0 \equiv R_{S0} + R_N + j\,\omega_0\,(\sigma_{S0}\,L_0 + L_N)\,, \tag{13.5}$$

with $R_{S0} + j\omega_0\sigma_{S0}L_0$ being the transient impedance of the machine proper. Eliminating the stator current $\underline{i}_S$ from Eqs. (13.3, 13.4) leads to

$$T_S \frac{d\underline{i}_{mS}}{dt} + \underline{i}_{mS} = \frac{1 + \sigma_S}{R_S}\,\underline{u}_S + \underline{i}_R\,e^{j\,\varepsilon}\,. \tag{13.6}$$

$T_S = L_S/R_S$ is the main time constant of the stator circuit. The expression for the electrical torque is, Eq. (10.52),

$$m_M = \frac{2}{3}\,L_0\,\mathrm{Im}\,[\underline{i}_S\,(\underline{i}_R\,e^{j\,\varepsilon})^*] = \frac{2}{3}\frac{L_0}{1 + \sigma_S}\,\mathrm{Im}\,[\underline{i}_{mS}\,(\underline{i}_R\,e^{j\,\varepsilon})^*]\,. \tag{13.7}$$

With the introduction of the rotor current vector in stator coordinates,

$$\underline{i}_R\,e^{j\,\varepsilon} = i_R(t)\,e^{j\,(\xi+\varepsilon)}\,, \tag{13.8}$$

the torque is

$$m_M = -\frac{2}{3}\,(1 - \sigma)\,L_R\,i_{mS}\,i_R\,\sin(\xi + \varepsilon - \mu) = -\frac{2}{3}\,(1 - \sigma)\,L_R\,i_{mS}\,i_{Rq}\,. \tag{13.9}$$

The position of the rotor current vector in field coordinates,

$$\delta = \xi + \varepsilon - \mu\,, \tag{13.10}$$

corresponds to the load angle. The angular relations of the various vectors are shown in Fig. 13.2. The instantaneous angular velocities are

$$\frac{d\xi}{dt} = \omega_2\,, \qquad \frac{d\varepsilon}{dt} = \omega\,, \qquad \frac{d\mu}{dt} = \omega_{mS}\,. \tag{13.11}$$

The real and imaginary parts of

$$\begin{aligned}
i_R\,e^{j\,(\xi+\varepsilon-\mu)} &= i_R\,e^{j\,\delta} = i_R\,\cos\delta + j\,i_R\,\sin\delta \\
&= i_{Rd} + j\,i_{Rq}
\end{aligned} \tag{13.12}$$

are defined as $d - q$-components of the rotor current vector in a stator flux-orientated reference frame or as direct and quadrature components of the rotor current "in field coordinates"; they are DC signals in steady state.

When assuming a symmetrical three-phase system of sinusoidal line voltages having the constant frequency ω_0, the voltage vector rotates on a circular path,

$$\underline{u}_S(t) = \frac{3\sqrt{2}}{2}\,U_S\,e^{j\,\omega_0\,t}\,, \tag{13.13}$$

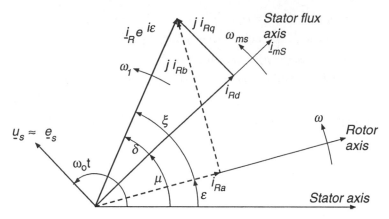

Fig. 13.2. Angular relationships of current vectors for doubly-fed induction machine

where U_S is the RMS-value of the line-to-neutral voltages.

Transforming Eq. (13.6) into field coordinates with the help of Eqs. (13.4, 13.8, 13.11–13.13) and splitting the result into real and imaginary components yields two real differential equations for the magnitude and angle of the magnetising current vector,

$$T_S \frac{di_{mS}}{dt} + i_{mS} = \frac{1 + \sigma_S}{R_S} u_{Sd} + i_{Rd}, \qquad (13.14)$$

$$\frac{d\mu}{dt} = \omega_{mS} - \frac{1}{T_S \, i_{mS}} \left[\frac{1 + \sigma_S}{R_S} u_{Sq} + i_{Rq} \right], \qquad (13.15)$$

where

$$u_{Sd} = \frac{3\sqrt{2}}{2} U_S \cos(\omega_0 t - \mu), \qquad (13.16)$$

$$u_{Sq} = \frac{3\sqrt{2}}{2} U_S \sin(\omega_0 t - \mu) \qquad (13.17)$$

are the field orientated direct and quadrature components of the line voltages.

The $d - q$-components of the rotor current vector can be derived from the measured rotor currents i_{R1}, i_{R2}, i_{R3} according to Eq. (13.12); the transition is again performed in two steps by first converting i_R to an orthogonal two phase AC system which is subsequently transformed into field coordinates. With an isolated neutral of the rotor winding we have with $\gamma = 2\pi/3$

$$i_R(t) = i_R \, e^{j\xi} = i_{R1} + i_{R2} \, e^{j\gamma} + i_{R3} \, e^{j2\gamma}$$

$$= \frac{3}{2} i_{R1} + j \frac{\sqrt{3}}{2} (i_{R2} - i_{R3}) = i_{Ra} + j \, i_{Rb}. \qquad (13.18)$$

The subsequent transformation into field coordinates yields

$$\underline{i}_R \, e^{j\,(\varepsilon-\mu)} = i_R \, e^{j\,(\xi+\varepsilon-\mu)} = i_R \, e^{j\,\delta}$$
$$= [i_{Ra} + j\,i_{Rb}][\cos(\varepsilon - \mu) + j\,\sin(\varepsilon - \mu)]$$
$$= i_{Ra}\,\cos(\varepsilon - \mu) - i_{Rb}\,\sin(\varepsilon - \mu)$$
$$+j\,[i_{Ra}\,\sin(\varepsilon - \mu) + i_{Rb}\,\cos(\varepsilon - \mu)] = i_{Rd} + j\,i_{Rq}\,;$$

$$(13.19)$$

the angle $\varepsilon - \mu$ may be interpreted as the rotor position in field coordinates.

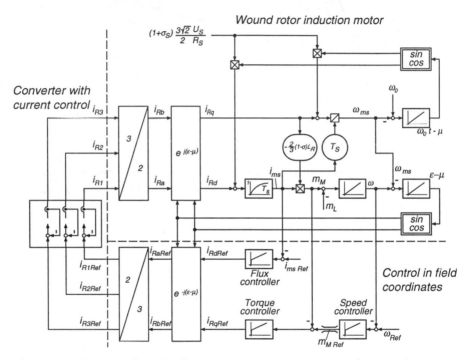

Fig. 13.3. Control of doubly-fed induction machine in stator flux coordinates

The block diagram of the doubly-fed machine is contained in the upper part of Fig. 13.3, showing the interactions described by the preceding equations. Considerable similarity exists with the block diagram of the induction motor, Fig. 12.12, except that there is an additional electrical input in the form of the line voltage U_S. The angle $\omega_0 t - \mu$ constitutes the advance of the voltage vector $\underline{u}_S$ against the magnetising vector $\underline{i}_{mS}$. Since the stator circuit is connected to a low impedance source, the angle $\omega_0 t - \mu \approx \frac{\pi}{2}$ experiences only small changes. As in the case of the stator-fed induction motor the magnitude of the flux can only be altered through the lag T_S, whereas the quadrature current i_{Rq} is available for rapidly controlling torque.

If the converter with current control is approximated by a three-phase current source with negligible lag, the idea of control in field coordinates can again be applied. In this case it means that the coordinate transformation $e^{j(\varepsilon-\mu)}$ inside the machine should be cancelled by an external transformation $e^{-j(\varepsilon-\mu)}$; with the subsequent phase split a balanced three-phase system of current references is generated. This is seen in the lower left portion of Fig. 13.3. The angle of transformation, $\varepsilon - \mu$, is readily available from measurements of rotor position as well as the stator and rotor currents. The transformation from field into rotor coordinates, in effect a modulation, is based on the following algorithm

$$
\begin{aligned}
\underline{i}_{R\,\text{Ref}} &= [i_{R1} + i_{R2}\,e^{j\,\gamma} + i_{R3}\,e^{j\,2\,\gamma}]_{\text{Ref}} \\
&= [i_{Ra} + j\,i_{Rb}]_{\text{Ref}} = [i_{Rd} + j\,i_{Rq}]_{\text{Ref}}\,e^{-j\,(\varepsilon-\mu)}\,,
\end{aligned}
\tag{13.20}
$$

which results in

$$
\begin{aligned}
i_{Ra\,\text{Ref}} &= i_{Rd\,\text{Ref}}\,\cos(\varepsilon - \mu) + i_{Rq\,\text{Ref}}\,\sin(\varepsilon - \mu)\,, \\
i_{Rb\,\text{Ref}} &= -i_{Rd\,\text{Ref}}\,\sin(\varepsilon - \mu) + i_{Rq\,\text{Ref}}\,\cos(\varepsilon - \mu)\,.
\end{aligned}
$$

The equivalent balanced three-phase system of current references is, Eq. (12.37),

$$
\begin{aligned}
i_{R1\,\text{Ref}} &= \tfrac{2}{3}\,i_{Ra\,\text{Ref}}\,, \\
i_{R2\,\text{Ref}} &= -\tfrac{1}{3}\,i_{Ra\,\text{Ref}} + \tfrac{1}{\sqrt{3}}\,i_{Rb\,\text{Ref}}\,, \\
i_{R3\,\text{Ref}} &= -\tfrac{1}{3}\,i_{Ra\,\text{Ref}} - \tfrac{1}{\sqrt{3}}\,i_{Rb\,\text{Ref}}\,.
\end{aligned}
\tag{13.21}
$$

Clearly the decoupled control scheme offers direct access to the $d - q$-current components. There are various possibilities for designing the higher level control as can be deduced from a discussion of the steady state characteristics of the drive.

At constant load torque and speed the rotor angle is $\varepsilon = \omega t$ and the flux vector rotates in synchronism with the line voltage vector, $\omega_{mS} = \omega_0$. Analogous to Eq. (13.13) the vector of the stator currents is, ignoring harmonics,

$$
\underline{i}_S(t) = \frac{3\sqrt{2}}{2}\,\underline{I}_S\,e^{j\,\omega_0\,t}\,, \qquad \underline{I}_S = I_S\,e^{j\,\varphi_S}\,,
\tag{13.22}
$$

where $\underline{I}_S$ is a complex phasor. Similarly the rotor current vector in stator coordinates is

$$
\underline{i}_R(t)\,e^{j\,\varepsilon} = \frac{3\sqrt{2}}{2}\,\underline{I}_R\,e^{j\,(\omega_2+\omega)\,t} = \frac{3\sqrt{2}}{2}\,\underline{I}_R\,e^{j\,\omega_0\,t}\,, \qquad \underline{I}_R = I_R\,e^{j\,\varphi_R}\,.
\tag{13.23}
$$

Inserting these definitions into Eq. (13.3) leads to an algebraic phasor equation

$$
R_S\,\underline{I}_S + j\,\omega_0\,L_S\,\underline{I}_S + j\,\omega_0\,L_0\,\underline{I}_R = U_{S0}\,,
\tag{13.24}
$$

which, with $L_S = (1 + \sigma_S)L_0$, is illustrated by the equivalent circuit in Fig. 13.4.

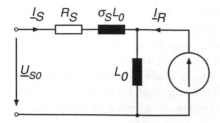

Fig. 13.4. Steady state equivalent circuit of doubly- fed induction machine with impressed rotor currents

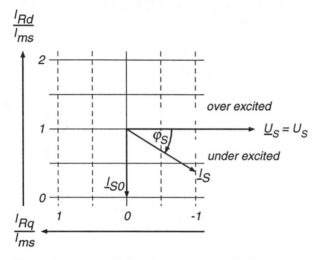

Fig. 13.5. Decoupled orthogonal control of stator current

The stator resistance may be neglected with larger machines,

$$R_S \ll \omega_0 L_S \, ,$$

so that the phasor of the magnetising current is determined by the impressed stator voltage,

$$\underline{I}_{mS} = (1 + \sigma_S) \underline{I}_S + \underline{I}_R \approx \frac{U_{S0}}{j\,\omega_0\,L_0} \, . \tag{13.25}$$

The no-load stator current ($I_R = 0$) is

$$\underline{I}_{S0} \approx \frac{U_{S0}}{j\,\omega_0\,L_S} \, , \tag{13.26}$$

hence Eq. (13.25) assumes the form

$$\underline{I}_S \approx \underline{I}_{S0}\left[1 - \frac{\underline{I}_R}{\underline{I}_{mS}}\right] = \underline{I}_{S0}\left[1 - \frac{I_R}{I_{mS}}\,e^{j\,(\xi+\varepsilon-\mu)}\right]$$

$$= \underline{I}_{S0}\left[1 - \frac{I_R}{I_{mS}}\,e^{j\,\delta}\right] = \underline{I}_{S0}\left[1 - \frac{I_{Rd}}{I_{mS}} - j\,\frac{I_{Rq}}{I_{mS}}\right]. \qquad (13.27)$$

This indicates that the decoupled control of the rotor current in $d - q$-components corresponds directly to orthogonal control of the active and re-active currents as is shown by the orthogonal grid in Fig. 13.5. Note that this operation in all four quadrants of the current plane is not limited to constant speed but applies to the whole torque/speed area, Fig. 13.1 b, accessible with the voltage, current and frequency limits of the converter. Depending on the operating condition of the drive, power is flowing from the rotor via the con-verter to the line or vice versa as is shown in Fig. 13.6. The phase sequence of the rotor currents is inverted below and above synchronism.

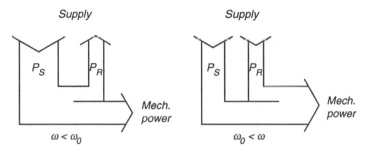

Fig. 13.6. Power flow of doubly-fed induction motor drive below and above synchronism

When assessing the reactive power balance, the reactive power drawn by the converter from the line must also be taken into account. If a cycloconverter is used, its lagging reactive input current may be substantial; it could impair the ability of the drive to operate with overall leading power factor. Measures for reducing the lagging reactive line current of the converter such as the reduction of its supply voltage with a transformer when operating the drive at low slip are frequently applied.

Starting the drive is accomplished with the help of resistors, as seen in Fig. 13.1 a; as soon as the motor has reached the operating speed range the current-controlled converter is switched in.

The orthogonal grid in Fig. 13.5 provides a clear indication of how the higher level control should be arranged. As the stator flux (i_{mS}) is essen-tially prescribed by the line voltage U_S, the d-component of the rotor current could be used to maintain the reactive stator current at a fixed value or pos-sibly a value dependent on stator voltage (reactive power bias). On the other hand, the q-component is the ideal input for torque control to which speed control can be superimposed. All reference quantities should be limited to

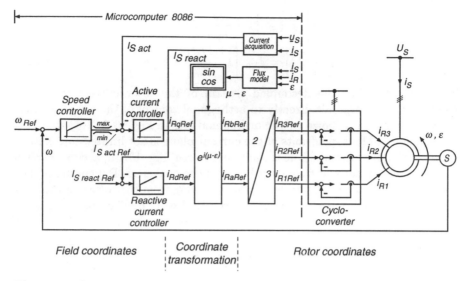

Fig. 13.7. Control scheme of doubly-fed induction motor drive with microcomputer

avoid undesirable operating conditions and to protect the equipment. The control section in the lower part of Fig. 13.3 represents only one of several possibilities; yielding torque/speed curves with torque limit as in Fig. 13.1 b.

Figure 13.7 depicts the control scheme of an experimental 22 kW drive with microprocessor control [A13]. The rotor winding is supplied from a 6-pulse cycloconverter with analogue current control; the converter was chosen to have adequate power rating up to 7.5 Hz corresponding to 15% slip. To improve the steady state accuracy of the current control loops, feed-forward voltage signals were added for counteracting the slip-proportional induced voltages in the rotor windings.

As was mentioned before, the angle μ of the magnetising current (stator flux) can either be computed from measurements of the currents, Eq.(13.4), or from stator voltages and currents; in combination with a measurement of the rotor position ε this provides the information for the modulation producing the reference signals for the rotor currents.

With field orientated control an excellent dynamic performance of the drive is achieved, even though this may not be essential for some of the applications mentioned, but the transparent structure of the control is beneficial as it offers a great degree of flexibility.

Figure 13.8 shows the effect of voltage feed-forward on the performance of a current loop in steady-state. The cycloconverter is operating at a frequency of 6.7 Hz corresponding to a speed of 1700 1/min of the 4-pole machine. The choice of the feed-forward scheme is based on the rotor voltage equation Eq. (13.1) which, introducing the magnetising current, is with $L_R/R_R = T_R$

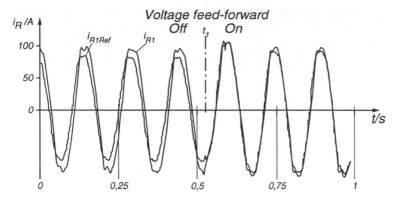

Fig. 13.8. Effect of voltage feed-forward on current control with a cycloconverter

$$\sigma\, T_R \, \frac{di_R}{dt} + i_R = \frac{1}{R_R}\, u_R - (1 - \sigma)\, T_R \, \frac{d}{dt}(i_{mS}\, e^{-j\,\varepsilon})\,. \tag{13.28}$$

The last term may be expanded in the form

$$
\begin{aligned}
\underline{e}_R &= (1 - \sigma)\, L_R \, \frac{d}{dt}(i_{mS}\, e^{-j\,\varepsilon}) \\
&= (1 - \sigma)\, L_R \, \frac{d}{dt}(i_{mS}\, e^{j\,(\mu-\varepsilon)}) \\
&= (1 - \sigma)\, L_R \left[\frac{di_{mS}}{dt} + j\,(\omega_{mS} - \omega)\,i_{mS}\right] e^{j\,(\mu-\varepsilon)}\,;
\end{aligned}
\tag{13.29}
$$

with $i_{mS} \approx$ const., this represents the slip-proportional voltage acting as a load disturbance in the current control loop; its effect is compensated by feed-forward with the necessary signals being generated in the microprocessor, Fig. 14.17.

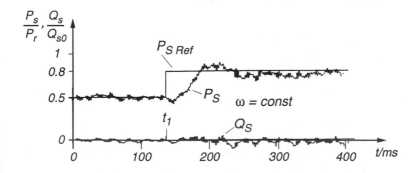

Fig. 13.9. Decoupled control of active/reactive power with a cycloconverter

The measured step response of the control loop for active stator power (with the speed control disabled) is seen in Fig. 13.9; it exhibits good dynamic performance and little cross coupling with the reactive power control loop. The ripple on the recorded traces is caused by harmonics of the cycloconverter and quantisation effects.

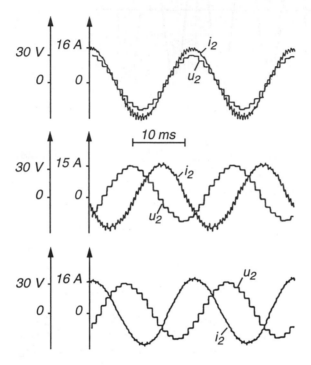

Fig. 13.10. Waveforms of total line-side current of a 22 kW doubly-fed drive with symmetrical GTO voltage source converter at different reactive power setpoints. Line current (**a**) in phase, (**b**) lagging, (**c**) leading

When employing a DC link PWM-converter instead of a cycloconverter for supplying the rotor-winding, the performance of the drive can be further improved in view of the higher switching frequency and the reduced harmonic interactions on the line- and motor-side. The reactive power drawn by the converter from the line can also be greatly reduced. Fig. 13.10 displays the total line current, comprising the stator as well as the converter input currents, when a symmetrical voltage source converter with GTO thyristors, as shown in Fig. 11.13, is used. By designing a decoupled orthogonal control structure also for the line side converter, based on the voltage vector, the active and reactive parts of the total line current can be controlled, as discussed in Sect. 13.2; this provides complete freedom with regard to the line-side characteristics of the drive, which may also be employed for power factor correction, apart from high dynamic performance drive control [A31, H36, H37, K17].

Even though this type of drive shows very interesting characteristics, the amount of hardware for the rotor side converter is considerable which is likely to limit the use to higher power and special applications, where the added flexibility provided by two-axes control is important. One prominent area of application is that of generators in large wind power stations operating on the constant frequency grid; the variable speed operation uses the high-inertia wind rotor effectively as a flywheel in gusty wind conditions. This helps to smooth the power fluctuations and reduces the rating of the generator as well as stresses on mechanical components such as shaft and gears. Another example are constant frequency ship- board generators, driven at variable speed by the main Diesel engine with its low fuel cost, so called shaft generators.

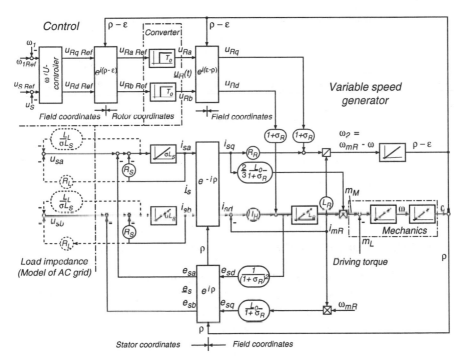

Fig. 13.11. Block diagram of a variable speed AC generator with impressed rotor voltages feeding an island grid at constant frequency

If the doubly-fed machine is normally acting as a generator, is is preferable to operate the rotor side converter as a variable frequency voltage source, i.e. without current control, in order not to impede transient currents in the rotor windings which are caused by disturbances on the stator side. This is illustrated in the block diagram in Fig. 13.11, where the converter is drawn in

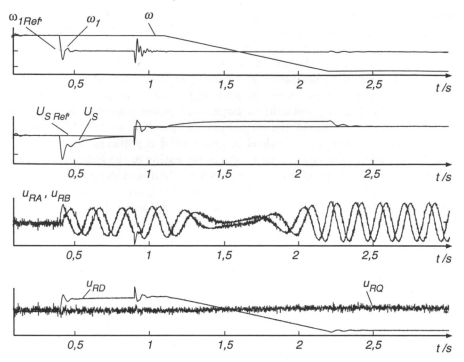

Fig. 13.12. Simulation results of a variable speed AC generator with impressed rotor voltages. Transients following step changes of the voltage reference and a ramp change of speed

rotor coordinates; the stator- fed AC grid is represented by a load impedance R_L, L_L in stator- and the electromagnetic interactions in rotor flux- coordinates. The structure of the diagram is similar to that shown in Fig. 12.38. A two- axes control is indicated for the stator voltage and stator frequency; with the help of two separate PI- controllers, the simulated results in Fig. 13.12 are obtained, showing the transients after step changes of the voltage reference with regard to frequency and amplitude and when the mechanical speed follows a ramp change; the sequence of the rotor voltages reverses as the rotor passes through the synchronous speed.

Future high power applications for this type of control are found in pumped storage hydro power stations for changing the speed of the turbine/pump when it operates with varying head, thus improving the hydraulic efficiency. Also, changing the speed of the motor-generator-set by small amounts allows to activate kinetic energy and thus offers the option of contributing to the dynamic stability of the power grid [H36, H52, H75, K17, T3].

13.2 Control of a Line-side Voltage Source Converter as a Reactive Power Compensator

The principle of field orientated control as discussed in Chap. 12 with the example of machine-side voltage source converters can equally be applied to line-side converters for achieving high dynamic performance. This is shown in Fig. 13.13 a, where a symmetrical voltage source PWM converter serves to connect the DC link of a doubly-fed machine through a transformer to the constant frequency grid. Line-side PWM converters are applicable whenever a DC bus is to be coupled with the AC grid, such as fuel cells or variable

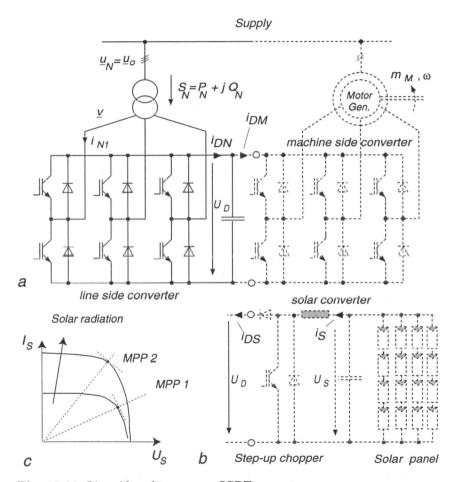

Fig. 13.13. Line-side voltage source IGBT converter
(a) in rotor circuit of a doubly-fed AC motor, **(b)** for solar generator;
(c) "Maximum Power Point" control of solar panel

speed wind energy plants employing a DC link; a solar generator is indicated in Fig. 13.13 b, where a step-up chopper raises the voltage U_S of the solar panels at the irradiation-dependent maximum power point to the level of the link voltage U_D.

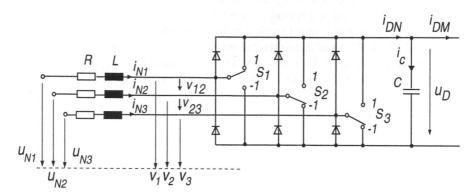

Fig. 13.14. Equivalent switching circuit of the line-side converter

An equivalent switching circuit of the line-side converter is seen in Fig. 13.14 corresponding to the circuit in Fig. 11.7 a; the transformer is represented by its line impedance. The time-dependent vector of the sinusoidal and symmetrical line voltages with constant frequency is with $\omega_N t = \omega_o t = \lambda$ and $\gamma = 2\pi/3$

$$\underline{u}_N(t) = u_{N1} + e^{j\gamma} u_{N2} + e^{j2\gamma} u_{N3} = 3\sqrt{2}/2\, U_N e^{j\lambda(t)} = u_{Na}(t) + j u_{Nb}(t);$$
(13.30)

similar definitions hold for the PWM voltages at the secondary of the transformer, i.e. at the input of the converter bridge,

$$\underline{v}(t) = v_1 + e^{j\gamma} v_2 + e^{j2\gamma} v_3 = v_{12} - v_{23}\, e^{-j\gamma} = v_a(t) + j v_b(t)$$
(13.31)

and the line-side currents

$$\underline{i}_N(t) = i_{N1} + e^{j\gamma} i_{N2} + e^{j2\gamma} i_{N3} = i_{Na}(t) + j i_{Nb}(t).$$
(13.32)

Control of the currents is again carried out in a moving frame of coordinates so that the feedback signals are DC quantities in steady state. According to Sect. 10.3 a suitable reference frame are now line voltage coordinates, where the vector $\underline{u}_N$ assumes the role of the induced motor voltages $\underline{e}_S$ and the pulse-width-modulated converter voltages $\underline{v}(t)$ serve as the actuating variables. Fig. 13.15 shows the vector diagram of the line-side converter. The transformed voltages and currents

$$\underline{v}(t)\, e^{-j\lambda} = v_d + j v_q, \quad \underline{i}_N(t) e^{-j\lambda} = i_{Nd} + j i_{Nq}$$
(13.33)

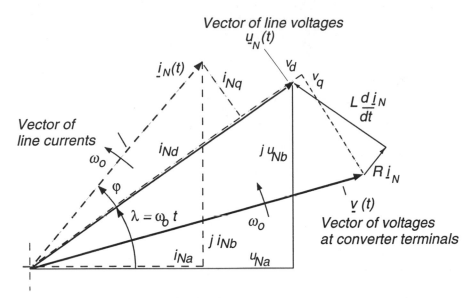

Fig. 13.15. Vector diagram of the line-side converter

may be called line-orientated variables.

The differential equation for the line-side impedance R, L

$$L \frac{di_N}{dt} + R i_N = u_N - v, \tag{13.34}$$

where $R << \omega_N L$ holds for higher power, is transformed into the moving coordinate system by multiplication with $e^{-j\lambda}$ and subsequently split into real and imaginary parts,

$$L \frac{di_{Nd}}{dt} + R i_{Nd} = 3\sqrt{2}/2 \, U_N - v_d + \omega_o L \, i_{Nq}, \tag{13.35}$$

$$L \frac{di_{Nq}}{dt} + R i_{Nq} = -v_q - \omega_o L \, i_{Nd}. \tag{13.36}$$

The instantaneous direct and quadrature components of the currents, i_{Nd} and i_{Nq}, are a measure for the active and reactive currents flowing from the line to the converter, as expressed by the instantaneous complex power,

$$u_N^* \, i_N = (3\sqrt{2}/2) \, U_N \, e^{-j\lambda} \, i_N = 3\sqrt{2}/2 \, U_N[i_{Nd}(t) + j i_{Nq}(t)]; \tag{13.37}$$

in steady state

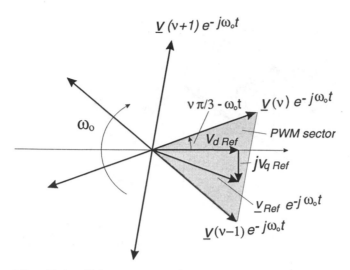

Fig. 13.16. Voltage vectors of converter in line-voltage coordinates

$$1/T_N \int_{t-T_N}^{t} (i_{Nd} + ji_{Nq})\, dt = 3\sqrt{2}/2\,(I_{Nact} + jI_{Nreact}) \tag{13.38}$$

holds. For the purpose of fast control, the instantaneous values $i_{Nd}(t), i_{Nq}(t)$ may be used as feedback signals in place of active and reactive currents, even though they are only identical as mean values and with high switching frequency, i.e. low harmonic content. The converter acts as a reactive compensator by producing a controllable reactive line current, whereas the active line current has the task of controlling the DC link voltage u_D.

The mesh equation for the DC link voltage is, combining the machine-side and the line-side link currents i_{DM}, i_{DN},

$$C\frac{du_D}{dt} = i_{DN} - i_{DM}; \tag{13.39}$$

i_{DM} is a disturbance in the voltage control loop. When the converter is used exclusively as a reactive compensator, i.e. without a machine-side converter, the constraint is $i_{DM} = 0$.

The available voltage vectors in moving coordinates at the terminals of the line-side converter are indicated in Fig. 13.16, from which the PWM converter selects by vectorial modulation in each interval T_o the two neighbouring vectors. The switching functions $S_i = \pm 1$ correspond again to the potential of the converter terminal i, as shown in Fig. 11.8; the magnitude of the voltage vectors is determined by the DC link voltage,

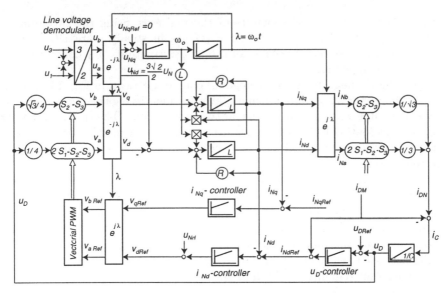

Fig. 13.17. Block diagram of the orthogonal control of the line-side converter

$$\underline{v} = v_a + j\,v_b = v_{12} - v_{23}\,e^{-j\gamma} = u_D/2[(S_1 - S_2) - (S_2 - S_3)e^{-j\gamma}]$$

$$= u_D/4[(2S_1 - S_2 - S_3) + j\sqrt{3}(S_2 - S_3)] = u_D\,e^{j\nu\pi/3}, \qquad (13.40)$$

where $\nu = 0, 1, 2...5$. The switching combinations result in a six-pointed star for the voltage vector $\underline{v}(t)$; in addition there are the two zero vectors $(S_1, S_2, S_3) = (-1, -1, -1)$ and $(1, 1, 1)$, when the converter terminals are short circuited at the upper or lower DC bus.

When neglecting the protective intervals and applying the equivalences given in Eqs. 13.21, the link current i_{DN} may be expressed by the line currents,

$$i_{DN} = S_1 i_{N1} + S_2 i_{N2} + S_3 i_{N3}$$

$$= 1/3\,[(2S_1 - S_2 - S_3)i_{Na} + \sqrt{3}\,(S_2 - S_3)i_{Nb}]; \qquad (13.41)$$

during the zero vector intervals the line currents are bypassing the DC-link, $i_{DN} = 0$.

The six-pointed star of the available converter voltages in line-voltage coordinates, Fig. 13.16,

$$\underline{v}(t)\,e^{-j\lambda} = v_d + jv_q = u_D\,e^{j(\nu\pi/3 - \omega_o t)} \qquad (13.42)$$

rotates with the angular velocity ω_o in clockwise direction and it is the task of the current controllers to generate the needed direct and quadrature voltage components $(v_d + jv_q)_{Ref}$ for transformation into the AC components $(v_a + jv_b)_{Ref}$; the vectorial modulator then determines for each switching period T_o the necessary signals (S_1, S_2, S_3). In view of the real voltage vector $\underline{u}_N(t)e^{-j\lambda(t)} = 3\sqrt{2}/2U_N$ and the rotating star $\underline{v}(t)\,e^{-j\lambda}$ the modulation is mainly performed in the vicinity of the real axis, using the two adjacent available voltage vectors. $v_d(t)$ and $v_q(t)$ are in steady state DC voltages with PWM-related noise; because of the transformation, they contain no line frequency components.

The block diagram of the orthogonal control is drawn in Fig. 13.17 showing in the upper left corner a demodulator for identifying the line vector from two of the line voltages and the usual transformations. The transformed currents, representing DC signals in steady state, are controlled in two channels, with the undesired interactions removed again by feed-forward at the output of the current controllers.

Experimental results obtained with a 20 kVA IGBT reactive compensator are depicted in Fig. 13.18 employing vectorial PWM with 4 kHz clock frequency [H36]. The line currents shown in Fig. 13.18 a are nearly sinusoidal and the transition from leading to lagging reactive current takes only a few ms. Changing the link voltage reference u_{DRef} in Fig. 13.18 b results in a brief active current component for charging the link capacitor, while the reactive current remains nearly undisturbed; clearly the decoupling control is effective.

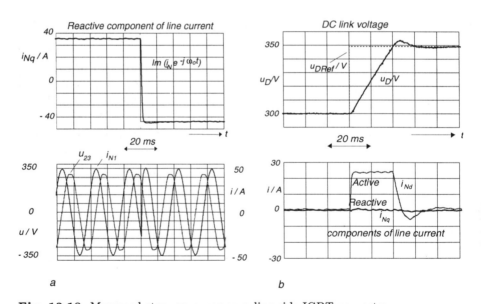

Fig. 13.18. Measured step responses on a line-side IGBT converter with vectorial PWM at 4 kHz clock frequency
(a) Reversal of the reactive current, **(b)** Change of DC link voltage

The example makes it evident that power electronics, often looked at as a cause of troublesome line-side interactions in the form of reactive currents and harmonics, can also provide a solution for removing such effects; this is due to the emergence of semiconductor devices with adequate switching frequency. The cost and complexity associated with the line-side converter will in future be tolerable even with lower power equipment as industrial drives. At higher power the flexibility will be important for the stable and economical operation of the supply grid.

13.3 Wound-Rotor Induction Motor with Slip-Power Recovery

The doubly-fed machine described in the Sect. 13.1 can operate as a motor or generator in a speed band above or below synchronism, drawing leading or lagging stator current from the line. Depending on the operating condition power flows from the converter to the rotor or in the reverse direction. This high degree of flexibility calls for a relatively costly converter but simplifications are possible when the specifications are relaxed. For example, when rectifying the three-phase rotor currents with diodes and feeding a DC link, the power flow becomes unidirectional. This is so because both voltage and current at the DC side of the rectifier are restricted to positive values; seen from the AC side the rectifier appears like a nonlinear, purely resistive network. The purpose of this scheme may therefore be to recover the slip power that would otherwise be lost in external rotor resistors; it can either be converted to mechanical power in a DC motor coupled to the shaft of the induction motor (Krämer-drive) or fed back to the AC line with the help of a line-commutated or other type of inverter (static Scherbius-drive) as is explained in Fig. 13.19 a.

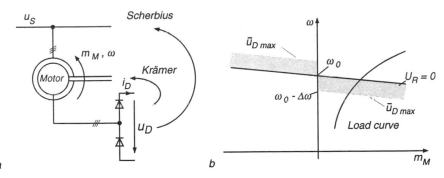

Fig. 13.19. Wound-rotor motor with slip power recovery scheme.
(a) Circuit; (b) Operating range

Again, slip power recovery is of particular interest at high power such as needed for rotating converters with flywheel storage in rolling mill drives or for boiler feed pumps extending into the 20 MW-range, Sect. 7.4. While the

Krämer-drive requiring an additional DC machine is no longer of interest, the Scherbius-drive with the associated solid state equipment is still used today.

In view of the diode rectifiers in the rotor circuit the operating range is now restricted to areas that would be available with external rotor resistors, i.e. motor operation below and regeneration above the torque/speed curve valid for short circuited rotor, $U_R = 0$, as shown in Fig. 13.19 b. Also, since the rotor currents cannot be impressed at will, the stator currents are always lagging the voltages as on a wound-rotor motor with secondary resistors. This excludes the control of the stator voltage which was possible with the more general scheme in Fig. 13.1 a.

Speed control is achieved by changing the back voltage u_D in the DC link. If $\bar{u}_D$ is increased, the speed must drop so that the rectified rotor voltage can maintain a current through the DC link; without current in the rotor windings the rotor produces no torque. Closed loop control is usual with this type of drive in to have access to torque and speed and to protect the equipment against overload.

Analytical modelling of the drive is made complicated by the rectifier-generated harmonics in the rotor windings which also cause distortion of the stator currents. The subsequent discussion is only an approximation, a more detailed analysis would have to be based on a digital simulation.

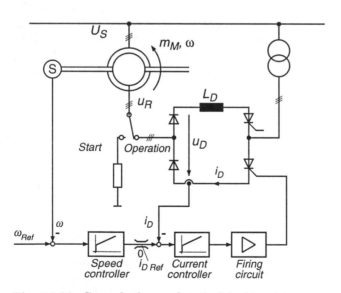

Fig. 13.20. Control scheme of static Scherbius-drive

The circuit of the Scherbius-drive and the usual control scheme employing an inner current- and an outer speed- loop is shown in Fig. 13.20. The starting resistor serves to bring the motor into the specified speed range, where the rotor voltage has sufficiently decreased to allow switching in the converter.

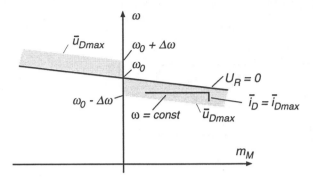

Fig. 13.21. Torque/speed curves of controlled Scherbius-drive

The DC link contains a line-commutated inverter for feeding the slip power back to the grid and a smoothing choke to produce continuous current over most of the load range.

A step-down transformer matches the supply voltage to the voltage in the DC link, which is quite low in the usual speed range, and reduces the reactive current at the line side of the inverter which is caused by the delayed firing around $\alpha \approx \frac{\pi}{2}$. Often the transformer ratio is changed in several steps to create optimal conditions over a wider speed range.

The rotor current and, hence, the current in the DC link are roughly proportional to torque; the inner current control loop, seen in Fig. 13.20, provides therefore a good substitute for torque control and torque limit, while the speed controller is superimposed, generating the necessary current reference. Further control functions may be added if the need arises; for example on a flywheel drive and with pulsating load, approximately constant line power could be specified resulting in a fluctuating speed level.

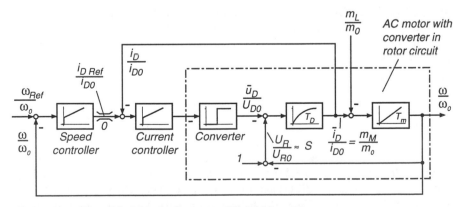

Fig. 13.22. Simplified block diagram of Scherbius-drive

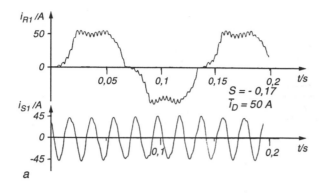

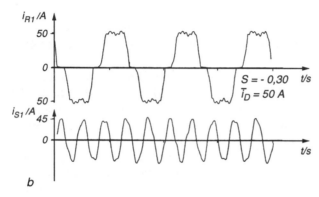

Fig. 13.23. Stator and rotor currents of Scherbius-drive in steady state operation above synchronous speed
(a) $S = -0.17$; (b) $S = -0.30$

The steady state torque/speed curves of the static Scherbius-drive are seen in Fig. 13.21 for $\omega_{Ref} = $ const. As long as the direct current is below the current limit, the speed is kept constant by the PI-controller; with increased load the speed controller eventually saturates causing the curve to bend down until the maximum voltage of the inverter is reached. Further drop in speed must be avoided since it would inevitably lead to a commutation failure of the inverter as explained in Fig. 8.17.

A simplified block diagram of the controlled drive is drawn in Fig. 13.22 with the control loops arranged according to Fig. 13.20; the normalising quantities U_{R0}, m_0, i_{D0} etc. are suitably chosen.

The inverter with the mean direct voltage $\bar{u}_D$ is represented by a control delay as described in Fig. 8.31; $\bar{u}_D$ is the back voltage opposing the rectified rotor voltage, with the difference driving the current through the DC link and the rotor windings. T_D is the filter time constant which is mainly due to the smoothing reactor. Clearly, the whole structure strongly resembles that of a controlled DC drive; therefore similar transients should be expected.

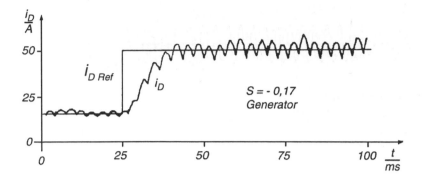

Fig. 13.24. Measured step response of link current control loop

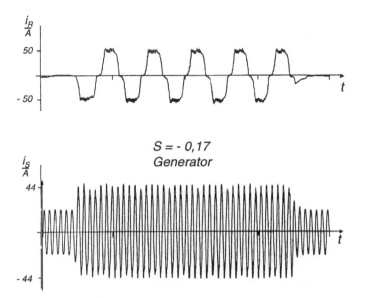

Fig. 13.25. Measured stator and rotor currents during step change of link current reference

As mentioned before, the details are of considerable complexity, in particular the current waveforms in the rotor and stator at different speed and load; this is shown by simulated results and measurements conducted on a 22 kW experimental Scherbius-drive [A13]. Measured waveforms of the stator and rotor currents at two operating points in the generating region (above synchronous speed) are seen in Fig. 13.23 with the same direct link current, $\overline{i}_D = 50$ A, in both cases; the speed-dependent distortion of the stator

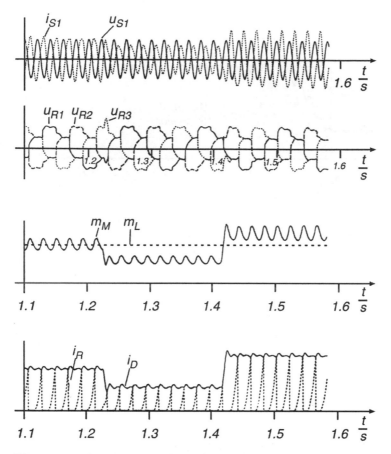

Fig. 13.26. Simulation results with a Scherbius drive

currents is clearly visible. A recorded step response of the inner current control loop is depicted in Fig. 13.24 showing good dynamic performance of the electronic torque control. The effect on the stator and rotor currents when changing of the link current between 0 and 50 A at constant speed (1725 1/min) is shown in Fig. 13.25.

As an example of the results obtainable with digital simulation, Fig. 13.26 displays some computed curves of voltages and currents as well as electrical torque. Clearly, the deviations from sinusoidal waveforms are substantial; of particular interest are the long commutation intervals of the uncontrolled rectifier, caused by the small slip-proportional driving voltage and the rotor leakage reactance. On the other hand, reducing the higher frequency harmonics tends to smooth the currents and the torque.

14. Variable Frequency Synchronous Motor Drives

The speed of a synchronous motor with DC excitation in the rotor is determined by the stator frequency and the number of poles. As long as efficient, variable frequency power supplies were unavailable this meant constant speed operation at fixed frequency. There are drive applications, where constant speed is desired or where the reactive power that can be generated with line- connected synchronous motors is an important feature; these are, besides electromechanical clocks, mainly high power drives, such as for piston compressors in the chemical industry. Another field of application exists in pumped storage power plants, where the synchronous generators serve as constant-speed-drives for pumps in periods of low demand for electrical power, feeding water into elevated reservoirs for later use during hours of peak demand. In this instance the type of motor is, of course, not a matter of choice but the synchronous machine is very well suited for this duty; it is, in fact, the only one that could be used at a power level of, possibly, several hundred MW.

Problems with large synchronous motors operating on a constant frequency supply may be caused by their inherent oscillatory response since the torque now depends on the load angle, and the required start-up procedure. Asynchronous starting at full or reduced line voltage with the help of the damper winding and the short circuited field winding as well as special starting motors are common practice; more recently, large synchronous machines are also started with variable frequency supplied from static converters. This can be extended to variable speed drives with many interesting features.

Historically, the synchronous motor fed with variable frequency from a DC link by thyratrons represented the first attempt at assigning the task of commutation to static equipment [A15, W29]. The thyratrons, gas filled discharge tubes with heated cathodes, that were available in those days, were of course hardly suitable for converter duty. Thereafter, the converter-fed synchronous motor did not receive much attention until this was completely changed by the advent of high power solid state switching devices available today.

As seen in the synopsis in Table 11.1 there are three main fields of application for adjustable speed synchronous motor drives,

- Large, low speed reversing drives, such as needed for gear-less rolling mills or mine hoists, with stringent requirements for high dynamic performance. In the past, these were mainly DC motors but they can now be built as AC synchronous motors supplied by cycloconverters, thereby eliminating all the design- and operating- restrictions inherent in DC machines [J1, L6, N4, R23, S4, S5, S67, S79].
- Large, high speed compressor drives up to 100 MW for pipe-lines or wind-tunnels, where the induced source voltages and, as a consequence, the capability of the synchronous machine of supplying reactive current, permits the use of simple DC link converters with natural or load commutation. DC motors could not be built at this power rating and speed because of the commutator [G25, H5, K63, N15, V11].

Since a synchronous motor does not have to be magnetised through the converter, as is the case with induction motors, a larger airgap is no disadvantage and can even be desirable for mechanical reasons. The slip-rings for supplying the rotating field winding with direct current can be avoided by employing an AC exciter with rotating rectifiers as is common on large turbo-generators.

- A third important and growing field of application are low power, high dynamic performance servo drives (usually < 10 kW) with permanent magnet excitation and transistor converter supply (MOSFET or IGBT). Their main advantage, when compared with induction motors, is the almost complete elimination of rotor losses; on the other hand, field weakening is more difficult [B29, G35, J2, L48, L51, P28]. After the initial problems of achieving a wider field weakening range have been solved with internal magnets built into the rotor, permanently excited synchronous motors are also an option for vehicle drives. Large synchronous motors with permanent magnets are of interest for the propulsion of military ships, because of the improved efficiency, Table 11.1; the stator phases may there be fed from separate four-quadrant converters to attain redundancy in case of faults.

The analysis and design of a control system for an electric drive calls for a dynamic model of the motor; with a synchronous machine this may be of considerable complexity if details such as saliency and unsymmetrical rotor windings are to be taken into account; as this is not the aim of the discussion here, substantial simplification may be achieved without much loss of validity by restricting ourselves to machines with cylindrical rotors, where the mathematical model of a symmetrical three-phase machine described in Sect. 10.1 remains applicable. With large machines and high speed a solid steel nonsalient rotor is the normal choice for mechanical reasons but even at low speed cylindrical rotor designs with symmetrical damper windings are often used; hence the model derived in Sect. 10.1 remains useful.

An exception is the switched reluctance motor with its salient construction of the stator and rotor; its synchronous function is based on sequentially feeding the separate windings on the stator poles. This calls for entirely different models and is not included here; a large specialised literature exists on this subject, e.g. [D10, L14, L19, P19].

In Table 10.1 it was demonstrated how the general model of AC motors can be adapted to various constraints. Larger synchronous motors usually have a symmetrical damper winding for reducing the effects of harmonics and negative sequence current components produced by the power converter; this also tends to lower the transient reactance of the machine and thus facilitates commutation. Hence, the single axis field winding of a synchronous machines may for simplicity be incorporated into the symmetrical damper winding by inserting an ideal DC voltage source u_F without internal impedance into a joint three-phase rotor winding, as seen in Table 10.1 d. This simplification is adequate for designing the control, it would of course be too inaccurate for the design of the machine.

Smaller motors, such as synchronous servo motors with permanent magnet excitation, Table 10.1 a,b do not normally possess damper windings but there is a slight electrical damping effect by eddy currents in the magnets; the mechanical damping of the drive can now be achieved by the control.

Since the control of permanent magnet excited machines is simpler than that of machines with field and damper windings, this case will be discussed first.

14.1 Control of Synchronous Motors with PM Excitation

Permanently excited synchronous motors are an attractive solution for servo drives in the kW-range. There are various designs, depending on the type of magnetic material used; best results are obtained with "rare earth" magnetic materials such as Samarium-Cobalt or Neodymium-Iron-Boron, which combine high flux density (beyond 1 T) and very large coercive force (up to 7000 A/cm). This makes it practical to magnetise the pole pieces separately rather than have to magnetise the complete assembly in place, as is necessary for instance with Alnico-magnets. Rare-earth magnets require also much less space and, with proper design, there is no danger of accidental demagnetisation by short circuit currents; unfortunately magnets of this type are still quite expensive.

Synchronous motors with rare earth permanent magnets have higher power density than comparable DC motors because the limiting effects of the mechanical commutator are absent; the power density exceeds also that of induction motors because there are no losses by torque-producing rotor currents that need to be removed by cooling. A comparison is seen in Fig. 14.1, showing the outline of an axial field DC disk motor, until a few years ago the ultimate choice for dynamic performance, and a synchronous AC servo motor with rare earth magnet excitation. The AC motor is not only more powerful, having lower mass and inertia, it exhibits also a much larger impulse torque $m_{max}\Delta t = 900~Nms$ as compared to 5.5 Nms for the disk motor; this, for instance, is important for quickly accelerating an inertia load through a given speed range $\Delta\omega$. The reason for this large difference is the thermal capacity of the AC stator winding which is embedded in slots, as compared to the open conductors of the printed armature winding in a DC disk motor. Due to their high power density and, consequently, the smaller size, permanent magnet synchronous motors have in recent years evolved as the preferred choice for positioning drives on machine tools and robots, where the motors are to be integrated into the mechanical load. The problems associated with backlash and elasticity of gears may be avoided by using linear synchronous motors with permanent magnets for precision servo drives [B78, G29, H33].

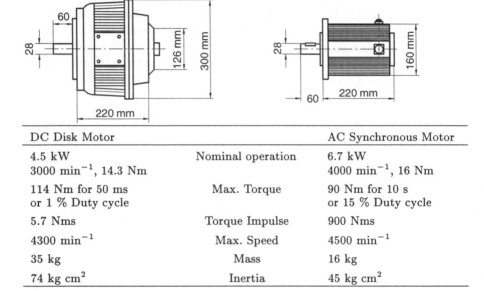

DC Disk Motor		AC Synchronous Motor
4.5 kW 3000 min^{-1}, 14.3 Nm	Nominal operation	6.7 kW 4000 min^{-1}, 16 Nm
114 Nm for 50 ms or 1 % Duty cycle	Max. Torque	90 Nm for 10 s or 15 % Duty cycle
5.7 Nms	Torque Impulse	900 Nms
4300 min^{-1}	Max. Speed	4500 min^{-1}
35 kg	Mass	16 kg
74 kg cm^2	Inertia	45 kg cm^2

Fig. 14.1. Comparison of DC and AC servo motors with permanent excitation

As an example, Fig. 14.2 shows the cross section of an experimental 1.2 kW, 4 pole synchronous motor (continuous rating) with a hollow steel rotor to which Sm-Co-Magnets produced from standard size bars are attached, thus reducing also the quadrature field. The whole rotor assembly is bandaged with Kevlar fibre after the hollow parts have been filled with light-weight nonconducting material. The stator is skewed by one slot width for reducing the torque ripple caused by the stator slots [L51, P28].

Rare earth permanent magnets can be considered as ampereturns sources; since they combine relatively high electrical resistivity and unity permeability for external magnetic fields, they may be viewed as part of the air gap. Hence the motor effectively features a constant wide air gap resulting in a relatively small synchronous reactance which minimises armature reaction.

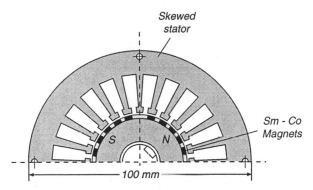

Fig. 14.2. Cross section of 4 pole synchronous servo motor with rare-earth permanent magnets mounted on the rotor surface

To adapt the mathematical model of the symmetrical two pole three-phase machine derived in Sect. 10.1 to this constraint, the rotor circuit, Table 10.1 a, is assumed to be fed from two current sources, Fig. 14.3. According to Eq. (10.10) this produces an impressed rotor current vector

$$ \underline{i}_R(t) = I_F - \frac{I_F}{2}\,c^{j\,\gamma} - \frac{I_F}{2}\,e^{j\,2\,\gamma} = \frac{3}{2}\,I_F\,, \qquad \xi = 0\,, \tag{14.1} $$

which is fixed to the rotor and creates a sinusoidal ampereturns distribution that moves synchronously with the rotor; it serves as a basis of reference for the control.

When further assuming that the stator currents of the servo motor are supplied by a current-controlled switched transistor converter of the type shown in Fig. 11.3, having adequate ceiling voltage and high clock frequency, the stator windings can again be considered as being fed by current sources following the commands from a set of reference values. More realistically, the response of the current control loops may be approximated by unity gain lag elements

$$ T_e\,\frac{di_{S\nu}}{dt} + i_{S\nu} = i_{S\,\nu\,\text{Ref}} \qquad \nu = 1,2,3 \tag{14.2} $$

with $\sum_1^3 i_{S\nu} = 0$, where in practice T_e may be in the order of 1 ms or less. Fig. 14.4 shows a recorded step response of a stator current control loop for the servo motor in Fig. 14.2, measured in standstill. In steady state condition the currents are approximately sinusoidal; due to the induced back voltage the lag is increasing at higher speed.

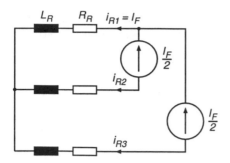

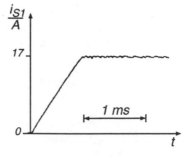

Fig. 14.3. Equivalent rotor circuit of synchronous motor with permanent magnet excitation

Fig. 14.4. Recorded step response of stator current control loop, employing a transistor converter

The stator voltage Eq. (10.50), while being of no consequence for the drive dynamics when the currents are impressed, is important for the converter design. With the angular rotational speed $\omega = d\varepsilon/dt$ it follows

$$R_S\, \underline{i}_S + L_S\, \frac{d\underline{i}_S}{dt} + j\,\omega\, \frac{3}{2}\, L_0\, I_F\, e^{j\varepsilon} = \underline{u}_S(t)\,. \tag{14.3}$$

The expression for the electric torque is

$$m_M(t) = \tfrac{2}{3}\, L_0\, \mathrm{Im}\left[\underline{i}_S\,(\underline{i}_R\, e^{j\varepsilon})^*\right] = L_0\, I_F\, \mathrm{Im}\left[\underline{i}_S\, e^{-j\varepsilon}\right] = \Phi_F\, i_{Sq} \tag{14.4}$$

where $\underline{i}_S\, e^{-j\varepsilon}$ is the stator current vector in a coordinate frame given by the rotor position. Since ε is easy to measure, this is the preferred orientation here; it corresponds to the $d-q$ or Park-Transformation commonly used with synchronous machines.

The mathematical model of the synchronous motor is contained in the right hand section of Fig. 14.5. The principle of controlling the motor is similar to the field orientated control of an induction motor except that now the rotor position serves as the angle of reference; hence there is no need for a flux model. The drive should of course be operated in such a way as to make best use of the motor and converter.

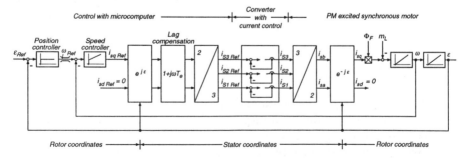

Fig. 14.5. Control of synchronous motor with permanent magnet excitation

The appropriate mode of control may be deduced from a phasor diagram, describing the voltage equation Eq. (14.3) in steady state. With $\omega_1 = \omega = $ const., $\varepsilon = \omega t$, the stator current and voltage vectors are

$$\underline{i}_S(t) = \frac{3\sqrt{2}}{2} \underline{I}_S e^{j\omega t}, \qquad \underline{u}_S(t) = \frac{3\sqrt{2}}{2} \underline{U}_S e^{j\omega t}. \tag{14.5}$$

After simplification, Eq. (14.3) assumes the form

$$(R_S + j\omega L_S)\underline{I}_S + \underline{E} = \underline{U}_S, \tag{14.6}$$

where

$$\underline{E} = j\frac{1}{\sqrt{2}}\omega \, \Phi_F \tag{14.7}$$

is the RMS-value of the source voltage induced in the stator winding by the permanent magnets, Fig. 14.6 a.

The expression

$$\underline{i}_S(t) e^{-j\omega t} = \frac{3\sqrt{2}}{2} \underline{I}_S = \frac{3\sqrt{2}}{2} I_S e^{j\delta} = \frac{3\sqrt{2}}{2}(I_{Sd} + j\, I_{Sq}) \tag{14.8}$$

represents the stator current in rotor coordinates, with δ being the current load angle; $\delta = 0$ is the no-load position. The phasor diagram is drawn in solid lines in Fig. 14.6 b for $\delta = \frac{\pi}{2}$ which is the optimal mode of operation, where the motor produces maximum torque for a given stator current. With a synchronous motor operating on the constant frequency supply, it would correspond to the static stability limit, but this is of no consequence here as the motor is self-controlled, similar to a DC machine where the commutator is fixed to the rotor.

Clearly, $\delta = \pm\frac{\pi}{2}$, $I_{Sd} = 0$ is the best choice as long as the converter can supply the necessary voltage, i.e. below base speed. Direct field weakening is not feasible with a permanently excited machine but a similar effect can be achieved by advancing the current vector beyond $\delta = \frac{\pi}{2}$, i.e. introducing a negative direct current component. This is indicated in Fig. 14.6 b in dashed lines resulting in a decrease of the terminal voltage U_{S1}. As a consequence the maximum value of the quadrature current I_{Sq} may have to be reduced in order not to exceed the limit for the total current,

$$I_S = \sqrt{I_{Sd}^2 + I_{Sq}^2} < I_{S\,\mathrm{max}}, \tag{14.9}$$

which leads to a corresponding reduction of torque, a characteristic feature of operation at reduced field.

The optimal control procedure is thus as follows: Operation with $I_{Sd} = 0$, $|I_{Sq}| < I_{Smax}$ up to the voltage limit, which is reached at the base speed; a further increase in speed can then be achieved by shifting the current into the

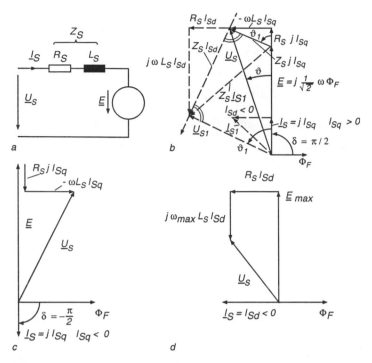

Fig. 14.6. Phasor diagrams of PM synchronous motor,
(a) Equivalent circuit diagram,
(b) Phasor diagram showing effect of field weakening,
(c) Phasor diagram during regeneration,
(d) Phasor diagram at maximum no-load speed

unstable region, $|\delta| > \frac{\pi}{2}$, while maintaining constant terminal voltage. In view of the large effective air gap with rare-earth permanent magnets, this method of "field weakening by armature reaction" calls for a large stator current in the direct axis and, hence, is a much less efficient procedure than would be possible on a synchronous machine with a DC field winding; however, as a short time measure it may be acceptable. As seen in Fig. 14.6 c the need for field weakening is likely to occur mainly in the motoring quadrants but, in principle, it is applicable to braking operation as well. A possible solution to the control problem is sketched in the lower part of Fig. 14.10, where field weakening is initiated by an auxiliary control loop with the aim of limiting the magnitude of the stator voltage.

The degree of field weakening that may be achieved by armature reaction can be estimated from the phasor diagram, Fig. 14.6 b. By initially assuming a purely quadrature current, the voltage components are

$$U_S \cos\vartheta = E + R_S I_{Sq},$$
$$U_S \sin\vartheta = \omega L_S I_{Sq},$$

where the voltage angle ϑ depends on speed and load,

$$\vartheta = \arctan \frac{\omega L_S I_{Sq}}{E + R_S I_{Sq}} . \tag{14.10}$$

By now adding a direct component, $I_{Sd} < 0$, while maintaining I_{Sq} and hence torque, the voltage phasor $\underline{U}_S$ moves along the dashed straight line in the indicated direction. The minimum value of the voltage is reached when the phasor $\underline{U}_S$ is orthogonal to this line, hence

$$U_{S1} = U_S \cos(\vartheta_1 - \vartheta) \tag{14.11}$$

where

$$\vartheta_1 = \arctan \frac{\omega L_S}{R_S} . \tag{14.12}$$

There is no point in advancing the voltage angle beyond ϑ_1, as the stator voltage then begins to rise again.

The additional direct component I_{Sd}, which appears as a reactive current component at the motor terminals, causes an increase of the stator current I_S; the current I_{S1} for the condition of minimum stator voltage is found from the related triangles

$$I_{S1} = \sqrt{\frac{E^2 + U_{S1}^2 - 2 E U_{S1} \cos \vartheta_1}{E^2 + U_S^2 - 2 E U_S \cos \vartheta}} \, I_{Sq} . \tag{14.13}$$

Since the stator current is limited by the converter, this calls for a corresponding reduction of the maximum quadrature current.

The following parameters, taken from the 1.2 kW servo motor (Fig. 14.2), may serve as an example of what can be expected with this method [L51]. At the base speed (2000 1/min) and at nominal torque (5.7 Nm), the relative voltage across the stator impedance is

$$\frac{\omega L_S I_{Sq}}{E} = 0.139 , \qquad \frac{R_S I_{Sq}}{E} = 0.066 ,$$

which results in $U_S/E = 1.075$ and $\vartheta = 7.5°$.

By adding a negative I_{Sd}-component and adjusting it for minimum stator voltage, the voltage angle becomes

$$\vartheta_1 = \arctan \frac{\omega L_S}{R_S} = 64.5° ,$$

and the stator voltage is reduced to

$$\frac{U_{S1}}{E} = \frac{U_S}{E} \cos(\vartheta_1 - \vartheta) = 0.584 .$$

At the same time, however, the stator current, Eq. (14.13), rises to

$$\frac{I_{S1}}{I_{Sq}} \approx 6.0 \,.$$

The torque and speed in this assumed operating condition remain unchanged with the effective stator impedance reduced by a factor $(U_{S1}/I_{S1})/(U_S/I_{Sq}) = 0.091$; this permits access to suitably defined operating points in the low torque/high speed region, as long as the voltage and current limits of the converter are not exceeded. In view of the high copper losses this procedure is of course restricted to short transients at light load which is common practice with DC or induction motor AC drives.

As a limiting case the situation depicted in Fig. 14.6 d may be considered, where the stator ampereturns are opposing the rotor excitation. This would result in the maximum speed obtainable by controlled armature reaction at zero torque, $I_S = I_{Sd} < 0$.

From the voltage equation

$$U_S^2 = (E_{\max} + \omega_{\max} L_S I_{Sd})^2 + (R_S I_{Sd})^2 \,, \tag{14.14}$$

follows with $E_{\max}/E = \omega_{\max}/\omega$ the maximum no-load speed, referred to base speed, at nominal current

$$\frac{\omega_{\max}}{\omega} = \frac{\sqrt{(U_S)^2 - (R_S I_{Sd})^2}}{E + (\omega L_S I_{Sd})} \,; \tag{14.15}$$

with the numbers listed above this results in

$$\omega_{\max}/\omega = 1.25 \,.$$

Of course, higher speeds can be obtained by allowing larger currents. With the particular servo motor described, the no load speed at five times nominal

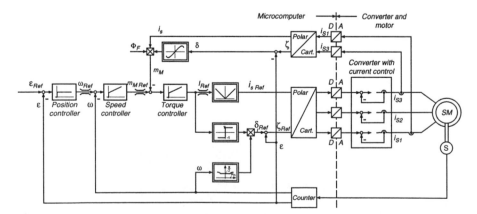

Fig. 14.7. Control of synchronous motor servo drive in polar coordinates

current is 8000 1/min. Better results in the field weakenig range are achieved when choosing different stator impedances in direct and quadrature axes.

Fig. 14.5 contains a block diagram of the rotor orientated control scheme in rectangular components but the control can of course also be presented in polar form. This is shown in Fig. 14.7, where a microcomputer performs all the control functions up to position control; operation at reduced field is achieved with the help of a suitably defined function $\delta_{Ref} = f(\omega)$.

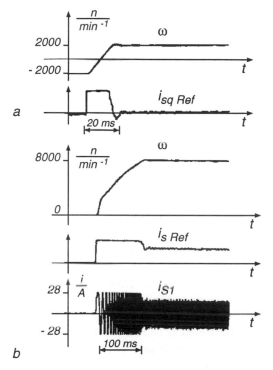

Fig. 14.8. Speed transients of microcomputer-controlled synchronous motor at no-load with six times nominal current
(a) speed reversal, (b) acceleration into field weakening region

The transformations from stator to rotor coordinates and vice versa now call for a simple modulo 2π addition of ε. All the inner control functions are executed in 0.6 ms-intervals which is adequate for stator frequencies of up to about 200 Hz but a further extension of the frequency range could be achieved with the help of signal processors. Transients that have been recorded with the 1.2 kW permanently excited synchronous motor at six times nominal current are displayed in Figs. 14.8 and 14.9. The control is performed by a microcomputer with a control structure similar to the one shown in Fig. 14.7; however, special adaptive software features have been

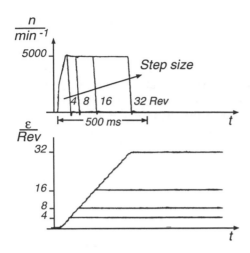

Fig. 14.9. Step response of time-optimal position control with synchronous motor servo drive at no-load

added to the speed- or position-control function to combine fast response with aperiodic damping [L51].

The speed response of the drive is limited, besides the inertia, by the maximum current which the converter can supply. Given sufficient stator current, the motors have very high short-time overload capacity which is important for servo drives designed for intermittent duty, Fig. 14.1.

The control circuits in Figs. 14.5 and 14.7 contain current control loops in stator coordinates, producing approximately impressed AC stator currents. The reference signals are sinusoidal in steady state which calls for careful tuning of the current controllers with regard to undesirable phase shifts. An alternative and usually preferable arrangement is shown in its basic form in Fig. 14.10, where the currents are controlled in moving rotor coordinates; this has the advantage that the controllers for i_{Sd}, i_{Sq} are processing DC signals in steady state. The effects of the induced voltages may again be compensated by feed forward at the outputs of the current controllers; the control can also be extended for field weakening.

As the phasor diagram in Fig. 14.6 b and the parameters of the experimental motor indicated, the increase of the stator currents in the field weakening region is due to the small stator inductance L_s, caused by the relatively large effective airgap as a consequence of the surface-mounted magnets. This can be improved with different motor designs, where the rotor is anisotropic, similar to a rotor with salient poles, thus increasing the longitudinal inductance L_{Sd} compared with the quadrature inductance L_{Sq} [B26,B28,B29,J2]; possible implementations of this idea are seen in Fig. 14.11 a,b. The magnets may either be surface mounted with the anisotropy caused by the internal shape of the rotor or they could be arranged internally [A8,B29,K89,M49]; in both

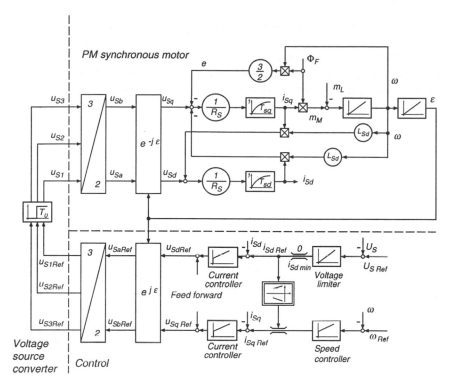

Fig. 14.10. PM-Synchronous motor drive with voltage source converter and current control in rotor coordinates

cases laminated rotor constructions would be used. These designs permit an easier displacement of the impressed flux by the direct stator current component i_{Sd} into geometrical leakage paths. Of course, a detailed modelling of the machine is much more complex but in priniple the result would correspond to the simplified model in Fig. 14.10 where different values of L_{Sd}, L_{Sq} in direct and quadrature axes are assumed, caused by varying cross sections of the magnetic paths and, possibly, saturation. In the design shown in Fig. 14.11 b1 a cage winding is included for self-starting of the motor on a constant frequency supply. With these measures PM synchronous motors may exibit a substantial degree of field weakening without excessive stator currents; this is necessary for instance on vehicle drives, reducing the need for a mechanical gear change.

Besides the PM synchronous motors having sinusoidal stator voltages and currents and smooth torque, i.e. low torque ripple in steady state, there are also more economical designs in the form of "brushless DC motors" , exibiting trapezoidal flux density distribution and phase voltages [S39, W14]. The stator currents have in principle a square waveform, as in Fig. 8.19 d, where one phase current is zero at any time; caused by the finite commutation intervals the currents are of course continuous, i.e. the corners are rounded off. The main advantage is that there is no need for a pulse-width-modulated converter and maximum power density is achieved; however, due to imperfect commutation there is considerable torque ripple making these motors less suited for high precision servo drives.

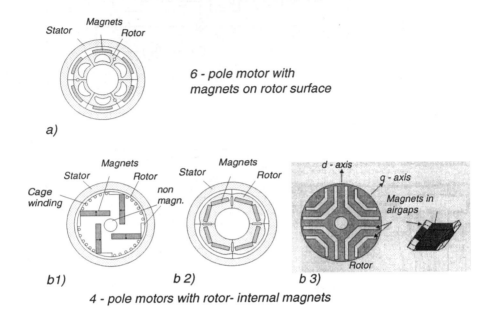

Fig. 14.11. Cross sections of PM synchronous motors with anisotropic rotor designs for extended field weakening range

14.2 Cycloconverter-fed Synchronous Motors with Field- and Damper-Windings

Another type of adjustable speed synchronous motor drives employs machines of conventional design having in the rotor a symmetrical damper winding and a single axis field winding fed with direct voltage through slip rings or from an AC exciter with rotating rectifiers. At low speed, for instance below 100 1/min, the stator can be supplied by a cycloconverter employing line commutation which makes this scheme particularly attractive for high power applications, as long as the frequency range of the cycloconverter suffices; the large number of thyristor branches is not objectionable at high power where parallel thyristors might otherwise be needed. The rotor has a single axis field winding and a symmetrical damper winding which serves for suppressing harmonics and negative sequence ampereturn waves, not for starting the motor because this is now achieved with the variable frequency supply. The basic circuit of this type of drive is shown in Fig. 14.12.

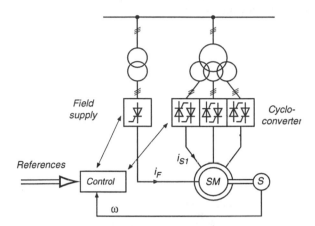

Fig. 14.12. Variable frequency synchronous motor drive with cycloconverter

Typical applications are large gearless drives, in the past dominated by DC machines supplied by dual converters (Fig. 9.4), for example for reversing rolling mills, mine hoists, ore mills and rotary cement kilns; at lower power, elevators or direct wheel drives on excavators in open-cast mines are also possible applications. The drive permits 4-quadrant operation with smooth transitions including standstill; hence it is perfectly suited for speed- as well as position- control requiring rapid response. The torque harmonics due to the cycloconverter are small and of relatively high frequency. An additional advantage is that cycloconverters with air- or liquid- cooling are proven designs from the DC era and are readily available up to MW rating.

A detailed analysis of the drive calls for an extended machine model including the converter that could not be dealt with manually or by intuition but would require a computer simulation [M17]. For qualitative purposes, however, the model derived in Sect. 10.1 still remains useful, as seen in Table 10.1 d. Hence we assume that the motor with a nonsalient cylindrical rotor has a symmetrical three-phase rotor winding into which, according to Fig. 14.13, an ideal source of direct voltage u_F is inserted as field power supply. Since this controllable voltage source contains no internal impedance, the damping effect of the rotor winding remains unimpeded and the symmetrical rotor winding is used for DC excitation [L42]. The simplified model is not offering the same freedom for choosing parameters as a separate winding arrangement but may be adequate for the design of the control system.

To adapt the mathematical model derived in Sect. 10.1 to this constraint, we again define with $\gamma = 2\pi/3$ a vector of externally applied rotor voltages

$$\underline{u}_R(t) = u_{R1} + u_{R2}\, e^{j\,\gamma} + u_{R3}\, e^{j\,2\,\gamma}$$

$$= u_{R1} - \tfrac{1}{2}\left(u_{R2} + u_{R3}\right) + j\,\frac{\sqrt{3}}{2}\left(u_{R2} - u_{R3}\right). \tag{14.16}$$

Because of the short circuit between phases 2 and 3,

$$u_{R2} = u_{R3}\,,$$

holds, hence

$$\underline{u}_R(t) = u_{R1} - u_{R2} = u_F(t)\,. \tag{14.17}$$

Thus the rotor voltage vector is real, being identical with the applied field voltage, Fig. 14.13.

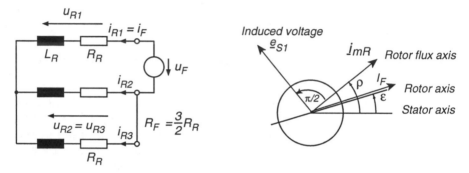

Fig. 14.13. Symmetrical rotor winding with DC excitation from a voltage source   **Fig. 14.14.** Definition of coordinate system

The rotor currents have mean values, to which transient terms i'_R are superimposed. Considering the isolated neutral of the winding, we find

$$\underline{i}_R(t) = (I_F + i'_{R1}) + \left(-\frac{I_F}{2} + i'_{R2}\right)e^{j\,\gamma} + \left(-\frac{I_F}{2} + i'_{R3}\right)e^{j\,2\,\gamma}$$

$$= \frac{3}{2}(I_F + i'_{R1}) + j\,\frac{\sqrt{3}}{2}(i'_{R2} - i'_{R3})\,. \tag{14.18}$$

The mean of the rotor current vector is also real,

$$\underline{i}_{R0} = \frac{3}{2}I_F = \frac{3}{2}\,\frac{U_F}{\frac{3}{2}R_R} = \frac{u_F}{R_R}\,, \tag{14.19}$$

but under dynamic conditions the vector can be deflected by induced current components.

Best results of the drive control may be expected if the stator currents are again current-sourced by the cycloconverter with the help of fast current control loops, as was shown in Sect. 13.1. This leaves the rotor voltage equation and the mechanical equations for modelling the drive dynamics, Eqs.(10.51–10.53),

$$u_F = R_R \, \underline{i}_R + L_R \frac{d\underline{i}_R}{dt} + L_0 \frac{d}{dt}\left(\underline{i}_S \, e^{-j\varepsilon}\right), \tag{14.20}$$

$$J \frac{d\omega}{dt} = \frac{2}{3} L_0 \, \mathrm{Im}\left[\underline{i}_S \, (\underline{i}_R \, e^{j\varepsilon})^*\right] - m_L, \tag{14.21}$$

$$\frac{d\varepsilon}{dt} = \omega. \tag{14.22}$$

The design of a control scheme in field coordinates calls for the selection of a suitable frame of reference. In the case of the synchronous motor with permanent magnets, Sect. 14.1, the rotor position ε was the natural choice as this was the orientation of the magnets producing the rotor ampereturns; however, the situation is different here, where the induced currents can shift the ampereturns-wave with respect to the rotor, Fig. 14.8. It is therefore appropriate to employ a flux vector as with the induction motor. A suitable choice is the rotor flux, defined in Eq. (12.22),

$$\underline{i}_{mR} = \underline{i}_S + (1 + \sigma_R)\underline{i}_R \, e^{j\varepsilon} = i_{mR} \, e^{j\varrho}. \tag{14.23}$$

This makes sense because the rotor winding in Fig. 14.13 has all the features of an induction machine; the only difference is that in steady state the magnetising current vector $\underline{i}_{mR}$ is in synchronism with the rotor, $\varrho - \varepsilon = \text{const.}$, whereas it was moving with slip frequency across the rotor of the induction motor, Fig. 12.11.

Eliminating $\underline{i}_R$ from Eqs. (14.20,14.23)

$$T_R \frac{d\underline{i}_{mR}}{dt} + (1 - j\omega\,T_R)\,\underline{i}_{mR} = \underline{i}_S + (1 + \sigma_R)\frac{u_F}{R_R}\,e^{j\varepsilon}, \tag{14.24}$$

and splitting in two real differential equations yields

$$T_R \frac{di_{mR}}{dt} + i_{mR} = i_{Sd} + (1 + \sigma_R)\frac{u_F}{R_R}\,\cos(\varrho - \varepsilon), \tag{14.25}$$

$$\frac{d}{dt}(\varrho - \varepsilon) = \omega_{mR} - \omega$$

$$= \frac{1}{i_{mR}\,T_R}\left[i_{Sq} - (1 + \sigma_R)\frac{u_F}{R_R}\,\sin(\varrho - \varepsilon)\right], \tag{14.26}$$

where

$$\underline{i}_S \, e^{-j\varrho} = i_S \, e^{j\,(\varsigma - \varrho)} = i_{Sd} + j\,i_{Sq} \tag{14.27}$$

is again the stator current vector in rotor flux coordinates.

Since the vector $\underline{i}_{mR}$ moves in steady state synchronously with the rotor, the right hand side of Eq. (14.26) vanishes,

$$i_{Sq}|_{t\to\infty} = (1 + \sigma_R)\, \frac{u_F}{R_R}\, \sin(\varrho - \varepsilon)|_{t\to\infty}\,. \tag{14.28}$$

The equation (12.27) for the driving torque remains unchanged,

$$m_M = \frac{2}{3}\, L_0\, \mathrm{Im}\,[\underline{i}_S\,(\underline{i}_R\, e^{j\,\varepsilon})^*] = \frac{2}{3}\,(1 - \sigma)\, L_S\, i_{mR}\, i_{Sq} = k\, i_{mR}\, i_{Sq}\,. \tag{14.29}$$

The equations describing the synchronous machine in field coordinates are graphically represented by the block diagram in the upper part of Fig. 14.15, containing once more the transformations from the stator- or rotor- orientated coordinate system into field coordinates. The diagram corresponds to that of the induction motor with the exception of the field voltage u_F which, after transformation into flux-orientated $d - q$-components, is added as an additional input signal at the appropriate points. In practice u_F may be supplied by a separate two-quadrant converter with closed loop voltage control, thus underlining the characteristics of a low impedance source. It would of course also be possible to introduce a control loop for the field current to eliminate the effect of changing field winding resistance due to temperature variations. The drive would then respond faster in the field weakening range, but the damping effect by transient currents induced in the low impedance field circuit might be impeded; also the model of the machine would have to be modified.

In contrast to the PM excited machine in Fig. 14.6 there is no need to apply stator current in the d-axis for field weakening, since this can now be achieved more efficiently by reducing the field voltage rather than increasing the stator current. The need to reduce the field becomes manifest again when the magnitude of the stator voltage approaches its limit as sensed by a limiting voltage controller; alternatively field weakening could be performed with an i_{mR}-control loop, the reference of which is generated as a nonlinear function of speed, $i_{m\,R\,\mathrm{Ref}}(\omega)$, Fig. 12.14.

The flux angle ϱ required for modulation and for computation of $i_{m\,R}$, $i_{S\,q}$ and m_M is again derived from a flux model, processed on-line by a microcomputer. As in case of the induction machine, the model may be defined in stator or in field coordinates, Fig. 12.16 a,b. Computational advantages are gained by taking the second approach, referring to Eqs. (14.25, 14.26), which describe the magnitude of the flux vector and its position relative to the rotor. This is represented by the block diagram in Fig. 14.16 which is a copy of the motor model in Fig. 14.15. Clearly, with $u_F = 0$, the flux model of the induction machine, Fig. 12.16 b, results. The angle $\varrho - \varepsilon$ is approximately constant in steady state.

The accuracy with which the stator currents are tracking the references may again be considerably improved by employing voltage feed-forward; its purpose is to cancel the voltage $\underline{e}_S$ behind the transient reactance which is induced by the magnetising current, i.e. the main flux.

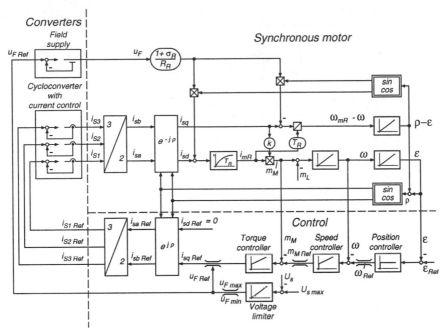

Fig. 14.15. Control of synchronous motor in field-coordinates

The stator voltage was derived in Eq. (12.45)

$$\underline{u}_S(t) = R_S\,\underline{i}_S + \sigma\,L_S\,\frac{d\underline{i}_S}{dt} + (1-\sigma)\,L_S\,\frac{d\underline{i}_{mR}}{dt}\,,\tag{14.30}$$

which in steady state, i.e. with

$$\omega_{mR} = \omega_1 = \omega = \text{const}, \quad i_{mR} = \text{const}, \quad i_S = \text{const}$$

assumes the form

$$\underline{u}_S(t) = (R_S + j\,\omega\,\sigma\,L_S)\,\underline{i}_S + \underline{e}_S(t)\,;\tag{14.31}$$

hence the induced voltage is

$$\underline{e}_S(t) = j\,\omega\,(1-\sigma)\,L_S\,\underline{i}_{mR}\,.\tag{14.32}$$

If the control functions are performed by a microcomputer, the sinusoidal three-phase components of $\underline{e}_S$ can be easily computed and fed into the current loops behind the current controllers as is shown in Fig. 14.17. Experimental results were seen in Fig. 13.8.

If the current controllers are of the analogue version, being part of the power converters, the scheme in Fig. 14.17 a would call for a second D/A converter that can be omitted by mixing the signals within the microcomputer, as shown in Fig. 14.17 b. Of course, perfect cancellation of $\underline{e}_S$ is not

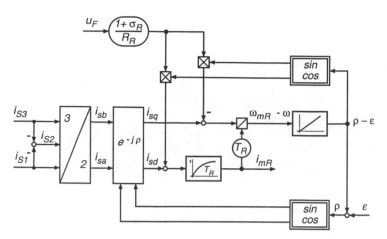

Fig. 14.16. Dynamic model for computing the rotor flux of synchronous machine

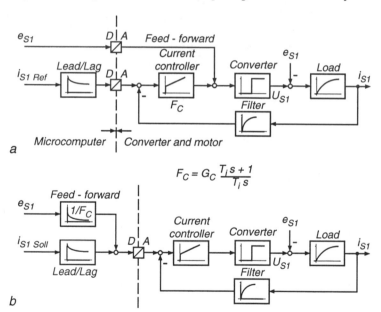

$$F_C = G_C \frac{T_i s + 1}{T_i s}$$

Fig. 14.17. Current control with voltage feed-forward

possible with any of the schemes since the nonlinear control dynamics of the converter separate the point of compensation from the actual entry point of the disturbance. Still, considerable improvements are possible, as was shown in Fig. 13.8. The residual lag of the current loops may be further reduced by inserting lead-lag filters at the reference inputs of the current controllers; this is indicated in Fig. 14.17.

An alternative approach to the control shown in Fig. 14.15 would again be to control the stator currents in field coordinates, according to Fig. 12.27 where the current controllers are processing DC signals in steady state.

14.3 Synchronous Motor
with Load-commutated Inverter (LCI-Drive)

The variable frequency synchronous motor drive with cycloconverter supply discussed in the preceding paragraph is suitable for large drives with high dynamic performance. Since the cycloconverter is line-commutated, only converter-grade thyristors are needed, although in a large quantity. On the other hand there is the frequency limitation of cycloconverters; with a 50 Hz supply and 6-pulse circuits the range extends to about $f_1 = 20\,Hz$. This is sufficient for a rolling mill motor having a base speed of 60 1/min and a maximum speed of 120 1/min; it could be realised with up to 20 poles at practically any power rating [S4, S79].

However, when large high speed drives are needed, for example 4500 1/min on a compressor for a pipe-line or a blast furnace, a maximum stator frequency of 75 Hz would be needed even with a two pole motor, so a cycloconverter can offer no solution; neither could a DC drive have been built for this duty. The rotor of the synchronous motor is made of solid steel, similar to that of a turbogenerator.

The simplest way to eliminate the restrictions imposed by the line frequency is a two-stage conversion, inserting a DC link for decoupling of the line- and the machine-side converters. Fig. 11.14 showed such a circuit containing only 12 thyristors; the machine-side converter can be further simplified with a synchronous motor because the stator winding contains a voltage source $\underline{e}_S$ in series with a transient impedance, Eq. (14.31). Hence the machine-side converter can be commutated by the load without resorting to forced commutation as was necessary with the induction motor; in other words the synchronous machine is operated in the overexcited region so that it can supply the reactive current neeeded by the phase-controlled machine-side converter. The basic circuit of a load-commutated inverter drive (LCI-drive) is seen in Fig. 14.18.

This leaves the problem of commutation at start-up, where the induced voltage is insufficient, but this can be circumvented by intermittently blanking the link current with the help of the line-side converter. This produces torque pulsations, but there are many applications where smooth operation in the very low speed range, say below 10% of nominal speed, is not necessary; examples are the turbo-compressors mentioned before, also gas-turbine or pumped storage sets, which are started with the synchronous generator acting as a motor.

Outside a very low speed range, where the current i_D must be made intermittent to allow commutation of the machine-side converter, the drive

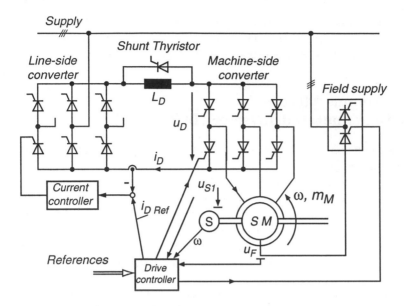

Fig. 14.18. Six-pulse converter with DC link
for supplying a synchronous machine

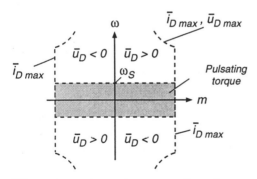

Fig. 14.19. Operating range of synchronous motor drive with direct current link

can operate smoothly in all four quadrants of the torque/speed plane, as is illustrated in Fig. 14.19. During regeneration the voltage $\bar{u}_D$ is reversed by delayed firing of the line-side converter whereas the direction of rotation is determined by the sequence of the firing pulses for the machine-side converter. The voltage $\bar{u}_D$ rises with the speed up to the limit given by the converter, a further increase of speed is then achieved by reducing the field current, i.e. by field weakening as with a DC drive. The line interactions of this type of drive are also similar to those of a converter-fed reversible DC drive, Chap. 8.

The basic mode of operation of the machine-side converter corresponds to that described in Sect. 8.3, except that the firing of thyristors must be synchronised to the variable frequency source voltage $\underline{e}_S$ which is performing the commutation. Outside the commutation intervals the stator current, flowing through two phases of the motor, is current-sourced by the DC link; its magnitude is maintained by closed loop control via the line-side converter.

The stator current has approximately the same waveform consisting of 120° blocks as seen in Fig. 11.15 a; one stator phase carries no current. The distorted waveform of the currents, which also produces torque pulsations, and the need for low impedance commutating paths make it desirable to provide continuous damper windings which assist the formation of a smoothly moving rotational flux wave, as was shown in Fig. 12.30 and 12.37.

The combination of

- current sources in the stator (except during commutation, where two terminals are short-cicuited through the converter), and
- low impedance rotor circuit

indicates that a detailed analysis of the system is complex and could only be handled by digital simulation [M17]; however, by approximation the same mathematical model of the machine can be employed as in the preceding section, Eqs. (14.20-14.22). The flux vector, Eq. (14.23), the induced voltage, Eq. (14.32) and the computational model for the rotor flux, Fig. 14.16, also remain valid.

Despite this similarity of the plant structure the field orientated control strategies illustrated in Figs. 12.33 and 14.15 are not directly applicable; this is so because the stator currents cannot be determined at will, as was possible with the current source converter employing forced commutation or the cycloconverter with line commutation and current control. Instead, firing of the thyristors in the machine-side converter must now be so timed as to

- permit safe commutation,
- result in minimum link current for a given torque; the line-side converter would otherwise operate at an unnecessarily low power factor.

These are overriding conditions which can best be fulfilled by always operating the machine-side converter at the limits of the control range, i.e.

- zero firing delay ($\alpha = 0$) when regenerating or
- maximum firing delay (α_{max}) in the motoring region.

The firing delay α is defined with respect to the induced voltages e_S behind the transient impedance.

The two modes of operation are sketched in Fig. 14.20 a,b. Clearly the commutation interval and the fact that the commutation must be completed before $\tau = \pi$, prevent operation with purely quadrature current. The situation is critical in motor operation with the machine-side converter serving as an inverter because the extinction angle γ, explained in Fig. 8.17, should be small in order not to reduce the torque unnecessarily but on the other hand it must be large enough to allow safe recovery of the outgoing thyristor. If

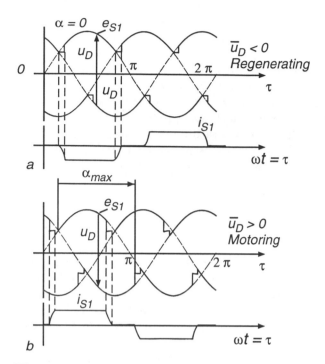

Fig. 14.20. Current and voltage waveforms of machine-side converter with natural commutation. **(a)** regenerating, **(b)** motoring

the extinction angle becomes too small, a commutation failure may be the consequence; while this would not constitute a major disturbance, with the link reactor preventing excessive rise of the current, it would still cause a severe torque transient until the current control loop has restored the link current and the inverter returns to its regular firing sequence; the incident is similar to the misfiring of a six-cylinder combustion engine.

Another problem is the synchronisation of the machine side converter; timing of the firing circuit with respect to the terminal voltages of the machine would be unsatisfactory since these voltages are highly distorted and out of phase with the nearly sinusoidal induced voltages e_S.

There are two possibilities for detecting e_S, the driving voltages of the commutating circuits. One is based on Eq. (14.31),

$$\underline{e}_S(t) = (1 - \sigma) L_S \frac{d\underline{i}_{mR}}{dt} = \underline{u}_S(t) - R_S \underline{i}_S - \sigma L_S \frac{d\underline{i}_S}{dt} , \qquad (14.33)$$

which could be evaluated with analogue circuitry or a microcomputer. A second option is to use the flux model in Fig. 14.16, where two orthogonal stator currents i_{Sa}, i_{Sb}, the angular rotor position ε and the field voltage u_F represent the input signals; this would definitely require a microcomputer for executing the various nonlinear functions. The first of the two approaches was chosen for a laboratory drive, as it seems to be more straight forward and less sensitive to uncertain parameters of the motor.

If the reconstruction of the induced motor voltages e_S is of sufficient accuracy, a closed loop control for the extinction angle could also be envisaged, as is common practice on high voltage DC transmissions (HVDC). However, the draw-back even of sophisticated nonlinear extinction angle control schemes is that they cannot entirely exclude commutation failures because there may always be late disturbances that are not taken into account when the firing instant is determined [F21, 77]. The problem with every extinction margin control is that only hindsight can tell with certainty, whether the previously calculated firing angle was adequate.

A major difference between the firing control schemes for the line- and the machine-side converters is that the latter has to function with both directions of sequence and over a wide frequency range; this is a strong incentive to employ digital methods, where a variable clock frequency may be used to change the slope of saw-tooth signals etc.

The commutation itself is nearly frequency independent, as long as the resistance of the commutation circuit can be neglected, because the driving voltages increase with speed and the time available for commutation is reduced inversely with speed, resulting in an approximately constant commutation angle τ_c over a wide speed range.

At very low speed, however, this is no longer true because the resistive voltage component consumes an increasing portion of the available voltage-time area; the minimal speed, where natural commutation ceases to function is usually below 10% of nominal speed. Operation in this speed range is still possible when reverting to intermittent link current; this is achieved by commanding zero link current, to which the current controller responds by temporarily shifting the firing pulses of the line-side converter to the commutation limit, i.e. applying maximum negative voltage to the DC-link. As soon as the current in the stator windings has decayed to zero and the recovery time for the thyristors has elapsed, the next thyristor-pair of the machine-side converter is made conducting and the link current can be restored.

The delay necessary for reducing and building up the link current can be considerably reduced if a shunt thyristor in parallel with the link inductor, Fig. 14.18, is fired at the time, when the current reference is blanked [L1]; this short-circuits the link inductor and allows the current to circulate freely without affecting the commutation; as soon as the machine-side converter is switched to the new conducting state and the line-side converter is reactivated, the shunting thyristor is quickly blocked by the voltages in the DC link. This allows a substantial reduction of the blanking time intervals in the low speed region and reduces the effects of pulsating torque.

It was mentioned above in connection with Fig. 14.20 that the machine-side converter, operating with natural commutation, invariably causes a certain amount of current in the d-axis, i.e. unintentional load dependent armature reaction. This can be offset by a corresponding change of field current. One possibility is to control the magnitude of the induced motor voltages e_S to a reference value rising with speed; when limiting the voltage reference above base speed, field weakening is initiated.

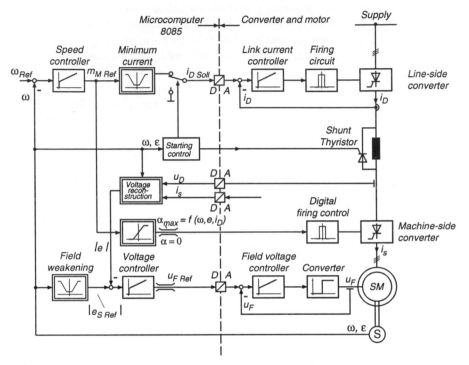

Fig. 14.21. Microcomputer based speed control of LCI synchronous motor drive with DC link converter

A control scheme incorporating these features is shown in Fig. 14.21. The power circuit contains four converters of different types and sizes.

- Line-side converter,
- Machine-side converter,
- Shunt thyristor,
- Field power supply,

for which the microcomputer is producing coordinated control commands. They are the reference values for the DC link current and field voltage as well as signals for the digital firing circuit of the machine-side converter and the shunting thyristor. The function generator prescribing the link current reference has the purpose of assuring continuous current; in the minimal current mode control is effected through the firing angle of the machine-side converter which quickly assumes its upper or lower limit $\alpha = 0$ or $\alpha = \alpha_{max}$ as the link current reference begins to respond to increased torque demand. The inverter limit angle α_{max} must be continuously adjusted to achieve minimal but safe extinction time; it can be determined either by open loop computation based on current, voltage and speed or by closed loop control, possibly with feed-forward compensation from the current. The control loop for the field voltage u_F is included in Fig. 14.21 which corresponds to the low impedance rotor model in Fig. 14.13. Also, control of the field voltage responds more favourably when transient currents are induced in the field winding when load is applied. A control loop for the field current i_F would tend to temporarily reduce the field voltage at a time, when it should be increased. On the other hand, control of the field current would eliminate the effect of changing field winding resistance due to temperature variations and improve the dynamic response in the field weakening region.

The sequence of the firing pulses for the machine-side converter can be based on the estimated angular position of the flux model, Fig. 14.16, or on the rotor position ε, if a voltage model according to Eq. (14.33) is used, whose accuracy is questionable at low speed.

An important feature is the starting control algorithm in the very low speed range by temporarily blanking the link current via the line-side converter and firing the shunt thyristor. This is a convenient way of quickly altering the mode of operation, such as torque reversal at low speed which could take too long when waiting for the next natural commutation.

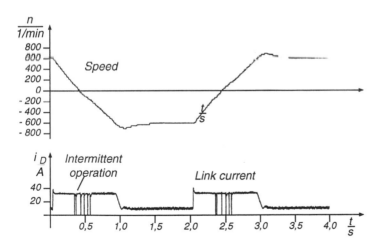

Fig. 14.22. Reversing transient of a 20 kW synchronous motor drive at half nominal speed

A 20 kW synchronous motor drive with microcomputer control has been designed in the laboratory and tested [R16, R17, R18]. Figure 14.22 depicts a recorded reversing transient at half nominal speed with the inertia of the drive increased by a dynamometer without load. The intermittent link current at low speed as well as the upper and lower current limits are clearly visible. The oscillogram demonstrates that the dynamic response is slower than could be expected with a field orientated type of control. However, there are many applications where fast response is not a primary concern.

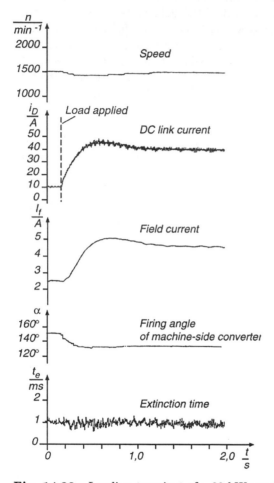

Fig. 14.23. Loading transient of a 20 kW synchronous motor drive

Further details are seen in Fig. 14.23 where load torque is applied; the rise of the field current is initiated by a transient and then maintained by the control loop for the induced motor voltages e_S, thus neutralising the effects of armature reaction. The trace of the extinction time indicates the good performance of this critical control function.

Finally, Fig. 14.24 shows the effect of a deliberately triggered commutation failure which causes a rapid rise of the link current. However, the current

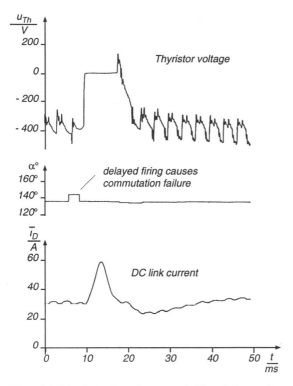

Fig. 14.24. Intentional commutation failure of machine-side converter

controller is able to quickly restore the link current so that the machine-side inverter can resume firing in the regular pattern.

As seen in Table 11.1 synchronous motor LCI-drives are being used up to the highest power ratings, mainly for compressors, gas turbines and pumped storage sets. Since the usual method of reducing resistive losses by increasing the motor voltages cannot be applied indefinitely because of the winding isolation of the motor (the usual limit is around 10 kV), converters in parallel connection are often chosen, as is shown in Fig. 14.25 for 12-pulse operation, at the same time reducing torque pulsations. For this purpose the motor with solid steel rotor is built with two separate stator windings, displaced by 30° electrically, which are fed from two separate DC link converters. At the line side, a corresponding phase shift may be produced by a star/delta connected transformer. Brushless excitation of the motor with an AC exciter and rotating rectifiers, common on large turbogenerators, is also indicated. Fig. 14.26 demonstrates the control scheme, where the speed controller supplies equal current references to the separate link current controllers; the back voltages u_{D1}, u_{D2} active in the DC links, are shifted according to the geometrical positions of the motor windings.

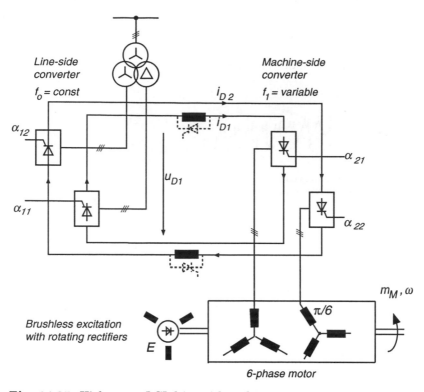

Fig. 14.25. High power LCI-drive with 6-phase motor

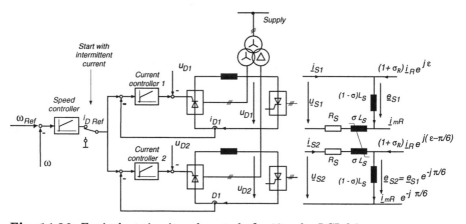

Fig. 14.26. Equivalent circuit and control of a 12-pulse LCI-drive

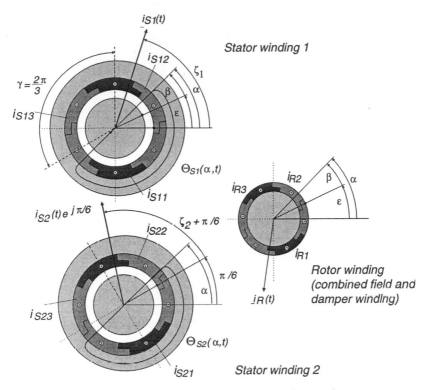

Fig. 14.27. Simplified winding scheme of a two pole, 6-pulse synchronous motor

The arrangement of the windings is demonstrated in Fig. 14.27 in simplified form; whereas in the rotor a symmetrical AC winding serves as a
combined field- and damper- winding, there are two electrically isolated and
by $\pi/6 = 30°$ displaced two pole AC windings in the stator; in accordance
with the simplifications introduced in Sect. 10.1 they are modelled as airgap-
windings. The pertinent ampereturns waves are, as in Eq. 10.2,

$$\Theta_{S1}(\alpha, t) = N_S[i_{S11}(t)\vartheta_S(\alpha) + i_{S12}(t)\vartheta_S(\alpha - \gamma) +$$
$$+ i_{S13}(t)\vartheta_S(\alpha - 2\gamma)], \tag{14.34}$$

$$\Theta_{S2}(\alpha, t) = N_S[i_{S21}(t)\vartheta_S(\alpha - \pi/6) + i_{S22}(t)\vartheta_S(\alpha - \pi/6 - \gamma) +$$
$$+ i_{S23}(t)\vartheta_S(\alpha - \pi/6 - 2\gamma)], \tag{14.35}$$

where $-1 < \vartheta_S(\alpha) < +1$ are dimensionless periodic functions describing the
distributions of the windings.

Two complex stator current vectors are defined for the two windings,

$$\underline{i}_{S1}(t) = i_{S1}(t)\, e^{j\zeta_1(t)}$$
$$= i_{S11}(t) + i_{S12}(t)\, e^{j\gamma} + i_{S13}(t)\, e^{j2\gamma} \tag{14.36}$$
$$= i_{S1a}(t) + j i_{S1b}(t)$$

and

$$\underline{i}_{S2}(t) = i_{S2}(t)\, e^{j\zeta_2(t)}$$
$$= i_{S21}(t) + i_{S22}(t)\, e^{j\gamma} + i_{S23}(t)\, e^{j2\gamma} \tag{14.37}$$
$$= i_{S2a}(t) + j i_{S2b}(t)\,.$$

When including the conditions for isolated neutral points and neglecting spatial harmonics, i.e. with $\vartheta_S(\alpha) = cos\alpha$, two travelling waves result

$$\Theta_{S1}(\alpha, t) = N_S/2\,[\underline{i}_{S1}(t)e^{-j\alpha} + \underline{i}_{S1}(t)^* e^{j\alpha}]$$
$$= N_S Re\,[\underline{i}_{S1}(t)e^{-j\alpha}] \tag{14.38}$$
$$= N_S i_{S1}(t) cos[\alpha - \zeta_1(t)],$$

$$\Theta_{S2}(\alpha, t) = N_S/2\,[(\underline{i}_{S2}(t)e^{j\pi/6})e^{-j\alpha} + (\underline{i}_{S2}(t)e^{j\pi/6})^* e^{j\alpha}]$$
$$= N_S Re\,[\underline{i}_{S2}(t)e^{j\pi/6}e^{-j\alpha}] \tag{14.39}$$
$$= N_S i_{S2}(t) cos[\alpha - (\zeta_2(t) + \pi/6)],$$

which are superimposed in the airgap,

$$\Theta_S(\alpha, t) = \Theta_{S1}(\alpha, t) + \Theta_{S2}(\alpha, t)$$
$$= N_S[i_{S1}(t) cos(\alpha - \zeta_1(t)) + i_{S2}(t) cos(\alpha - (\zeta_2(t) + \pi/6))]. \tag{14.40}$$

A corresponding definition holds for the rotor with the mechanical angle $\varepsilon(t)$,

$$\Theta_R(\alpha, \varepsilon, t) = N_R Re\,[\underline{i}_R(t)e^{j\varepsilon}e^{-j\alpha}] = N_R i_R(t) cos[\alpha - \xi(t) - \varepsilon(t)], \tag{14.41}$$

where

$$\underline{i}_R(t) = i_R(t)e^{j\xi(t)} \tag{14.42}$$

is the vector of the rotor current. Linear superposition of the ampereturns produces, assuming the leakage factor σ_S, a radial flux density wave B_S at the stator side of the airgap with length h

$$B_S(\alpha, \varepsilon, t) = (\mu_o/2h)[(1 + \sigma_S)(\Theta_{S1}(\alpha, t) + \Theta_{S2}(\alpha, t)) + \Theta_R(\alpha, \varepsilon, t)]. \tag{14.43}$$

The leakage flux cannot be accurately modelled with the simplifications applied; the two stator windings are assumed to be firmly coupled but this is of minor importance since they are fed by current sources.

With the definitions in Sect. 10.1 the flux vectors of the two stator windings are obtained

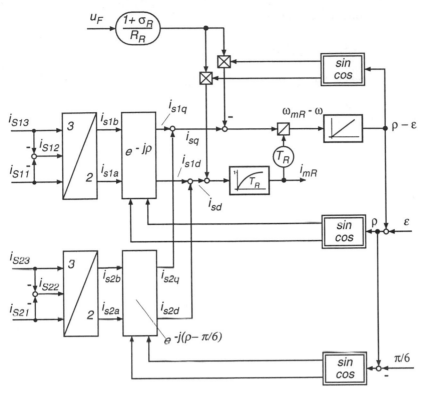

Fig. 14.28. Flux model of a 6-phase synchronous motor

$$\underline{\Psi}_{S1}(t) = L_o \left[(1 + \sigma_S)(\underline{i}_{S1} + \underline{i}_{S2}e^{j\pi/6}) + \underline{i}_R e^{j\varepsilon} \right] \tag{14.44}$$

and

$$\underline{\Psi}_{S2}(t) = L_o \left[(1 + \sigma_S)(\underline{i}_{S1}e^{-j\pi/6} + \underline{i}_{S2}) + \underline{i}_R e^{j(\varepsilon - \pi/6)} \right]$$
$$= \underline{\Psi}_{S1}(t)\, e^{-j\pi/6} \tag{14.45}$$

resulting in the flux model in Fig. 14.28. The pertinent voltage equations are

$$\underline{u}_{S1}(t) = R_S \underline{i}_{S1} + \frac{d}{dt}\underline{\Psi}_{S1}$$
$$= R_S \underline{i}_{S1} + L_S \frac{d}{dt}\left[\underline{i}_{S1} + \underline{i}_{S2}e^{j\pi/6} \right] + L_o \frac{d}{dt}\left[\underline{i}_R e^{j\varepsilon} \right] \tag{14.46}$$

and

$$\underline{u}_{S2}(t) = R_S \underline{i}_{S2} + \frac{d}{dt}\underline{\Psi}_{S2} \tag{14.47}$$
$$= R_S \underline{i}_{S2} + L_S \frac{d}{dt}\left[\underline{i}_{S1}\, e^{-j\pi/6} + \underline{i}_{S2} \right] + L_o \frac{d}{dt}\left[\underline{i}_R e^{j(\varepsilon - \pi/6)} \right],$$

The rotor voltage equation is

$$u_F(t) = R_R \underline{i}_R(t) + \frac{d}{dt}\underline{\Psi}_R$$

$$= R_R \underline{i}_R + L_o \frac{d}{dt}\left[(1+\sigma_R)\underline{i}_R + (\underline{i}_{S1} + \underline{i}_{S2}e^{j\pi/6})e^{-j\varepsilon}\right], \qquad (14.48)$$

where $u_F(t)$ is again an assumed impressed voltage source serving for DC excitation. The design of the control corresponds to that shown with the 6-pulse drive.

15. Some Applications
of Controlled Electrical Drives

The preceding chapters were dealing mainly with the different types of electrical drives and their control; applications were only mentioned as they affected the operation of the machines and the associated equipment. Also, the driving specifications of a mechanical load are normally not met by just one type of electrical drive and the variety of applications can be bewildering; in this chapter some problems associated with controlled electrical drives will be explained in more detail. For this we begin with a 4-quadrant drive, be it DC or AC, the basic structure of which is contained within the dashed lines in Fig. 15.1. The moving masses are at first assumed to be rigidly coupled, forming a lumped inertia. The inner loop which comprises the power converter and part of the electrical machine assures fast torque control; with an integrating controller it exhibits unity gain and serves for linearisation. Once the torque loop with the equivalent lag T_e is closed, there is little difference between a DC and an AC drive.

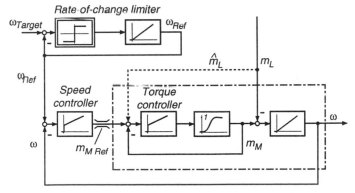

Fig. 15.1. Speed control loop with rate-of-change-limiter and load feed-forward

Torque control is mandatory on high performance drives except for the smallest power ratings, because torque serves as the controlling input to any mechanical plant. Whenever load torque must be counteracted or the speed is to be changed, it is only possible through the torque reference; hence the response of the torque control loop limits the control bandwidth of the complete drive. In view of the practical difficulties of measuring mechanical torque over a wide frequency range it is most desirable to have an electrical substitute available as feedback signal; since torque control is usually embedded in higher level controls offering a corrective input, steady state precision of the torque signal is not a primary concern as long as the dynamic behaviour of the real motor torque is properly reflected. By limiting the torque reference, protection of the power converter, the motor and the mechanical load is achieved.

Sometimes it is possible to extract an estimate of the load torque, $\hat{m}_L$, from a load model and feed it to the input of the torque controller as a disturbance feed-forward. This is called inverse load modelling; it results in the best possible dynamic response of the drive that could only be improved with a faster actuator, i.e. an increased control band-width.

For the following the closed torque loop is considered to be part of the drive equipment. For safety reasons it is normally not accessible to the user because its internal functions may be too involved and subject to critical adjustments.

15.1 Speed Controlled Drives

By superimposing to the torque control a speed control loop the most frequently used drive control scheme results; its steady state characteristics areS drawn in Fig. 2.6, where the torque limit may be adjusted to twice nominal torque. When the speed reference ω_{Ref} is manually set, it is normally specified on larger drives that speed changes should take place with less than maximum torque; on the other hand the control should respond to load changes as quickly as possible and with full torque. These at first sight contradictory demands are met by making the speed loop as fast as is practical but including a rate-of-change limiter at the reference input, outside the control loop. This is seen in Fig. 7.11, also showing transients of the speed reference following a step change of the speed target-point. The rate-of-change limiter may be interpreted as a simple dynamic model having a desirable and safe response, no matter how quickly an impatient operator may be changing the speed target speed. Similar provisions are of course also necessary if the target speed is generated automatically; for example on programmed rolling mill or machine tool drives where the various controlled motions have to be coordinated to achieve the desired end result, for instance an accurate spatial trajectory of a robot hand.

When the mechanical plant is of large physical dimension, it is often desirable to share the load between several motors. Fig. 15.2 shows the example of a multiple motor drive on rotary printing presses, where eight or more individual printing stations may be mechanically coupled by a long drive shaft.

To avoid the transmission of large fluctuating torques through the shaft which would result in undesirable torsional motions, each section is driven by an individual motor that is supposed to carry most of the sectional load so that only small synchronising torques are transmitted through the drive shaft; the load in the different sections and, correspondingly, the motor sizes may be different. Similar problems arise with multiple motor drives on paper machines of earlier design having a mechanical drive shaft or on trains with multiple drives.

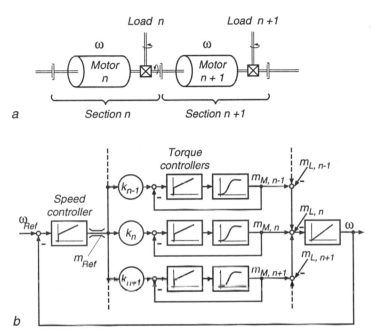

Fig. 15.2. Multiple motor drive with mechanical drive shaft
(a) Mechanical system; (b) Block diagram

Since the motors in Fig. 15.2 a are semi-rigidly coupled, only one speed controller is needed for which the feedback signal is taken from a suitably placed speed sensor. (Each motor is usually equipped with a tachometer because the drive system may have to be rearranged by opening and closing the clutches.) Sharing of the total torque is achieved by the torque controllers which receive their shares (k_n) of the total torque reference from the common speed controller; a simplified block diagram of this control scheme is drawn in Fig. 15.2 b.

When the motors of a multiple drive system are only loosely connected, a different situation exists, where each motor needs its own speed controller. This problem arises for instance on large paper machines without mechanical drive shaft extending over a length of 100 m and containing dozens of individual drive motors; similar structures are found in continuous hot strip rolling mills and many production processes in the fibre and textile industry.

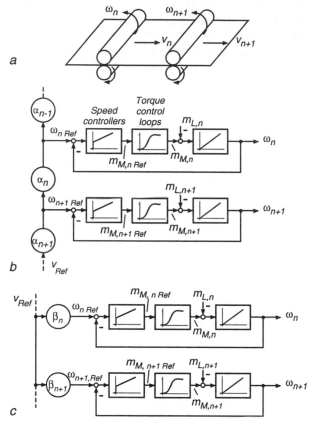

Fig. 15.3. Sectional drive system for continuous production process
(a) Mechanical system;
(b) Block diagram with serial reference chain;
(c) Block diagram with parallel reference structure

   A characteristic of such drives is that the individual sections are only lightly coupled through a strip of red-hot steel or a continuous web of thin material that cannot sustain any appreciable forces; on the other hand, the material must be kept sufficiently taut to avoid creases or folds. Fig. 15.3 shows a schematic example of a continuous process with a multiple drive system. The velocity of the material is usually increasing as it proceeds through the different stations; this is obvious with the continuous rolling mill, where the metal sheet becomes thinner and longer when passing through each pair of rolls but applies equally to a paper machine, even though the speed increase may only be a fraction of one percent in each section. The speed of each motor must therefore be maintained in precise relative synchronism with the preceding and subsequent sections (ignoring constant or slowly varying scale factors such as gear ratios or roll diameters). The speed ratios must be accurately maintained independent of the speed level that may be changing during the process and with the type of production [K28].

A proven strategy for this type of drive system is to derive the speed references in a serial form, so that the ratios of reference speeds in neighbouring sections,

$$\alpha_n = \left. \frac{\omega_n}{\omega_{n+1}} \right|_{\text{Ref}} ,$$

remain constant even if the speed ratio may have been changed in another section of the drive. This control principle, called "progressive draw" is indicated in Fig. 15.3 b; a change of the overall velocity reference v_{Ref} or of a preceding ratio simultaneously affects all following drives in the desired way. The idea of achieving "progressive draw" by deriving the speed reference $\omega_{n\,\text{Ref}}$ from the measured speed ω_{n+1} of the preceding drive can be dismissed because after a few repetitions this would result in poorly damped and even unstable transients, similar to a queue of motor cars where the attention of each driver is solely fixed on the rear of the leading vehicle.

If all the speed references $\omega_{n\,\text{Ref}}$ were derived in parallel from the common velocity reference v_{Ref}, a change of one speed ratio would require all subsequent settings to be changed as well; this scheme which has merits in other applications is sketched in Fig. 15.3 c.

The control variables are usually processed with analogue means including operational amplifiers or precision potentiometers but digital methods are becoming more widely used; the main reasons are

- Better accuracy,
- Possibility of storing reference data and
- Ease of recording and evaluating measurements.

The improved accuracy of digital methods is due to the fact that speeds can be measured to high resolution by counting angular increments and that very precise reference frequencies are readily available from crystal oscillators ($\Delta f / f_0 < 10^{-6}$); also digital controllers exhibit no drift effects. An example of a digital speed control is described in Sect. 15.3; typically, the specified error of speed ratios on a large paper machine producing 10 m wide paper at a velocity of 30 m/s may be $< 10^{-4}$/day, which would be very difficult to attain with analogue methods.

As was mentioned in Chap. 2 the assumption of a lumped inertia may not apply to some drives where the mechanical construction results in appreciable elasticity. This could be due to a long drive shaft or gears separating the motor from the load; the same is true for cranes or hoists where the elasticity of the rope is not negligible [J18, R2]. Another example are robots, where elasticity is the consequence of a lightweight construction. A model of this type of drive is depicted in Fig. 15.4 together with a simplified linear block diagram. K is the torsional spring constant of the flexible transmission; its inertia as well as its inherent damping are neglected, $\varepsilon_1 - \varepsilon_2$ is the torsional displacement of the shaft. The load torque may again consist of various components including lift torque and friction, it is represented here by a variable independent torque m_L. In practice part of the load torque would be speed-dependent and, hence, could exert a damping effect.

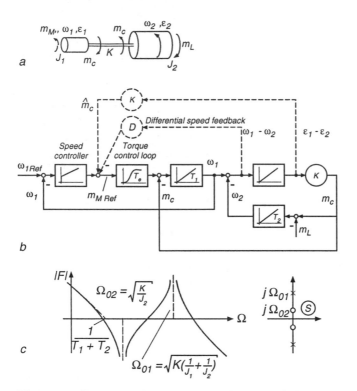

Fig. 15.4. Two-mass drive system with torsional elasticity
(a) Mechanical scheme, **(b)** Block diagram, **(c)** Bode diagram

The inner torque control loop of the motor, which in the case of a DC drive would be the armature current loop, is represented by an equivalent lag having unity gain and time constant T_e; it must react sufficiently fast to effectively control this type of plant. The reason is again that corrective action can only be applied at the reference side of the torque controller, i.e. through the torque control loop. The oscillatory mechanical part of the plant contains two inner loops with two integrators each. The open loop transfer function between torque and motor speed is,

$$F_1(s) = \frac{L\left(\omega_1/\omega_0\right)}{L\left(m_M/m_0\right)} = \frac{1}{(T_1 + T_2)\,s}\,\frac{(s/\Omega_{02})^2 + 1}{(s/\Omega_{01})^2 + 1},$$ (15.1)

with the following abbreviations

$$T_1 = \frac{J_1\,\omega_0}{m_0}, \quad T_2 = \frac{J_2\,\omega_0}{m_0}, \quad \Omega_{01} = \sqrt{K\left(\frac{1}{J_1} + \frac{1}{J_2}\right)}. \quad \Omega_{02} = \sqrt{\frac{K}{J_2}}.$$ (15.2)

It contains imaginary pairs of poles and zeros which make this a difficult plant to control; the Bode-diagram is sketched in Fig. 15.4 c.

When attempting to control the motor speed ω_1, which is the usual practice because the speed sensor is normally attached to the free end of the motor shaft, it is realised that a reasonably damped speed loop with a PI-controller can only be achieved under the condition

$$\frac{J_1}{J_2} = \frac{T_1}{T_2} \gg 1,$$

i.e. if poles and zeros are close together and are nearly cancelling. This is seen in Fig. 15.5, where measured step responses of the speed control loop are shown for different cases following changes of the speed reference $\omega_{1\,\text{Ref}}$ and of the load torque m_L. The transients were recorded on a DC drive, with the resonant load being simulated by another controlled DC drive. The parameters of the PI-controller were in each case so chosen as to obtain optimal results [B72, B73].

It is realised that, even when the motor speed follows the reference speed in a swift and well damped transient, the load speed, being coupled to the motor speed through the transfer function

$$F_2(s) = \frac{L\left(\omega_2/\omega_0\right)}{L\left(\omega_1/\omega_0\right)} = \frac{1}{(s/\Omega_{02})^2 + 1}$$ (15.3)

remains outside the control loop and can exhibit practically undamped oscillations. Obviously, this calls for a different approach to the control of the plant.

When assuming for a moment that in an ideal case both speeds ω_1, ω_2 as well as the torsional displacement $\varepsilon_1 - \varepsilon_2$ of the shaft are measurable, then a very effective scheme for decoupling the mechanical plant could be devised, which is indicated by the dashed signal paths in Fig. 15.4 b: By feeding back to the torque controller a signal which is proportional to differential speed, damping is achieved because the two masses are kept aligned, thus removing the cause for subsequent oscillations. Also, when adding an estimate of the coupling torque $\hat{m}_c$ to the torque reference, the actual loading on the motor is in effect compensated by feed-forward and one of the two internal feed-back paths is opened [T30]. Again, a precondition for success by these

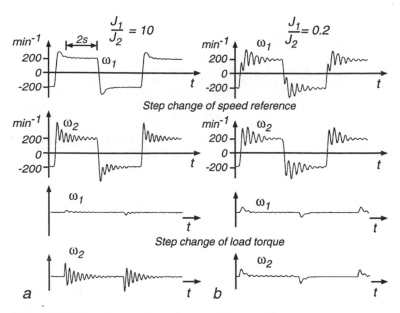

Fig. 15.5. Transients of speed control loop with two-mass system, recorded with a 40 kW DC drive **(a)** $J_1 = 10 J_2$; **(b)** $J_1 = 0.2 J_2$

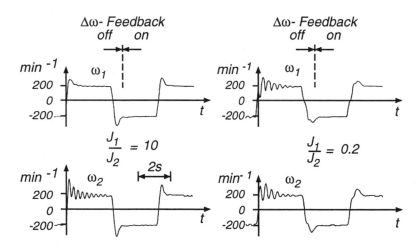

Fig. 15.6. Effect of auxiliary $\Delta\omega$-feedback on speed control of a two-mass system, recorded with a 40 kW DC drive

measures is that the torque loop is responding sufficiently fast, $T_e \ll T_1, T_2$. The effectiveness of $\Delta\omega$-feedback for the speed control of the two mass system is apparent from the recorded transients in Fig. 15.6.

Unfortunately, feedback signals for speed and position of the load, ω_2 and ε_2, are usually not available. There may be practical reasons, for example that the load is driven though a gear and rotates at a very low speed so that mounting a speed sensor would be difficult and expensive. Similar problems exist if the motion of the load inertia is translational, for instance on an elevator or mine hoist.

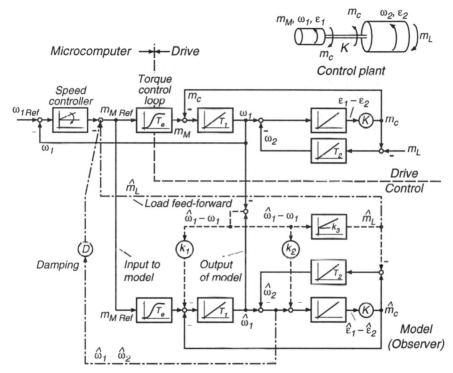

Fig. 15.7. Speed control of a two-mass drive system with auxiliary feedback signals derived from an observer

In such a situation one could attempt to derive the necessary information from a dynamic model, called an observer [W17, W18], which is shown in Fig. 15.7 in simplified form. The observer is a mathematical model of the plant that is evaluated in real time in parallel with the actual transients of the plant. Establishing this model requires the structure and the parameters of the plant to be known fairly accurately, which is usually no great problem with mechanical systems. The model is driven by the measurable inputs of the plant, in this case the torque reference which is already available in the microcomputer; the remaining inputs, particularly the load and frictional torques, are unknown and have to be estimated. Keeping the model in line with the actual plant is achieved by comparing some accessible output variable from the plant, for example the motor speed ω_1, with its estimated counterpart $\hat{\omega}_1$ and correcting the model by additional inputs so that the error vanishes. If the model parameters are correctly representing the plant, it can be assumed that the remaining estimated variables in the model are also tracking the real variables, i.e. that the model serves as an accurate observer. Clearly, designing an observer calls for a detailed analysis of the system to be modelled [L80].

The structure of the observer seen in Fig. 15.7 gives an idea of how the correction of the model can be achieved; each of the integrators in the model is influenced by the error $\hat{\omega}_1 - \omega_1$ in order to obtain maximum flexibility in selecting the eigenvalues of the observer by suitable choice of the gain factors k_i. This is so because the observer represents itself a feedback system, the dynamic properties of which are the subject of a separate analysis.

Realisation of observers for complex applications has only become practical since distributed digital signal processing with microelectronic components is possible; neither analogue techniques nor a main-frame process computer could provide an economic solution.

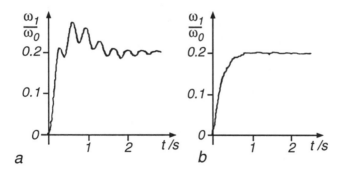

Fig. 15.8. Step response of speed control of a two-mass system, recorded with a 1 kW DC drive.
(a) PI-speed controller; **(b)** P-speed controller and observer

The estimate of the load torque, $\hat{m}_L$, is generated by integrating the speed error $\hat{\omega}_1 - \omega_1$; this permits the use of a proportional instead of a proportional-plus-integral (PI) speed controller thus reducing the speed overshoot following a change of the reference without sacrificing steady state accuracy. Results for a 1 kW DC drive with microcomputer control are shown in Fig. 15.8 [W12, W13]. The step response of the speed control loop with an optimally adjusted PI-speed controller for $J_1 = 5J_2$ and a resonant frequency $f_{01} = 5$ Hz is shown in Fig. 15.8 a. As seen in Fig. 15.8 b, it is possible to improve this rather unsatisfactory transient considerably by applying differential speed feedback and damping signals from the observer; the oscillations are now all but eliminated. The algorithms for the speed controller and the observer have been implemented in a single board 8086-microcomputer; the sampling period for the complete algorithm was 2 ms, the storage requirement < 1 k byte of ROM.

The characteristics of the plant and its parameters must be known for the computed variables produced by the observer to be trustworthy. While the dynamic structure of mechanical systems is usually known from the geometry, the distribution of masses with their linkages etc. may not be known accurately or may change under varying operating conditions. Examples are mine hoists, where the winding rope at different length acts like a variable spring, or robots with changing geometry when positioning a mass by a rotary movement at a varying radius. If the parameter changes are known or can be derived from other sources of information, it is of course possible to include them in the model, forming a time varying observer; this is feasible if the observer is implemented in a microcomputer.

However, if the variations of plant parameters are unknown or even the structure of the plant is uncertain, the problem of estimating internal variables becomes much more complex because it involves the identification of the plant as well as the adaptation of the observer and the controllers. This task has also come now within reach of a microcomputer but it would require a more detailed presentation than is possible in the present context [B73, M21, W13].

15.2 Linear Position Control

In many applications of electrical drives it is specified that the shaft of the motor or the load should be controlled to a constant or variable reference angle or that a machine part, possibly the tool of a milling machine, should follow a prescribed trajectory; this rotational or translational movement may be part of a multi-dimensional motion. It is generalised on a robot having six or more position-controlled drives to be able to place the tool in any position of the workspace and point it in any direction. While position-controlled electrical servo drives at lower power rating (usually < 10 kW) are of particular importance as feed drives on machine tools or robots, this is by no means the sole area of application. Servo drives are found wherever mechanical motion is required for controlling technical equipment in industry or transportation, be it the positioning of servo-actuated valves in chemical plants and power stations or the deflection of control surfaces on an aircraft. Indeed, position control is also found on higher power drives, such as elevators, mine-hoists or automatic commuter trains.

Depending on the application there are of course different requirements with regard to accuracy or dynamic response; some are discussed in the following sections. At the same time they demonstrate again the immense variability of electrical drives.

Let us assume at first that the position reference and the feedback signal representing the measured position are continuous or sampled variables and that a linear position control scheme is desired. Specifications of this type apply for example to elevators, servo drives for radar or satellite antennae or to feed drives on machine tools for contour milling. If the reference input is a mechanical signal, such as the position of a pick-up sensing the contour of a mechanical model to be duplicated by the machine tool in a different scale, the control system is also called a position follower, but in most cases the position reference and the feedback signals are represented electrically.

It is further assumed for the subsequent arguments that the rotation of the motor shaft is converted by an ideal gear into the translational motion of a machine part having the velocity v and the position x. Transmission errors such as nonlinearity of the lead screw or ambiguity due to backlash can be reduced either by employing a precise spindle or by a direct position measurement instead of a rotary sensor attached to the motor shaft. Position sensors are available in great variety, operating on different principles and covering a wide range of accuracy; in some cases a simple potentiometer or resolver may suffice, while digital encoders or high resolution incremental sensors may be required in others; ultimate precision can be assured, if necessary, by optical methods employing laser interferometry; the cost and complexity rise of course steeply with the specified accuracy.

The principle of cascaded control was explained in Fig. 7.5, showing a system of several superimposed control loops for torque, speed and position. An acceleration control loop which has the purpose of eliminating the effects of load torque, is occasionally added as an option, usually on the basis of an approximate feedback signal derived from the speed measurement. Acceleration control is no substitute for torque control, however, because it cannot prevent static overload, for instance when the drive is mechanically jammed.

The advantages of nested control loops in a cascaded system have been described before; they are

- Transparent structure, employing all measurable state variables
- Step-by-step design, beginning with the innermost loop, thereby solving the stability problem in several smaller steps,
- Use of standard controllers, P, PI, PID etc.
- The effects of nonlinearities, for instance in the power actuator, are restricted to the inner loop, resulting in a quasi-linear, unity gain response,
- Disturbances acting on the plant are quickly couteracted by inner loops,
- Intermediate variables can be limited by clamping the pertinent reference variables,
- Commissioning is greatly simplified by closing one control loop after the other, from inside out,

- Opening of outer loops permits simple procedures for diagnostic and field tests.

At first sight there is only one serious drawback of the cascade control structure as seen in Fig. 7.5, that is caused by the fact that the response to the reference input becomes progressively slower as more outer loops are added; it can be shown that under simplifying assumptions the closed loop equivalent time constant T_e increases by a factor of at least two for each additional loop [K25, 38]; hence the response to the reference input of a multiple loop control system may be slower than the response of a corresponding single loop system – provided, of course, that a stable single loop control can be devised. As a consequence of these accumulating delays the position control scheme in Fig. 7.5 may exhibit unacceptable dynamic errors when the reference position is time-varying.

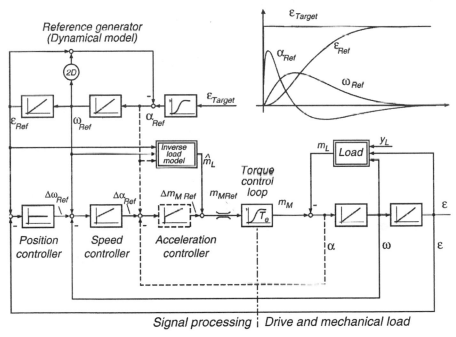

Fig. 15.9. Multiple-loop position control with feed-forward from a dynamic model of the reference trajectory and the load

Fortunately, this disadvantage can easily be removed by employing feed-forward to the inner loops, as is shown in Fig. 15.9, where a reference generator produces reference variables that otherwise would have to be generated by controllers in the outer loops responding to dynamic errors. The feed-forward signals must, of course, be coordinated for compatibility, otherwise there could be contradictory demands with the result of saturating individual controllers.

On multiple-axes servo drives such as used on continuous milling machines, where the tool should accurately follow a given spatial contour, the set of reference signals $(x, v, a)_{\text{Ref}}$ is usually computed off-line for all axes and stored in a memory to be fed in parallel to the various drive controllers.

Another solution is seen in Fig. 15.9 where the reference variables for each axis are generated on-line by a dynamic model representing the desired response of the drive; the effect of this scheme is that temporary saturation of controllers is avoided which could occur if unrealistic (for instance discontinuous), reference signals were applied; by employing a suitably chosen reference model the dynamic behaviour of the control system becomes reproducible and dependable because all the controllers remain in their linear high-gain mode and keep the respective errors small. Naturally provision must be made for unexpected disturbances such as load torque or line voltage fluctuations which means that the torque control loop must have sufficient margin for absorbing the additional load, should it occur.

The existence of a dynamic model can also be very useful in other applications; for example, controlling a high-speed elevator swiftly and still agreeably for the passengers requires that certain limits of acceleration $a(t)$ and jerk da/dt should not be exceeded; this can be achieved by a dynamic model producing a set of smooth reference variables. Since the feed-forward signals improve the precision of the trajectory of motion, it also becomes possible to accurately predict the stopping distance given any initial condition $(x, v, a)_0$; this is important for being able to decide instantly whether a call from one of the forward floors or a late stopping demand by one of the passengers can be answered immediately or whether the caller has to wait for the next chance, as speed reversal between scheduled stops is ruled out. Also, position overshoot is not allowed on an elevator because the passengers would dislike it and on a machine tool as it could leave marks on the work piece.

The ultimate in dynamic performance is obtained by feeding an estimate of the load torque $\hat{m}_L$ directly to the input of the torque control loop as is indicated in Fig. 15.9, thus relieving the speed controller from generating the main part of the torque reference. The torque estimate must be based on a model of the load and the reference trajectory, definitely not on measured values of acceleration, speed or position, as this would imply a danger of instability due to unavoidable inaccuracies and computational delays. Since forming an estimate of the load torque for a given trajectory of motion effectively involves the inversion of the equation of motion, this is also called "feedforward by inverse load modelling" [F28, L41]. With multiple axes robots the inverse load model may be of considerable complexity; a simple example was shown in Fig. 2.4.

When controlling the motion of a machine tool to high precision, for example with a maximum position error of 10^{-5} m of a total travel distance

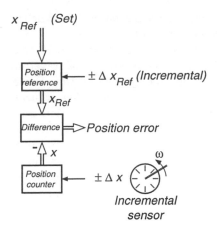

Fig. 15.10. Digital position control

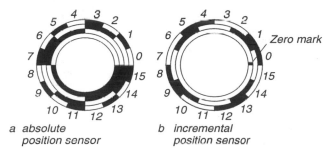

a absolute
position sensor

b incremental
position sensor

Fig. 15.11. Absolute and incremental position sensors

of 1 m, it is necessary to employ digital signal processing not only for the
position reference but also for the feedback signal and the error detection,
Fig. 15.10. This is the only way to avoid drift effects which would occur with
analogue signal processing; only a digital controller is capable of determining
with certainty minute differences between two large numbers, for example
25,000 and 24,999. Since a digital speed controller does not exhibit drift, and
position is the true integral of speed, the position controller can be of the
proportional type; there is no need for an integrating channel for achieving a
high steady state accuracy. Naturally this calls for comparable precision also
of the digital sensors for rotary and translational motion, combining high
accuracy and resolution.

An encoder, Fig. 15.11 a, produces an absolute measurement of the distance from a fixed position, for example by supplying n bit of a dual word
covering the range of $0 \leq x \leq (2^n - 1)\Delta$, where Δ is the resolution. It is
preferable to employ codes where only one bit changes at any one time, to

exclude ambiguity (Gray code); a similar effect is achieved with a synchronising signal from the channel with the highest resolution.

Another somewhat simpler sensor is of the incremental type, Fig. 15.11 b,
which generates a forward or reverse pulse for every increment of travel; by
counting these pulses, the actual position is obtained, provided the counter
has been initially set by a separate calibrating pulse or some other accurate
position measurement. Incremental sensors which normally function on a
magnetic or optical basis, are sometimes considered unsatisfactory because
pulses might be lost, causing a corresponding undetected position error until
the calibrating position is passed again. However, with todays sensors, using
LED light sources and the associated integrated circuitry, incremental sensors
can be regarded as very reliable components; if desired, battery-buffering of
the counter can be provided.

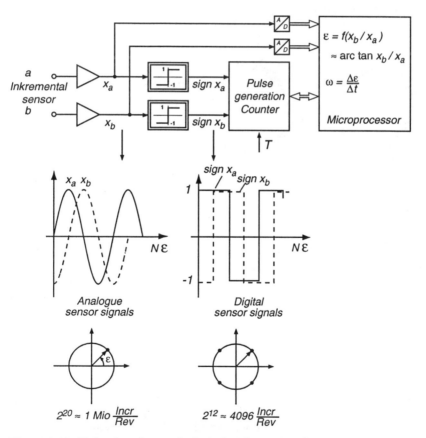

Fig. 15.12. Enlarging the resolution of an incremental encoder
by analogue interpolation

The function of an incremental sensor for angle measurements is explained with the help of Fig. 15.12. There are two optical or magnetic measuring channels producing out-of-phase signals, approximating $x_a = \hat{x} \sin N\varepsilon$ and $x_b = \hat{x} \cos N\varepsilon$, where ε is the angle to be measured and N is the number of periods per revolution of the sensor shaft; a typical value is $N = 2^{12}$. By clipping these signals, two orthogonal square wave functions sign x_a and sign x_b are created. Whenever one of these functions is changing sign, an incremental motion of $\Delta\varepsilon = \pm 2\pi/4N$ is registered, where the sense of rotation is derived from the direction of the signal change and the value of the other signal. With $N = 2^{12} = 4096$, this corresponds to an angular resolution of 2^{-14} or about $1/16\,000$ of one revolution of the sensor shaft.

By analogue interpolation, also shown in Fig. 15.12, this resolution can be greatly improved [S45]. The signals x_a and x_b are not be exactly sinusoidal and the amplitude $\hat{x}$ may depend on the speed of rotation $\omega = d\varepsilon/dt$, but by forming the ratio x_a/x_b and computing $\arctan x_a/x_b$, several lower bits of angular position measurement are generated, increasing the total resolution to 18 to 20 bit, or 250 000 to 1 Mio increments per revolution. While the dependability of the additional bits may be somewhat in doubt, the analogue interpolation serves as a most welcome increase of the measurement bandwidth when detecting slow rotational speeds.

With both types of position sensors it is easy to perform a digital velocity measurement. By sampling in short and precise time intervals T the position signal available at the output of the encoder or the counter and subtracting subsequent readings, the average velocity in the last sampling interval is obtained.

$$v(\nu + 1) - \frac{1}{T} [x(\nu + 1) - x(\nu)] ; \tag{15.4}$$

correspondingly, an acceleration signal can be detected from subsequent velocity measurements,

$$a(\nu + 1) = \frac{1}{T} [v(\nu + 1) - v(\nu)] = \frac{1}{T^2} [x(\nu + 1) - 2\,x(\nu) + x(\nu - 1)] . \tag{15.5}$$

With a given sampling interval T the resolution of a digital speed measurement is reduced at low speed because fewer angular increments are accumulated in each period T; the limit is reached when only a few counts are providing a very coarse representation of speed with questionable value. This then calls for either increasing the resolution of the sensor, for instance by analogue interpolation, or enlarging the sampling period. This could be done either by choosing a multiple period mT, thus reducing the bandwidth of the measurement, or by converting from a frequency- based measurement, as described, to a time- based measurement, where the intervals between successive angular pulses are counted with a stable high frequency reference and the result is inverted for obtaining a speed signal. With a microcomputer, frequency- and time- based sensing may be carried out in parallel, selecting whichever result offers a better resolution.

The resolution of the speed sensing can also be improved without analogue interpolation by forming in each interval the integer difference of the incremental counts and simultaneously measuring with a high frequency reference the differential time between the first and last increments [K10].

In the past, when digital signal processing equipment was bulky and expensive, it was only used when it was absolutely necessary to achieve the required accuracy, i.e. in the outer position control loop of a servo drive, but with todays microelectronic hardware this is changing rapidly. All the control functions in Fig. 15.9, including the control of an induction motor and the reference model, can now be incorporated as a modular algorithm in a microcomputer which finds room on a postcard size printed card. If necessary, the sampling time in the order of 1 ms can be reduced by a factor of ten when employing signal processors processors; hence DC servo motors are rapidly supplanted by digitally controlled and compact AC drives. This is a remarkable development made possible by the joint advances of microelectronics and power electronics.

It was mentioned before that backlash in gears and couplings may cause problems with position controlled drives because the contact between two adjoining parts of the drive is temporarily lost when the transmitted force is reversed; as a consequence the surfaces formerly exerting opposite forces are moving apart, until contact is again made at the other side of the gear tooth. Clearly this discontinuity of the transmitted force, possibly combined with Coulomb friction, is in conflict with a smooth and precise position control [B65]. Backlash can be avoided by employing high quality mechanical components, but wear cannot be excluded. A mechanical solution would be the use of spring-loaded gears with split wheels which maintain a bilateral contact force irrespective of the sign of the transmitted load force.

Another solution requiring only standard gears is to split the drive into several units and to preload the gears electrically; this is of particular interest if the type of load makes a distribution of the driving force desirable anyway. An example could be the azimuth motion of a large satellite antenna, i.e. the rotation around the vertical axis; antennae of this type require very accurate positioning, for example with a final tolerance of less than $1/100\,°$. This corresponds to a resolution of the full circle to one part in 36,000 which definitely calls for direct measurement by a digital sensor mounted at the pivot bearing of the main construction. The two or more servo motors could be arranged at symmetrical points around the circumference and exert torque through pinion drives to the central gear. A simplified schematic is drawn in Fig. 15.13 assuming two motors, with the main structure of the antenna and the two drive motors being lumped into three separate rigid bodies. The backlash assumed in both gear trains is made ineffective by adding opposite offset signals to the torque references; this results in preloading the gears in the low torque region, where backlash would be most troublesome. As the required load torque $m_{M\ Ref}$ increases, the motor that initially opposes the motion, eventually reverses its torque and comes to the help of the other motor until there is nearly equal load sharing at full torque.

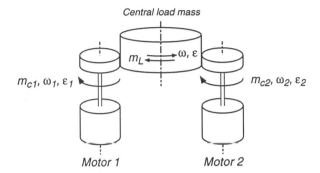

Fig. 15.13. Position drive with two servo motors and electrically preloaded gears for counteracting the effects of backlash

When assuming symmetry between the two motors, neglecting all frictional torques and referring all angles, angular velocities, inertias and torques to the central axis, the block diagram in the right hand side of Fig. 15.14 describes the mechanics of the drive.

The torque control loops are again replaced by equivalent lags having unity gain. In addition to inherent damping effects not shown in Fig. 15.14, such as speed dependent frictional torques, active damping will definitely be needed with this complicated mechanical structure. Two kinds of oscillations are likely to require countermeasures

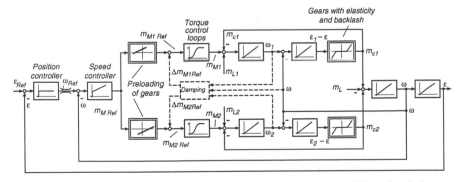

Fig. 15.14. Simplified block diagram of three-mass twin drive with preloaded gears for counteracting backlash

- oscillations of the two drive motors in phase opposition while the central mass remains at rest,
- oscillations of the two motors in phase against the central mass.

Provided the torque control loops respond sufficiently fast, both oscillations can be damped by adding suitable signals to the torque references,

$$\Delta m_{M1\,Ref} = -D_1\left(\omega_1 - \omega_2\right) - D_2\left(\omega_1 + \omega_2 - 2\omega\right),$$
$$\Delta m_{M2\,Ref} = D_1\left(\omega_1 - \omega_2\right) - D_2\left(\omega_1 + \omega_2 - 2\omega\right).$$

The signals for the motor speeds ω_1, ω_2 can be derived from sensors attached to the motor shafts. It is noted that the steady state characteristics of the drive system are not altered by these auxiliary feedback loops because the signals vanish in steady state under symmetrical conditions [T31].

Fig. 15.15 demonstrates another pair of nonlinear functions that would be advantageous for the preloading of gears because they cause the motors to produce equal torque at full load while maintaining constant overall gain in the speed control loop. Again these nonlinear functions would be cumbersome to realise with analogue circuitry but they present no problems when microcomputers are employed for the control.

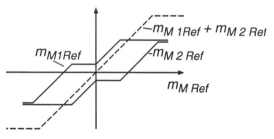

Fig. 15.15. Torque functions for electrically preloading of of gear train

Another example of a dual axis position controlled drive is shown in Fig. 15.16 a. It could be part of a machine tool or a robot with a mechanical structure in cylindrical coordinates r, ϵ, z, where each axis is activated by a separate position controlled drive having an inner torque control loop with the equivalent lag T_e. This is a simplification of the example discussed in Chap. 2. For the particular application it could be necessary to generate the reference trajectory in x, y, z-coordinates, while only machine orientated r, ϵ, z-measurements are available. Clearly the motion in the z-direction is decoupled if it is vertically orientated so that gravitational effects are excluded, but a transformation between (r, ϵ) and (x, y) is needed. It can be procured by transforming the x, y-reference data into polar coordinates or inversely by converting the r, ϵ-feedback signals into cartesian coordinates; the first possibility is chosen for the microcomputer based control scheme in Fig. 15.16 b because all the inner control functions are then performed in the actual

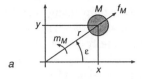

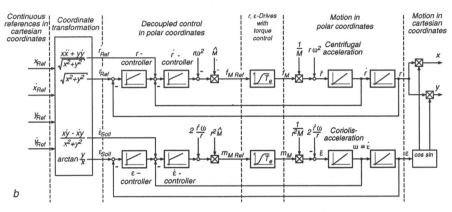

Fig. 15.16. Dual axis position control in polar coordinates
with reference signals supplied in cartesian coordinates
(a) Mechanical model; **(b)** Block diagram

drive coordinates. The set $(x, y, z)_{Ref}$ would again be produced by a suitable reference generator.

Dynamic coupling effects between the r, ϵ motions caused by centrifugal and Coriolis-forces can be approximately compensated by adding suitable corrective terms to the torque references, as was mentioned before; the effectiveness of this method of compensation, called inverse load modelling, depends of course on a fast response of the inner control loops.

With the present state of microelectronics all the control functions seen in Fig. 15.16 b can easily be executed with 1 kHz sampling frequency, which would be adequate in most cases.

15.3 Linear Position Control
with Moving Reference Point

In Sect. 15.1 it was mentioned that there are cases, for example with paper mills, where the speed of each drive section must be controlled with high precision in relative synchronism to the speed of a neighbouring section. This task can also be considered as a position control problem, where the reference position changes in steady state at constant velocity. As an example the digital speed control scheme in Fig. 15.17 which had been developed prior to the advent of microelectronics may be briefly discussed.

It is based on a conventional analogue speed control loop using, for example, a DC tachometer as a speed sensor. Due to unavoidable drift of the analogue components there is a normal control error, usually below 10^{-2} of nominal speed, which could, of course, be reduced by one or two orders of magnitude, but only with considerable effort, such as thermal stabilisation of the critical components, and at high cost.

The idea of the scheme in Fig. 15.17 is to add a digital corrective channel for compensating the steady state error of the analogue speed control loop; rapid response is not important since a well-tuned analogue speed control leaves little to be desired in this respect.

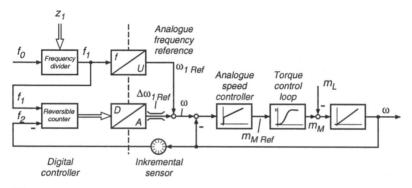

Fig. 15.17. Quasi-continuous digital speed control

The digital controller compares the pulse train f_2 from an incremental sensor, where each pulse corresponds to an angular increment $\Delta\epsilon$, with a reference pulse train f_1 derived from a high accuracy frequency source. A reversible counter registers the forward and reverse pulses; if none is lost, the content of the counter is a measure of the angular error of the drive shaft with respect to the reference angle. With the help of a digital-to-analogue converter it is transformed into a small corrective signal $\Delta\omega_1$ for the analogue speed loop. Hence the digital channel may be understood as an integral speed controller without drift or as a proportional position controller [J24, L31].

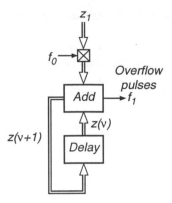

Fig. 15.18. Programmable frequency divider

In order to generate a finely adjustable reference frequency f_1 having the same steady state accuracy as the fixed clock frequency f_0 derived from a Quartz oscillator, the programmable frequency divider shown in Fig. 15.18 might be used. It contains an arithmetic unit for executing a recursive addition according to

$$z(\nu + 1) = [z(\nu) + z_1] \bmod z_{\max} , \tag{15.6}$$

where z_{max} is the capacity of the register while $z_1 < z_{max}$ corresponds to the desired reference frequency f_1 established by the overflow pulses. The addition is synchronised with the clock frequency f_0. The mean balance of the continued addition is

$$f_1 \, z_{\max} = f_0 \, z_1$$

so that a pulse train results, the mean frequency of which is adjustable in fine steps and has the same precision as the constant clock frequency f_0,

$$f_1 = \frac{z_1}{z_{\max}} \, f_0 . \tag{15.7}$$

The overflow pulses are not quite equidistant but when operating the frequency divider at a sufficiently high clock frequency, this inherent phase jitter has no effect on the speed of the motor [L32].

Similar methods are used in the process industry for in-line blending, where the flow rates of liquids to be mixed in prescribed ratios are measured by turbine wheel flow meters whose speed is transmitted and processed in the form of pulse trains [R12].

An adjustable reference frequency may also be formed with the help of a phase-locked loop (PLL) having a scaler with capacity z_1 in the feedback path; this produces at the output of a voltage controlled oscillator (VCO) a multiple of the constant clock frequency, $f = z_1 f_0$, which may then be reduced by a scaler with capacity z_{max}.

In various production lines where continuous strips or webs of material are being processed in subsequent stations it is often specified that the stations should be decoupled for preventing tension in the material. For this purpose a limited storage of material is accumulated in the form of a loop between the stations which, by taking up the difference of incoming and outgoing material, represents a measure of the synchronism between the two adjoining drives. Naturally, one should attempt to maintain the loop at approximately constant mean length to have adequate reserves in both directions, should the speeds of the two drives exhibit temporary fluctuations.

This is illustrated in Fig. 15.19 a showing the principle of a magnetic tape drive as used on computers. The incoming tape is fed from a reel and pressed against the drive rolls A_1, A'_1 to be transported with high acceleration and velocity past the read-write head S; subsequently the tape is rewound on another reel driven by a separate motor. By keeping the tape between the reels taut, the maximum acceleration would be severely limited by the inertia of the reels; also, excessive force in the tape could not be excluded. In order to avoid these problems, loops are formed on both sides of the drive rolls which are kept staight with moderate forces by mechanical or pneumatic equipment; this greatly reduces the effective inertia and the applied forces, when the tape is accelerated. The more slowly accelerating reel drives must of course produce sufficient maximum speed for preventing the storage loops from reaching their upper or lower limit.

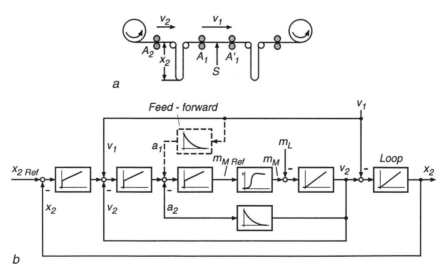

Fig. 15.19. Feed drive for magnetic tape   (a) Schematic; (b) Block diagram

Only the left half of the tape drive in Fig. 15.19 a is considered in the following, with the tape velocity v_1 at the read-write head S assumed to be an independent (impressed) variable. The speed v_2 of the feed rolls should be so controlled as to maintain the length x_2 of the storage loop at a prescribed value. Because of

$$\frac{dx_2}{dt} = \tfrac{1}{2}\,(v_2 - v_1) \tag{15.8}$$

the block diagram in Fig. 15.19 b results, where the storage represents an integrator with two inputs cancelling in steady state. Normally the feeding reel is controlled to maintain constant tension of the unwinding tape.

The quality of the control can be further improved by introducing to the speed controller of the feed rolls A_2 a feed-forward signal from the velocity v_1, which acts as a disturbance. In view of the quickly changing operating conditions it could even be appropriate to provide acceleration feed-forward to reduce the necessary length of the storage loop. It is noted that the inertia of the reel drive depends on whether the spool is full or nearly empty; also, most magnetic tape drives are reversible, i.e. they can operate in either direction.

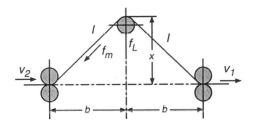

Fig. 15.20. Geometry of loop in a continuous rolling mill

On multi-stand continuous rolling mills the storage loops between successive stands, which are necessary to trim the coordinated speed control described in Sect. 15.1, have a different geometry, as is shown in Fig. 15.20; an electrically or hydraulically powered looper roll pushes the material with a force f_L upwards and thereby keeps the strip taut. Due to the geometry of the loop there is a nonlinearity between the deflection x and the length of the loop, $l = \sqrt{b^2 + x^2}$ leading to a nonlinear differential equation

$$\frac{2\,b}{v_0}\,\frac{x}{l}\,\frac{d\,(x/b)}{dt} = \frac{v_2}{v_0} - \frac{v_1}{v_0}, \tag{15.9}$$

where v_0 could be the rated velocity; the integrating time constant

$$T_i = \frac{2\,b}{v_0}\,\frac{x}{l} \tag{15.10}$$

depends on the ratio x/l, i.e. the shape of the loop. $x/l = 1$ corresponds to the special case in Fig. 15.19 a. Operating the loop at small values of x/l becomes increasingly difficult because the integrating time constant is reduced; this calls for fast response of the looper and the adjoining main mill drive used for controlling the loop.

The force f_m exerted on the material is also a function of the loop geometry; in relation to the vertical force f_L exerted by the looper roll, it is

$$\frac{f_m}{f_L} = \frac{1}{2}\frac{l}{x},\tag{15.11}$$

also indicating a critical range at small values of x/l, where the gain becomes very large.

A similar problem as with loop control arises with automated tracked vehicles developed in recent years for urban transportation, Fig. 15.21 a. The distance x of each vehicle to the leading vehicle must be controlled to a speed-dependent safe value x_{Ref} by applying driving or braking force $f_M \gtrless 0$. The distance can be measured, for example, by radar or by sensing equipment installed in the guideway.

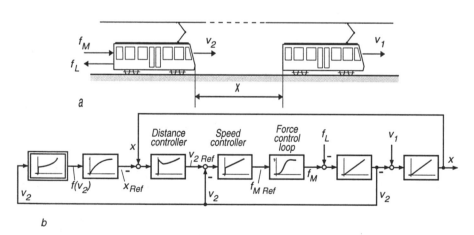

Fig. 15.21. Distance control of tracked vehicles
(a) Schematic; (b) Block diagram

When again regarding the velocity of the forward vehicle as an independent disturbance, the control system for the distance between the vehicles could have the form shown in Fig. 15.21 b. It represents a cascaded scheme containing inner loops for the driving force and velocity. The desired distance x_{Ref} to the leading vehicle is generated as a delayed nonlinear function of velocity, for example a parabola approximating a safe braking distance.

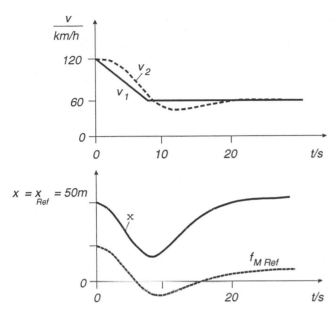

Fig. 15.22. Braking transients of distance control scheme

A computed transient of the distance control scheme is shown in Fig. 15.22; it may be caused by a sudden constant deceleration of the leading vehicle; a PID-controller and a linear function $x_{Ref}(v_2)$ were assumed for the example. Transients of this sort and their dangers are well known to the motorist; the task of maintaining a safe distance can be alleviated by employing advance signals, such as feed-forward from the deceleration of the leading vehicle or the rate of distance change, as is common practice in road traffic by use of braking lights or warning signs.

15.4 Time-optimal Position Control with Fixed Reference Point

Some applications of drives call for the discontinuous motion of a machine part from one steady state position to another. Examples are boring machines where the tool must be positioned for the next drilling operation or reversing rolling mills where the upper roll is lowered by a screw-down drive as soon as the slab of material has left the roll gap. Since the load is disengaged during the travel, there is no specified trajectory $x(t)$; the only requirement is that the drive should follow the discontinuous position reference in minimum time – given the power rating of the drive – and without overshooting the target position.

With the assumption that

- the response time of the torque control loop can be neglected,
- the jerk, i.e. the derivative of acceleration, need not be restricted (which would be necessary if a transportation system with on-board passengers or a weakly damped mechanical structure were involved),
- the frictional torques are ignored,

the drive with the associated mechanical linkages can be approximated by a double integrator having the acceleration $a(t)$ as input and the position $x(t)$ as output, as seen in Fig. 15.9.

The well known time-optimal solution that could be derived with the calculus of variations, consists of two intervals of maximum acceleration/deceleration and possibly an intermediate interval with constant maximum velocity. This is shown in Fig. 15.23 for two different step changes of the reference position [38,44]. Velocity and position are plotted versus time as well as in state space which, because x and v are the only state variables, reduces to a state plane. Beginning at $t = 0$ in steady state condition, $v(0) = x(0) = 0$, a new position reference is commanded, $x_{Ref} = x_{01}$, calling for a new target state $v(t_e) = 0$, $x(t_e) = x_{01}$ with $t_e \to$ min. The time optimal transient consists of maximum initial acceleration, operation at maximum speed and controlled deceleration to the prescribed target point; the maximum values of acceleration and deceleration are assumed equal in Fig. 15.23. If the amplitude of the position step is reduced, $x_{02} < x_{01}$, the maximum velocity may not be reached; the acceleration is then immediately followed by deceleration with the switch- over taking place at half the distance travelled.

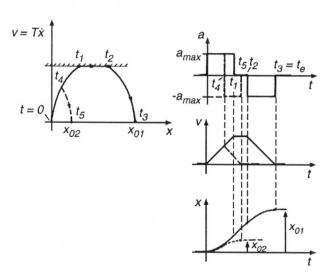

Fig. 15.23. Idealised time-optimal positioning transient with limited acceleration (a) State plane; (b) Response vs. time

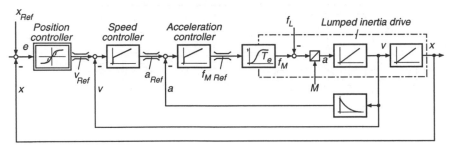

Fig. 15.24. Time-optimal position control scheme

In the state plane, the intervals of constant acceleration or deceleration are characterised by parabolae marked with time, the interval at constant velocity appears as a straight line parallel to the x-axis.

The multi-loop position control scheme in Fig. 15.24 closely models the idealised behaviour shown in Fig. 15.23 [A24]. The nonlinear position controller may be realised digitally to achieve reproducible accuracy at high resolution, for example 0.1 mm with a maximum distance of 1 m on a screw-down positioning drive. The scheme differs from the linear one in Fig. 15.9 by the nonlinear function in the position controller,

$$v_{\text{Soll}} = \sqrt{2\, a_{\max} |e|} \ \text{sign}\ e\,, \tag{15.12}$$

which generates the velocity reference as a function of the remaining position error e; its purpose is to guide the drive with constant maximum deceleration into the reference point. The mode of operation, after a step change of the position reference has occurred, is then as follows: When x_{Ref} changes by a value exceeding twice the maximum braking distance, the velocity reference v_{Ref} initially assumes its maximum value v_{max} causing the velocity controller to become saturated and the drive to start at maximum acceleration. When the velocity reaches the reference value, the velocity controller reverts to linear operation and maintains maximum velocity. As the position error is gradually reduced, $v_{\text{Ref}}(e)$ follows the parabolic function, resulting in constant deceleration until the reference position is reached. The drive may then be disconnected by disabling the power supply and arrested with a mechanical brake. This is necessary because of instability due to the infinite gain of the position controller at $e \approx 0$. In case the position reference changes by less than twice the maximum braking distance, the drive will switch from acceleration to deceleration before the maximum velocity is reached.

The torque control loop in Fig. 15.24 serves only for protection in case of overload; the maximum current should be adequate to produce maximum acceleration at full load torque.

Another more elegant method for realising minimum time position control is shown in Fig. 15.25, where the linear reference signal generator of Fig. 15.9 has been modified to generate the feed-forward signals required

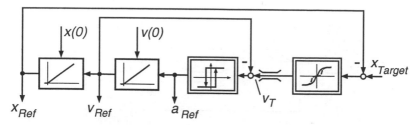

Fig. 15.25. Dynamic reference model for minimal-time position control

for time-optimal transients. The advantage of this arrangement is that the control system proper remains unchanged, i.e. linear and optimally tuned. The choice, whether the transients should be linear or time-optimal is solely determined by the reference model; this structure is particularly suited for implementation on a microcomputer.

As a further variation, the nonlinear function in Fig. 15.25 is modified around zero (in dashed lines) to prevent overshoot and to achieve a stable control mode at the reference position. The advantages of both control strategies are then combined: For large position error, the drive performs nearly time-optimally while it becomes linear once the position error has entered a narrow range around steady state condition; the same could be applied to the position controller in Fig. 15.24.

It has been mentioned that the second order approximation is only applicable as long as the residual lag of the torque control loop can be neglected. Thus the application of this control scheme is restricted to converter supplied

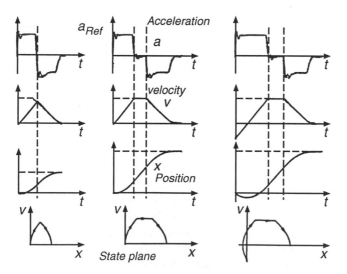

Fig. 15.26. Minimal-time positioning transients with different initial conditions

DC or AC drives having an effective lag of the torque loop of a few *ms*. A time-optimal transient of an AC servo drive was shown in Fig. 14.9; some simulated transients, assuming different initial conditions, are plotted in Fig. 15.26.

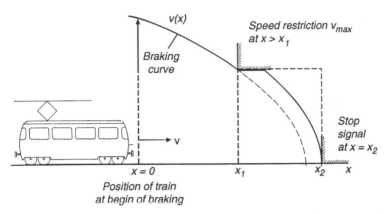

Fig. 15.27. Controlled deceleration of automatic train along a veocity profile

Controlled deceleration profiles are also used on other drives such as elevators or automatic trains, where at any time a fictitious braking curve $v(x, t)$ must be known, along which the vehicle could be decelerated or brought to standstill with the maximum specified deceleration. By continuously comparing this projected velocity profile with possible forward speed restrictions, it can be determined, when braking should begin to meet the constraint. Fig. 15.27 illustrates the situation, where an automatic train with its projected velocity profile approaches a restricted speed zone.

Whenever passengers on board vehicles are involved, it is usually specified that the accelerating forces must change gradually to avoid discomfort or anxiety (as every passenger standing in a bus knows); a similar situation exists, when complex mechanical structures such as satellite aerials or robots are to be moved, because abruptly changing acceleration or deceleration creates unnecessary forces and oscillations. This calls for smooth but swift positioning, still using all the available drive power. Some idealised time-optimal positioning transients with limited rate-of-change of acceleration (jerk) are drawn in Fig. 15.28, beginning with large change of position reference (a) which is gradually reduced to case (e); (b) and (d) are limiting cases separating different modes of operation. Clearly, the condition that da/dt should be finite amounts to the definition of acceleration $a(t)$ as another continuous state variable; hence, the state vector $[a(t), v(t), x(t)]$ could be represented as a continuous trajectory in a three- dimensional space.

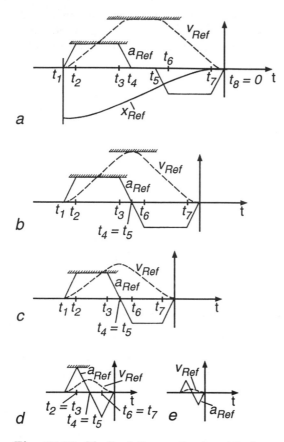

Fig. 15.28. Idealised time-optimal positioning transients with limited jerk

As this is difficult to visualise, some characteristic curves are plotted in Fig. 15.29 in the x, v-plane, showing how the controlled deceleration phase of the transient could be implemented; the starting phase is no particular problem because it follows the simple strategy of applying maximum da/dt until the peak values of acceleration and, possibly, velocity have been reached.

Figure 15.30 shows yet another modification of the dynamic model to be used for generating the reference functions $[a(t), v(t), x(t)]_{\mathrm{Ref}}$ for time-optimal positioning. The acceleration reference is now the output of another integrator, making it a continuous state variable of the dynamic model. The reference variables are again to be used as inputs for an optimally tuned linear multi-loop position control scheme as seen in Fig. 15.9. The nonlinear function

$$a_T(t) = f(e, v, a) \tag{15.13}$$

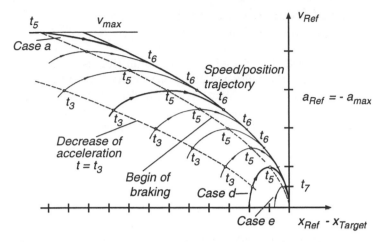

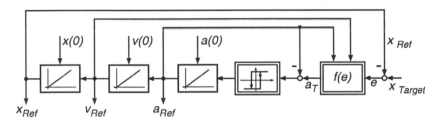

Fig. 15.29. Switching curves and trajectories in state-plane

Fig. 15.30. Dynamic model for minimum time, limited jerk, positioning transients

is so chosen as to generate in each interval a target acceleration a_T which the acceleration reference then approaches with constant slope.

Clearly, the main advantage of control with the help of a dynamic reference model is that the computation of the complex set of reference functions can be performed with an idealised mathematical model that is not affected by the unavoidable approximations when tuning the real drive control system.

As is indicated in Fig. 15.30 it is, of course, necessary to bring the initial conditions of the model in line with the plant; this would best be done in standstill, i.e. with $x(0), v(0) = a(0) = 0$.

15.5 Time-optimal Position Control with Moving Reference Point

In some applications the task of positioning in minimum time is made more complicated by the fact that the reference point itself is moving. This could assume the form of a rendez-vous-problem, where the drive is supposed not only to reach the reference point in minimum time but to assume also the same final velocity as the target. The minimum time requirement could be due to the condition that the initial state of the drive should be subject to least restrictions for reaching the rendez-vous position in time.

Examples for this type of problem are the control of a robot with the task of gently picking up a fragile object on a moving conveyer or of a flying shear that is to cut a strip of metal, emerging from a rolling mill at full speed, into various custom-lengths sections. Similar problems exist also with rotary printing presses, when a new supply roll of paper must be attached to the end of the outgoing roll while the press is running; the situation vaguely resembles a relay race, where the new runner must start in time for accepting the token from his predecessor in a specified space, or driving on the ramp to a motorway, where the entering vehicle is merging into a moving line of cars.

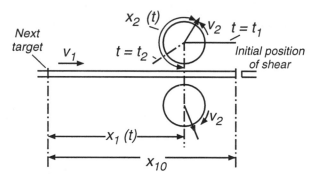

Fig. 15.31. Principle of a flying shear

The basic layout of a rotating shear is sketched in Fig. 15.31. The strip- or rod-shaped material coming from the last stand of a rolling mill moves with velocity v_1 through the shear, which consists of two mechanically coupled drums bearing opposing cutting blades. Initially the shear may be at rest; at a predetermined time t_1 is is accelerated, so that the tools separate the material at a prescribed point. The blades and the material should move in approximate synchronism during the separation, but a slightly higher velocity of the tools could also be specified for separating the material. Following the cut, the shear is decelerated and usually set at rest at a selected waiting position. It is noted that the accelerating and decelerating phase must be completed in less than two revolutions of the shear which calls for high dynamic performance of the drive; with DC machines this condition was sometimes met with twin motors for reducing the effective inertia. There is no limiting specification on the jerk in this application, except the finite rise time of the motor torque.

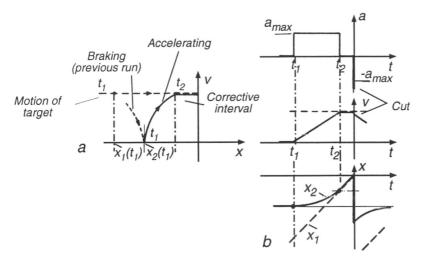

Fig. 15.32. Discontinuous motion of flying shear

Beside this discontinuous mode of operation there is also a continuous mode, when, due to lack of time, the shear moves at a pulsating speed without coming to standstill; the reason may be increased velocity of the material or the need for shorter cuts of material.

An idealised accelerating, cutting and subsequent positioning cycle is shown in Fig. 15.32 with the variables plotted vs. time and in the state plane; the circular motion of the shear is represented by the circumferential path of the blades. Time and position are defined relative to the cut, i.e. the acceleration phase takes place in the negative range of time and position; the material velocity is assumed to remain constant after the start of the shear, $v_1 = $ const. The shear begins to accelerate with $a = a_{max}$ at $t = t_1$; at t_2 it reaches cutting speed and then continues in synchronism with the material until the cut occurs at $t = 0$. The interval $t_2 < t < 0$ serves for correction of velocity and position to eliminate errors caused by tolerances and the finite response time of the speed loop.

Because of the assumed constant velocity of the material the position $x_1(t)$ of the reference point changes linearly with time, while the shear position $x_2(t)$ follows in the interval $t_1 < t < t_2$ a parabolic function. During the acceleration phase

$$t_a = t_2 - t_1 = \frac{v_1}{a_{\max}} \qquad\qquad (15.14)$$

the reference point moves by

$$x_1(t_2) - x_1(t_1) = v_1 t_a = \frac{v_1^2}{a_{\max}} . \qquad\qquad (15.15)$$

The shear velocity rises linearly with time in this interval

$$v_2(t) = (t - t_1) a_{\max} = v_1 \frac{t - t_1}{t_2 - t_1} , \quad t_1 \le t \le t_2 , \qquad\qquad (15.16)$$

which results in a necessary distance for acceleration

$$x_2(t_2) - x_2(t_1) = \tfrac{1}{2} a_{\max} (t_2 - t_1)^2 = \frac{1}{2} \frac{v_1^2}{a_{\max}} ; \qquad\qquad (15.17)$$

this is half the distance covered by the target point during the same interval. The remaining corrective interval is travelled in approximate synchronism,

$$x_1(t_2) = x_2(t_2) = v_1 t_2 < 0 . \qquad\qquad (15.18)$$

While it would also be possible to determine the instant when the shear should start at maximum acceleration by a time-based criterion, it seems more appropriate to use a closed loop control scheme for velocity and position which remains in operation during the complete cycle until the shear comes to rest again; this has the advantage that tolerances such as a slightly varying target velocity can be compensated during the final part of the approach [B16, K69, L39, O7].

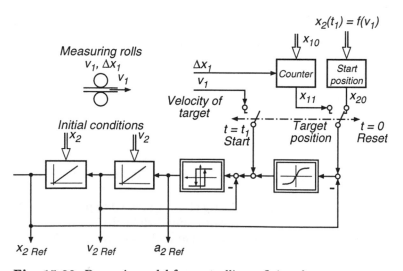

Fig. 15.33. Dynamic model for controlling a flying shear

The control of the shear drive is facilitated by the use of a suitable dynamic model in combination with a high performance multiloop position control scheme as in Fig. 15.9. The dynamic model in Fig. 15.33 is of the type shown in Fig. 15.25 producing the reference signals for minimal time positioning transients; the only difference is a feed-forward signal of the target velocity v_1, for example coming from measuring rolls. There are two modes of operation, intermittent and continuous motion.

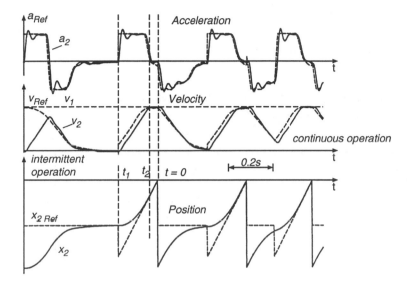

Fig. 15.34. Simulated transients of time-optimal shear drive

- In reset of the intermittent mode the feed-forward signal is removed and the position reference corresponds to the starting position x_{20} which is a function of target velocity

$$x_{20} = -\frac{v_1^2}{2\,a_{\max}} + v_1\,t_2 . \tag{15.19}$$

This condition (Reset and Wait) is maintained until time t_1 when the target position counter reaches the starting position count

$$x_1(t_1) = -\frac{v_1^2}{a_{\max}} + v_1\,t_2 . \tag{15.20}$$

- At this time t_1 the velocity feed-forward signal is added. As a consequence, the drive starts with maximum acceleration. The position reference input for the model is now derived from the target position counter which is counted towards zero by angular increments of the measurements rolls.

- As soon as the target position counter reaches zero and the cut occurs, the model is again put into reset while the desired length x_{10} of the next section is entered into the target position counter as an initial condition for the next run.

Changing the operating mode is indicated in Fig. 15.33 by mechanical switches; in reality this would of course be done electronically or by software.

It can be shown that the model in Fig. 15.33 produces time-optimal positioning transients as long as the physical restrictions imposed by the drive are not exceeded. Some simulated transients are depicted in Fig. 15.34 where the operating conditions are changing due to a reduction of the desired section length; this results in a transition from discontinuous to continuous mode. The transients prove that the control scheme can in principle meet these difficult specifications; it is of course important to select a drive that is able to rapidly follow the reference functions.

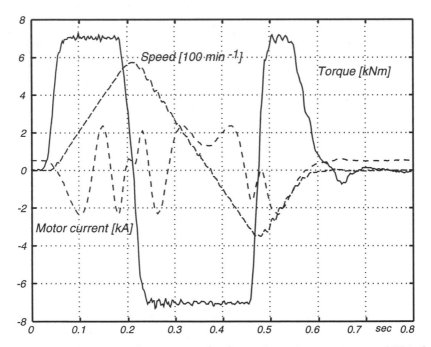

Fig. 15.35. Measured transients of a flying shear drive with a 500 kW induction motor supplied by voltage source GTO- inverter and field orientated control

Even on intermittent drives of this kind, calling for large overload capacity and fastest possible response, controlled AC motors are now successfully supplanting DC motors; their reduced inertia is an added advantage. An example is seen in Fig. 15.35 showing recorded transients of a flying shear drive during an acceleration and repositioning cycle. The 500 kW induction motor is supplied from a 800 kVA GTO voltage source inverter and has a control structure as shown in Fig. 12.28 [K75].

The accuracy of the material lengths to be expected from this type of drive control depends of course on many factors such as accuracy and response of the drive control loops, accuracy and resolution of the measurements from the moving material as well as the length of time t_2 available for correction during the final approach. Typically, some cm tolerance could be expected, when cutting material of 10 m length. The task of controlling a flying shear is quite complex; on the other hand, the gain in productivity is substantial, when comparing it with a shear that can cut only when the material is at rest.

The next step towards fully automatic operation is on-line minimisation of scrap by computing an optimal combination of different custom lengths sections that can be produced from a given size ingot. All these control functions, the flying shear control with a dynamic reference model as well as the optimisation can be executed today on a single board standard microcomputer.

Bibliography

Books and Proceedings

[1]   Amin, B., Induction motors, Analysis and torque control. 2nd Ed., Document Service, Paris, 2000

[2]   Bedford, B.D., Hoft, R.G., Principles of inverter circuits. Wiley, New York 1964

[3]   Bird, B.M., King, K.G., An introduction to power electronics. Wiley, New York 1983

[4]   Bödefeld, Th., Sequenz, H., Elektrische Maschine. 8th Edition, Springer, Berlin-Heidelberg-New York 1971

[5]   Bose, B.K., Power electronics and AC drives, Prentice Hall, Englewood Cliffs, N.J. 1986

[6]   Bühler, H., Einführung in die Theorie geregelter Gleichstromantriebe. Birkhäuser, Basel-Stuttgart 1962

[7]   Bühler, H., Einführung in die Theorie geregelter Drehstromantriebe, Bd. 1 Grundlagen, Bd. 2 Anwendungen. Birkhäuser, Basel-Stuttgart 1977

[8]   Bühler, H., Regelkreise mit Begrenzungen. VDI Fortschrittsberichte, Reihe 828, Bd.8, VDI Verlag Düsseldorf 2000

[9]   Buxbaum, A., Schierau, K., Berechnung von Regelkreisen der Antriebstechnik. Elitera, Berlin 1974

[10]   Concordia, C., Synchronous machines - Theory and performance. Wiley, New York 1951

[11]   Davis, R.M., Power diode and thyristor circuits. Peter Peregrinus, 1971

[12]   De, G., Electrical drives and their controls. Acad. Books Ltd, Bombay-New Delhi 1970

[13]   De Almeida, A., Bertoldi, P., Leonhard, W. (Eds.), Energy efficiency improvements in electric motors and drives, Springer Verlag Berlin Heidelberg New York 1997

[14]   Dewan, S.B., Straughen, A., Power semiconductor circuits. Wiley, New York 1975

[15]   Eckhardt, H., Grundzüge elektrischer Maschinen. Teubner, Stuttgart 1981

[16]   Ernst, D., Ströle, D., Industrieelektronik. Springer, Berlin 1973

[17]   Finney, D., The power thyristor and its applications. Mc Graw Hill, New York 1980

[18]   Fröhr, F., Orttenburger, F., Technische Regelstreckenglieder bei Gleichstrom-Antrieben. Siemens AG, Berlin-München 1971

[19]   Groß , H., Stute, G., Elektrische Vorschubantriebe für Werkzeugmaschinen. Siemens AG, Berlin-München 1981

[20]   Gyugyi, L., Pelley, B.R, Static power frequency changers. Wiley, New York 1976

[21]   Habinger, E., Die Technik der elektrischen Antriebe. VEB Technik, Berlin 1979
[22]   Hamming, R.W., Numerical methods for scientists and engineers. Mc Graw Hill, New York 1962
[23]   Heumann, K., Grundlagen der Leistungselektronik, 2. Aufl. Teubner, Stuttgart 1978
[24]   Heumann, K., Stumpe, A.C., Thyristoren. Teubner, Stuttgart 1969
[25]   Hofmann, A., Stocker, K., Thyristor-Handbuch, 4. Aufl. Siemens AG, Berlin-München 1976
[26]   Hoft, R.G., Semiconductor power electronics. Van Nostrand Reinhold 1986
[27]   Jötten, R., Leistungselektronik I. Vieweg, Braunschweig 1977
[28]   Jordan, H., Weis, M., Asynchronmaschinen, Vieweg, Braunschweig 1969
[29]   Kessler, C. (Ed.), Digitale Signalverarbeitung in der Regelungstechnik. VDE-Buchreihe Bd. 8, VDE, Berlin 1962
[30]   Kleinrath, H., Stromrichtergespeiste Drehfeldmaschinen. Springer, Wien 1980
[31]   Kovacz, K.P., Racz, J., Transiente Vorgänge in Wechselstrommaschinen. Ung. Akad. d. Wiss., Budapest 1959
[32]   Keuchel, U., Stephan, R.M., Microcomputer-based adaptive control applied to thyristor-driven DC-motors. Springer, Berlin-Heidelberg-New York 1993
[33]   Kümmel, F., Elektrische Antriebstechnik. Springer, Berlin-Heidelberg-New York 1971
[34]   Laible, Th., Die Theorie der Synchronmaschine im nichtstationären Betrieb. Springer, Berlin 1952
[35]   Langhoff, J., Raatz, E., Geregelte Gleichstromantriebe in der Praxis der Stromrichter- und Regelungstechnik. Elitera, Berlin 1977
[36]   Lappe, R., Thyristorstromrichter für Antriebsregelungen. VEB-Technik, Berlin 1970
[37]   Leonhard, A., Elektrische Antriebe, 2. Aufl. Enke, Stuttgart 1959
[38]   Leonhard, W., Einführung in die Regelungstechnik, 6. Aufl. Vieweg, Braunschweig 1991
[39]   Leonhard, W., Regelung in der elektrischen Antriebstechnik. Teubner, Stuttgart 1974
[40]   Leonhard, W. (Ed.), Control in power electronics and electrical drives. 1st IFAC Symp., Düsseldorf 1974, VDI/VDE, 1974
[41]   Leonhard, W. (Ed.), Control in power electronics and electrical drives. 2nd IFAC Symp., Düsseldorf 1977, Pergamon Press, Oxford 1977
[42]   Leonhard, W. (Ed.), Microelectronics in power electronics and electrical drives. ETG-Fachberichte Vol. 11, VDE, Berlin 1982
[43]   Leonhard, W. Digitale Signalverarbeitung in der Mess- und Regelungstechnik. 2. Aufl. Teubner, Stuttgart, 1990
[44]   Lerner, A.J., Schnelligkeitsoptimale Regelungen. (Translation from Russian), Oldenbourg, München 1963
[45]   Lyon, W.V., Transient analysis of alternating-current machinery. MIT Press and J. Wiley & Sons, New York 1954
[46]   Marti, K., Winograd, O., Stromrichter. Oldenbourg, München 1933
[47]   Mazda, F.F., Thyristor control. Wiley, New York 1973
[48]   Meyer, M., Selbstgeführte Thyristor-Stromrichter, 3. Aufl. Siemens AG, Berlin-München 1974
[49]   Möltgen, G., Netzgeführte Stromrichter mit Thyristoren, 3. Aufl. Siemens AG, Berlin – München 1974
[50]   Möltgen, G., Stromrichtertechnik, Einführung in Wirkungsweise und Theorie. Siemens AG, Berlin 1983

[51]   Mohan, N., Undeland,T.M., Robbins, W.P., Power electronics, Converters, applications and design, J. Wiley, New York 1989
[52]   Mohr, O. (Ed.), Steuerungen und Regelungen elektrischer Antriebe. VDE-Buchreihe Bd. 4, VDE, Berlin 1959
[53]   Murphy, J.M.D., Thyristor control of AC motors. Pergamon Press, Oxford 1973
[54]   Murphy, J.M.D., Turnbull, F.G., Power electronic control of AC motors. Pergamon Press, Oxford 1988
[55]   Novotny, D.W., Lipo, T.A., Vector control and dynamics of AC drives. Clarendon Press 1996
[56]   Orttenburger, F., Das regelungstechnische Verhalten der Gleichstrommaschine. Siemens AG, Berlin-München 1972
[57]   Pearman, R.A., Power electronics: Solid state motor control. Prentice Hall, 1980
[58]   Pelly, B.R., Thyristor phase-controlled converters and cycloconverters. Wiley, New York 1971
[59]   Pfaff, G., Regelung elektrischer Antriebe. Oldenbourg, München 1971
[60]   Pfaff, G., Meier, Ch., Regelung elektrischer Antriebe II, (Geregelte Gleichstromantriebe). Oldenbourg, München 1982
[61]   Puchstein, A.F., Lloyd, T.C., Conrad, A.G., Alternating current machines. Wiley, New York 1954
[62]   Rajashekara, K., Kawamura, A., Matsuse, K., Sensorless control of AC motor drives. IEEE Press, 1996
[63]   Say, M.G., Alternating current machines, 5th Ed. Pitman, London 1983
[64]   Say, M.G., Taylor, E.O., Direct current machines, Halsted Press 1980
[65]   Schönfeld, R., Digitale Regelung elektrischer Antriebe. Hüthig, Heidelberg 1990
[66]   Schröder, D., Elektrische Antriebe, Bd. 1-4, Springer Verlag 1995
[67]   Seefried, E., Müller, G., Frequenzgesteuerte Drehstrom- Asynchronantriebe. Verlag Technik, München 1992
[68]   Sen, P.C., Thyristor DC-drives. Wiley, New York 1981
[69]   Sen, P.C., Principles of electrical machines and power electronics. Wiley, New York 1989
[70]   Simon, W., Die numerische Steuerung von Werkzeugmaschinen. Hanser, München 1971
[71]   Slemon, G.R., Straughen, A., Electric machines. Addison Wesley, 1980
[72]   Späth, H., Steuerverfahren für Drehstrommaschinen. Springer, Berlin-Heidelberg-New York-Tokyo 1983
[73]   Steimel, K., Jötten, R., (Hrsg.), Energieelektronik und geregelte elektrische Antriebe. VDE-Buchreihe B 11, VDE, Berlin 1966
[74]   Steven, R.E., Electrical machines and power electronics. Van Nostrand Reinhold, 1983
[75]   Taegen, P., Einführung in die Theorie der Elektrischen Maschinen I, II. Vieweg, Braunschweig 1971
[76]   Tschilikin, M.G., Elektromotorische Antriebe. (Translation from Russian), VEB-Technik, 1957
[77]   Uhlmann, E., Power transmission by Direct Current. Springer, Berlin-Heidelberg-New York 1975
[78a]   Vas, P., Vector control of AC machines. Clarendon Press, 1990
[78b]   Vas, P., Sensorless vector and direct torque control. Oxford University Press, Oxford, New York, Tokyo 1998
[79]   VEM-Handbuch – Leistungselektronik. VEB-Technik, Berlin 1979

[80]   VEM-Handbuch – Die Technik der elektrischen Antriebe. VEB-Technik, Berlin 1979
[81]   Vogel, J., Elektrische Antriebstechnik. 5th Ed. Hüthig, Heidelberg, 1991
[82]   Wasserrab, Th., Schaltungslehre der Stromrichtertechnik. Springer, Berlin-Göttingen-Heidelberg 1962
[83]   Watzinger, H., Stromrichter-Gleichstromantriebe. Hüthig, Heidelberg 1980
[84]   Weber, W., Adaptive Regelsysteme. Oldenbourg, München 1971
[85]   Weh, H., Elektrische Netzwerke und Maschinen in Matrizen-Darstellung. Teubner, Stuttgart 1968
[86]   Wood, P., Switching power converters, Van Nostrand Reinhold 1981
[87]   Yamamura, S., AC motors for high performance applications, analysis and control. Marcel Dekker, New York 1986
[88]   Zadeh, L.A., Desoer, C.A., Linear system theory. McGraw Hill, New York 1963

Papers, Surveys and Dissertations

[A1]   Abbondanti, A., Method of flux control in induction motors driven by variable frequency, variable voltage supplies. Proc. Int. Semiconductor Power Converter Conf. Rec. 1977, pg. 177
[A2]   Abbondanti, A., Brennen, M.B., Variable speed induction motor drives use electronic slip calculation based on motor voltages and currents. IEEE Trans. Ind. Appl. 1975, pg. 483
[A3]   Abraham, L., Transient characteristics and limits of a squirrel-cage motor fed from a frequency-converter with DC current link. 2nd IFAC Symp., Control in Power Electronics and Electrical Drives, Düsseldorf 1977, pg. 475
[A4]   Abraham, L., Heumann, K., Koppelmann, F., Wechselrichter zur Drehzahlsteuerung von Käfigläufermotoren. AEG Mitteilungen 1964, pg. 89
[A5]   Abraham, L., Koppelmann, F., Käfigläufermotoren mit hoher Drehzahldynamik. AEG Mitteilungen 1965, pg. 118
[A6]   Ahrens, D., Günther, H., Elektrische und mechanische Einflußgrößen bei Lageregelkreisen. AEG Mitteilungen 1976, pg. 249
[A7]   Acarnley, P.P., French, C.D., Torque estimation and control in PM motor drives. Proc. EPE 95, Sevilla, pg. 3.411-3.416
[A8]   Ackva, A., Binder, A. Greubel, K. Piepenbreier, B., Electric vehicle drive with surface-mounted magnets for wide field weakening range. Proc. EPE 97, Trondheim, pg. 1.548-1.553
[A9]   Ahrens, D., Raatz, E., Regelungstechnische Untersuchungen von Antrieben mit Kupplungs-oder Getriebelose. AEG Mitteilungen 1973, pg. 210
[A10]   Aime, M.L., Degner, M.W., Lorenz, R.D., The effects of saturation induced saliency movement on flux angle estimation. Proc. Workshop on Advanced Motion Control (AMC) 98, Coimbra, pg. 369-374
[A11]   Ainsworth, J.D., Harmonics instability between controlled static convertors and AC-networks. Proc. IEE 1967, pg. 949
[A12]   Akagi, H., Nabae, A., A new control scheme for compensating the torque transfer function of a self-controlled synchronous motor. IEEE Trans. Ind. Appl. 1984, pg. 795
[A13]   Albrecht, P., Die geregelte doppeltgespeiste Asynchronmaschine als drehzahlvariabler Generator am Netz. Diss. TU Braunschweig 1984

[A14] Albrecht, P., Vollstedt, W., Microcomputer control of a variable speed doubly-fed induction generator operating on the fixed frequency grid. Microelectronics in Power Electronics and Electrical Drives, Darmstadt 1982, ETG-Fachberichte 11, pg. 363

[A15] Alexanderson, E.F.W., Mittag, A.H., The thyratron motor. AIEE Trans. 1934, pg. 1517

[A16] Aller Castro, J.M., Montilla, A.B., On-line parameter estimation of the induction machine model using active and reactive power balance for spatial vectors of field oriented drives. Proc. EPE 97, Trondheim, pg. 4.609–4.614

[A17] Ambrozic, V., Klaassen, H., A modified method for the determination of the rotor time constant of an induction motor in a cold state. Proc. 8th EDPE, Pula, 1994, pg. 66

[A18] Amler, G., Ein neues Steuer-und Regelverfahren für einen Stromzwischenkreis-Umrichter mit abschaltbaren Leistungshalbleitern. Diss. TU Braunschweig, 1991

[A19] Amler, G., A new control method for current source inverters with self-extinction devices, combining low distortion of voltage and current with fast dynamic response. Proc. EPE 91, Firenze, pg. 3–211

[A20] Andresen, E. C., Pfeiffer, R., Werth, L., Fundamentals for the design of high speed induction motor drives with transistor inverter supply. Proc. EPE 89, Aachen, pg. 823

[A21] Andresen, E. C., Gediga, S., Schwartz, H. J., A continuous PWM-square wave transition method for voltage vector control of induction motors. Proc. EPE 91, Firenze, pg. 2–156

[A22] Andresen, E. C., Haun, A., Influence of the pulse width modulation control method on the performance of frequency inverter induction motor drives. Europ. Trans. on Electr. Power Engng. 1993, pg.151

[A23] Andria, G., Salvatore, L., Signal processing for determining flux and torque of induction motors. Proc. IEE-Power Electronics and Variable Speed Drives, London 1988, pg. 296

[A24] Anke, K., Ertel, K., Sinn, G., Digitale Wegregelung. Siemens Zeitschrift 1960, pg. 664

[A25] Anke, K., Kessler, G., Müller, H., Digitale Drehzahlregelung. Siemens Zeitschrift 1960, pg. 660

[A26] Antic, D., Klaassens, J.B., Deleroi, W., An integrated boost-buck and matrix converter topology for low speed drives. Proc. EPE 93, Brighton, pg. 5-21

[A27] Appt, D., Baisch, R., Brömer, G., Wegmessung und Wegregelung an numerisch gesteuerten Werkzeugmaschinen. Siemens-Zeitschrift 1974, Beiheft Steuerung und Antriebe zur Automatisierung der Werkzeugmaschine, pg. 12

[A28] Appun, P., Control of rotating and linear induction motors for vehicle drives. 2nd IFAC Symp., Control in Power Electronics and Electrical Drives, Düsseldorf 1977, pg. 809

[A29] Armstromg, G.J., Atkinson, D.J., A comparison of model reference adaptive system and extended Kalman filter estimators for sensorless vector drives. Proc. EPE 97, Trondheim, pg. 1.424–1.429

[A30] Arnet, B., Jufer, M., Torque control on electric vehicles with separate wheel drives. Proc. EPE 97, Trondheim, pg. 4.659–4.664

[A31] Arsudis, D., Doppeltgespeister Drehstromgenerator mit Spannungszwischenkreis-Umrichter im Rotorkreis für Windkraftanlagen. Diss. TU Braunschweig 1989

[A32] Arsudis, D., Vollstedt, W., Sensorlose Regelung einer doppeltgespeisten Asynchronmaschine mit geringen Netzrückwirkungen. Archiv für Elektrotechnik, 1991, pg. 89

[A33] Atkinson, D.J., Acarnley, P.P., Finch, J.W., Parameter identification tech-
       niques for induction motor drives. Proc. EPE 89, Aachen, pg. 307
[A34] Atkinson, D.J., Acarnley, P.P., Finch, J.W., Method for the estimation of
       rotor resistance in induction motors. Proc. EPE 91, Firenze, pg. 3–338
[A35] Atkinson, D.J., Acarnley, P.P., Finch, J.W., Observers for induction motor
       state and parameter estimation. IEEE Trans. Ind. Appl. 1991, pg. 1119
[A36] Attaianese, C., Damiano, A., Marongiu, I., Perfetto, A., A DSP-based non-
       linear controller for induction motor drives. Proc. EPE 95, Sevilla, pg. 3.914–
       3.918
[A37] Attaianese, C., Damiano, A., Marongiu, I., Perfetto, A., Identification of
       rotor parameters and speed estimation in induction motor drives. Proc. EPE
       97, Trondheim, pg. 4.615–4.619
[A38] Attaianese, C., Tomasso, G., Perfetto, A., Position control of induction mo-
       tor by means of vectorial torque control (VTC) strategy. Proc. EPE 99,
       Lausanne
[B1]   Bachmann, G., Engel, D., Geregelte Gleichstrom-Hauptantriebe bis 400 kW
       für Werkzeugmaschinen. Siemens Energietechnik 1981, pg. 269
[B2]   Bader, W., Theorie der Schaltungen für die Widerstandsbremse von selbster-
       regten Gleichstrom-Reihenschlußmotoren bei elektrischen Fahrzeugen. Wiss.
       Veröff. Siemens Konzern 1930, pg. 209
[B3]   Balestrino, A., De Maria, G., Sciavicco, L., Adaptive control design in ser-
       vosystems. 3rd IFAC Symp. Control in Power Electronics and Electrical
       Drives, Lausanne 1983, pg. 125
[B4]   Bannack, A., Nguyen Hong Ha, Riefenstahl, U., Speed and angular synchro-
       nism control of machines with flexibly coupled drives. Proc. EPE 95, Sevilla,
       pg. 3.464-3.468
[B5]   Barrenscheen, J., Flieller, D., Kalinowski, D., Louis, J.P., A new sensorless
       speed and torque control for permanent magnet synchronous motors: Real-
       isation and modelling. Proc. EPE 95, Sevilla, pg. 3.839-3.844
[B6]   Bassi, E., Benzi, F., Scattolini, R., Design of a digital adaptive controller for
       electrical drives in industrial applications. IEEE Trans. Ind. Electr. 1992,
       pg. 120
[B7]   Bassi, E., Benzi, F.P., Bolognani, S., Buja, G.S., A field orientation scheme
       for current-fed induction motor drives based on the torque angle closed-loop
       control. IEEE Trans. Ind. Appl. 1992, pg.1038
[B8]   Bauer, F., Heinle, G., Eliminating the harmonics from measured current
       values in PWM drives. Proc. EPE 91, Firenze, pg. 3-343
[B9]   Bayer, K.H., Waldmann, H., Weibelzahl, M. Die TRANSVEKTOR-Re-
       gelung für den feldorientierten Betrieb einer Synchronmaschine. Siemens
       Zeitschrift 1971, pg. 765
[B10] Bayer, K.-H., Blaschke, F., Stability problems with the control of induction
       machines using the method of field orientation. 2nd IFAC Symp., Control in
       Power Electronics and Electrical Drives, Düsseldorf 1977, pg. 483
[B11] Becker, H., Dynamisch hochwertige Drehzahlregelung einer umrichterge-
       speisten Asynchronmaschine. Regelungstechnische Praxis 1973, pg. 217
[B12] Becker, H., ROBOT CONTROL, eine Mikroprozessorsteuerung für Indus-
       trieroboter. Siemens Energietechnik 1980, pg. 137
[B13] Beineke, S., Wertz, H., Schütte, F., Grotstollen, H., Fröhleke, N., Identifica-
       tion of nonlinear two-mass systems for self-commissioning speed control of
       electrical drives. Proc. IECON 98, Aachen, pg. 2251–2256
[B14] Bellini, A., Figalli, G., Tosti, F., Linearised model of induction motor drives
       via nonlinear feedback decoupling. Proc. EPE 91, Firenze, pg. 3-36

[B15] Belmans, R.J.M., Verdyck, D., Geysen, W., Findlay, R.D., Electromechanical analysis of the audible noise of an inverter-fed squirrel-cage induction motor. IEEE Trans. Ind. Appl. 1991, pg. 539

[B16] Bender, K., Pandit, M., Weber, W., Verlustoptimale Regelung von rotierenden Scheren. Regelungstechnik 1970, pg. 540; 1971,S. 8

[B17] Bender, K., Beschleunigungsgeregelte Antriebe in optimalen Regelkreisen. Regelungstechnik 1972, pg. 423

[B18] Ben Ammar, F. Pietrzak-David, M., de Fornel, B., Mirzaian, A., Field-oriented control of high-power induction motor drives by Kalman filter observation. Proc. EPE 91, Firenze, pg. 2-182

[B19] Ben Ammar, F. Pietrzak-David, M., de Fornel, B., Mirzaian, A., Power range extension of an induction motor speed drive by using a three-level GTO inverter with space vector modulation. Proc. EPE 93, Brighton, pg. 5-219

[B20] Ben Brahim, L., Kawamura, A., Digital current regulation of field-oriented controlled induction motor based on predictive flux observer. IEEE Trans. Ind. Appl. 1991, pg. 956

[B21] Ben Brahim, L., Kawamura, A., A fully digitised field-oriented controlled induction motor drive using only current sensors. IEEE Trans. Ind. Electr. 1992, pg. 241

[B22] Ben-Brahim, L., Improvement of the stability of the V/f controlled induction motor drive system. Proc. IECON 98, Aachen, pg. 859-864

[B23] Ben-Brahim, L., Tadakuma, S., Practical considerations for sensorless induction motor drive system. Proc. IECON 98, Aachen, pg. 1002-1007

[B24] Bergin, T. J., Cholewka, J., The direct load angle control of a CSI drive. Proc. IEE-Power Electronics and Variable Speed Drives, London 1988, pg. 305

[B25] Bernot, F., Gonthier, L., Bocus, S.D., Elbaroudi, S., Berthon, A., Kauffmann, J.M., High efficiency drive for electric vehicles permanent magnet motor fed by a two-stage converter. Proc. EPE 97, Trondheim, pg. 1.554-1.559

[B26] Bianchi, N., Bolognani, S., Parasiliti, F., Villani, M., Prediction of overload and flux-weakening performance of an IPM motor drive: analytical versus finite element approach. Proc. EPE 99, Lausanne

[B27] Bichler, U., Synchronmaschinen mit aktiver Dämpfung. Diss. TU Braunschweig 1985

[B28] Bilewski, M., Fratta, A., Giordano, L., Vagati, A., Villata, F., Control of high-performance interior permanent magnet synchronous drives. IEEE Trans. Ind. Appl. 1993, pg. 328

[B29] Binns, K.J., Wong, T.M., Analysis and performance of a high-field permanent-magnet synchronous machine. IEE Proceedings 1984, pg. 252-258

[B30] Blaabjerg, F., Pedersen, J.K., Digital implemented random modulation strategies for AC and switched reluctance drives. Proc. IECON 93, Maui, pg. 676

[B31] Blaschke, F., Ripperger, H., Steinkönig, H., Regelung umrichtergespeister Asynchronmaschinen mit eingeprägtem Ständerstrom. Siemens Zeitschrift 1968, pg. 773

[B32] Blaschke, F., Hütter, G., Scheider, U., Zwischenkreisumrichter zur Speisung von Asynchronmaschinen für Motor-und Generatorbetrieb. ETZ-A, 1968, pg. 108

[B33] Blaschke, F., Das Prinzip der Feldorientierung, die Grundlage für die TRANSVEKTOR-Regelung von Asynchronmaschinen. Siemens Zeitschrift 1971,pg. 757

[B34]   Blaschke, F., The principle of field orientation as applied to the new TRANSVEKTOR closed loop control system for rotating field machines. Siemens Review 1972, pg. 217

[B35]   Blaschke, F., Das Verfahren der Feldorientierung zur Regelung der Asynchronmaschine. Siemens Forschungs-und Entwicklungsberichte 1972, pg. 184

[B36]   Blaschke, F., Das Verfahren der Feldorientierung zur Regelung der Drehfeldmaschine. Diss. TU Braunschweig 1973

[B37]   Blaschke, F., Böhm, K., Verfahren der Flußerfassung bei der Regelung stromrichtergespeister Asynchronmaschinen. 1st IFAC-Symp., Control in Power Electronics and Electrical Drives, Düsseldorf 1974, Vol. 1, pg. 635

[B38]   Blaschke, F., Ströle, D., Transformationen zur Entflechtung elektrischer Antriebsregelstrecken. Regelungstechnik 1973, pg. 105

[B39]   Blasco-Giminez, R., Asher, G.M., Sumner, M., Rotor time constant identification in sensorless vector control drives using rotor slot harmonics. Proc. EPE 95, Sevilla, pg. 1.083–1.088

[B40]   Bodson, M., Chiasson, J., Novotnak, R., A systematic approach to selecting flux references for torque maximization in induction motors. IEEE Trans. Control Syst. Techn. 1995, pg. 388

[B41]   Bodson, M., Chiasson, J., Novotnak, R., Nonlinear speed observer for high-performance induction motor control. IEEE Trans. Ind. Electr. 1995, pg. 337

[B42]   Böcker, J., Janning, J., Discrete-time flux observer for PWM inverter fed induction motors. Proc. EPE 91, Firenze, pg. 2–171

[B43]   Böhm, E., Führung und Regelung von Antennenantrieben für Satelliten-Bodenstationen. Siemens Zeitschrift 1974, pg. 828

[B44]   Böhm, K., Wesselak, F., Drehzahlregelbare Drehstromantriebe mit Umrichterspeisung. Siemens Zeitschrift 1971, pg. 753

[B45]   Boehringer, A., Knöll, H., Transistorschalter im Bereich hoher Leistungen und Frequenzen. ETZ 1979, pg. 664

[B46]   Boehringer, A., Stute, G., Ruppmann, C., Vogt, G., Würslin, R., Entwicklung eines drehzahlgesteuerten Asynchronmaschinenantriebs für Werkzeugmaschinen. Zeitschrift für industrielle Fertigung 1979, pg. 463

[B47]   Böning, B., Leonhard, W., Propulsion control of a track-powered linear synchronous motor for a magnetically levitated vehicle on the basis of power measurements in the inverter station. Proc. Int. Conf. on Electrical Machines, Brussel 1978, paper L 5/ 1

[B48]   Boersting, H., Knudsen, M., Vadstrup, P., Standstill estimation of electrical parameters in induction motors using an optimal input signal. Proc. EPE 95, Sevilla, pg. 1.814–1.819

[B49]   Bogachev, Y.P., Izosimov, D.B., New generation of digital controlling electrical drives: high-precision synchronous servo drive for machine tool axes - robotics - mechatronic modules. Proc. EPE 95, Sevilla, pg. 3.863-3.868

[B50]   Boidin, M., de Fornel, B., Reboulet, C., Commande en couple sans capteur mèchanique d'une machine asynchrone d'induction. Conumel 83, Toulouse pg. IV-43

[B51]   Borojevic, D., Robust nonlinear control algorithm for fast positioning in servo drives. Proc. EPE 89, Aachen, pg. 1375

[B52]   Bose, B.K., Adjustable speed AC drives – a technology status review. Proc. IEEE 1982, pg. 116

[B53]   Bose, B.K.(Ed.), Microcomputer control of power electronics and drives. IEEE, New York 1987

[B54]   Bose, B.K., Power electronics and motion control– Technology status and recent trends. IEEE Trans. Ind. Appl. 1993, pg. 902

[B55] Bottura, C.P., Augusto Filho, S.A., Robust torque tracking control for the induction machine. Proc. IECON 98, Aachen, pg. 1533–1537

[B56] Bowes, S.R., Bird, B.M., Novel approach to the analysis and synthesis of modulation processes in power converters. Proc. IEE 1975, pg. 507

[B57] Bowes, S.R., Midoun, A., Microprocessor implementation of new optimal PWM switching strategies. Proc. IEE 1988, Pt. B. pg. 269

[B58] Bowes, S.R., New regular-sampled harmonic elimination PWM techniques for drives and static power converters. Proc. EPE 91, Firenze, pg. 3-241

[B59] Bowes, S.R., Advanced regular-sampled PWM control techniques for drives and static power converters. Proc. IECON 93, Maui, pg. 662

[B60] Bowler, P., Carter, R.C., Jones, D.M., The development and performance of a digital subharmonic modulator for AC drives. Proc. IEE-Power Electronics and Variable Speed Drives, London 1988, pg. 403

[B61] Braga, G., Farini, A., Fuga, F., Manigrasso, R., Synchronous drive for motorised wheels without gearbox for light rails system and electric cars. Proc. EPE 91, Firenze, pg. 4–78

[B62] Brandes, J., Greubel, K., Winter, U., Antriebe für Elektrofahrzeuge - ein Konzeptvergleich. etz 1994, H. 15, pg. 846–851

[B63] Brandenburg, G., Tröndle, H.-P., Das dynamische Verhalten des Registerfehlers bei Rotationsdruckmaschinen. Siemens Forschungs-und Entwicklungsberichte 1976, pg. 17-20 und pg. 65-71

[B64] Brandenburg, G., Wolfermann, W., State observers for multi-motor-drives in processing machines with continuously moving webs. Proc. EPE 85, Brussels, pg. 3.203-3.210

[B65] Brandenburg, G., Schäfer, U., Compensation of Coulomb friction in industrial elastic two-mass-system through adaptive control. Proc. EPE 89, Aachen, pg. 1409-1415

[B66] Brandenburg, G., Papiernik, W., Feedforward and feedback strategies applying the principle of input balancing for minimal tracking errors in CNC machine tools. Proc. Workshop on Advanced Motion Control (AMC) 96, Tsu City, pg. 612-618

[B67] Brandenburg, G., Geißenberger, S., Kink, C., Schall, N-H., Schramm, M., Multi-motor line shafts for rotary printing presses - a revolution in printing machine techniques. Proc. Workshop on Advanced Motion Control (AMC) 98, Coimbra, pg. 457-462

[B68] Brandstetter, P., Cermak, T., Chlebis, P., Kozina, J., Stefanik, M., Microcomputer system control in field coordinates of synchronous machine. Proc. EPE 91, Firenze, pg. 4-560

[B69] Brandstetter, P., Mech, M., Control methods for permanent magnet synchronous motor drives with high dynamic performance. Proc. EPE 95, Sevilla, pg. 3.805-3.810

[B70] Bressani, M., Odorico, A., Sica, M., High-speed, large-power induction motors for direct coupling to variable-speed gas compressors. Proc. EPE 93, Brighton, pg. 5–366

[B71] Brickwedde, A., Head, D., Graham, H., Microprocessor controlled 50 kVA PWM inverter drive. IEEE Ind. Appl. Ann. Meet. 1981, pg. 666

[B72] Brickwedde, A., Microprocessor-based adaptive control for electrical drives. 3rd IFAC Symp. Control in Power Electronics and Electrical Drives, Lausanne 1983, pg. 119

[B73] Brickwedde, A., Selbsteinstellender, on-line adaptiver Regler auf Mikrorechnerbasis für elektrische Antriebe. Diss. TU Braunschweig 1985

[B74] Brösse, A., Henneberger, G., Klepsch, Th., Positioning acuracy of a sensorless controlled servo drive system. Proc. EPE 95, Sevilla, pg. 3.167-3.172

[B75]   Brösse, A., Henneberger, G., Sensorless control of a switched reluctance mo-
        tor using a Kalman filter. Proc. EPE 97, Trondheim, pg. 4.561–4.566
[B76]   Brösse, A., Henneberger, G., Schniedermayer, M., Lorenz, R.D., Magel, N.,
        Sensorless control of a SRM at low speeds and standstill based on signal
        power evaluation. Proc. IECON 98, Aachen, pg. 1538-1543
[B77]   Brown, G.M., Szabados, B., Belmans, R.J.M., High power cycloconverter
        drive for double- fed induction motors. Proc. EPE 91, Firenze, pg. 2- 282
[B78]   Brunotte, C., Schumacher, W., Detection of the starting rotor angle of a
        PMSM at standstill. Proc. EPE 97, Trondheim, pg. 1.250-1.253
[B79]   Brunotte, C., Regelung und Identifizierung von Linearmotoren für Werk-
        zeugmaschinen. Diss. TU Braunschweig, 2000
[B80]   Bühler, E., Eine zeitoptimale Thyristor-Stromregelung unter Einsatz eines
        Mikroprozessors. Regelungstechnik 1978, pg. 37
[B81]   Bühler, H., Investigation of a rectifier regulating circuit as a sampled data
        system. 4th IFAC-Congress, Warsaw 1969
[B82]   Bühler, H., Umrichtergespeiste Antriebe mit Asynchronmaschinen. Neue
        Technik 1974, pg. 121
[B83]   Bünte, A., Grotstollen, H., Parameter identification of an inverter-fed induc-
        tion motor at standstill with a correlation method. Proc. EPE 93, Brighton,
        pg. 5-97
[B84]   Bünte, A., Grotstollen, H., Off-line parameter identification of inverter-fed
        induction motor at standstill. Proc. EPE 95, Sevilla, pg. 3.493-3.496
[B85]   Buja, G.S., Indri, G.B., Optimal pulse width modulation for feeding AC
        motors. IEEE Trans. Ind. Appl. 1977, pg. 38
[B86]   Buja, G.S., Fiorini, P., Microcomputer control of PWM inverters. IEEE
        Trans. Ind. Electr. 1982, pg. 212
[B87]   Buja, G.S., Menis, R., Valla, M.I., Disturbance torque estimation in a sen-
        sorless DC drive. Proc. IECON 93, Maui, pg. 977
[B88]   Busse, A., Holtz, J., A digital space vector modulator for the control of
        a three-phase power converter. Microelectronics in Power Electronics and
        Electrical Drives, Darmstadt 1982 ETG-Fachberichte 11, pg. 189
[B89]   Buxbaum, A., Regelung von Stromrichterantrieben bei lückendem und
        nichtlückendem Ankerstrom. AEG Mitteilungen 1969, pg. 348
[B90]   Buxbaum, A. Aufbau und Funktionsweise des adaptiven Ankerstromreglers.
        AEG Mitteilungen 1971, pg. 371
[B91]   Buxbaum, A., Einsatz von adaptiven Reglern bei geregelten Stromrichter-
        Stellgliedern. VDE-Aussprachetag Freiburg 1973, pg. 99
[B92]   Bystron, K., Strom-und Spannungsverhältnisse beim Drehstrom-Drehstrom-
        Umrichter mit Gleichspannungs-Zwischenkreis. ETZ-A, 1966, pg. 264
[B93]   Bystron, K., Meissen, W., Drehzahlsteuerung von Drehstrommotoren über
        Zwischenkreisumrichter. Siemens Zeitschrift 1965, pg. 254
[C1]   Cambronne, J.P., Semail, B., Rombaut, C., Vector control of a PWM current
        source inverter-fed induction motor. Proc. EPE 91, Firenze, pg. 2-177
[C2]   Capolino, G.A., Du, B., Extended Kalman observer for induction machine
        rotor currents. Proc. EPE 91, Firenze, pg. 3-672
[C3]   Cardoletti, L., Jufer, M., Sensorless position detection using the supply volt-
        age for a programmable current drive for synchronous motors. Proc. EPE
        91, Firenze, pg. 4–123
[C4]   Cardoso, F.D.S., Martins, J.F., Pires, V.F., A comparative study of a PI,
        neural network and Fuzzy genetic approach controllers for a AC drive. Proc.
        Workshop on Advanced Motion Control (AMC) 98, Coimbra, pg. 375-380

[C5]   Castillo-Castañeda, E., Okazaki, Y., Load variation compensation in real-time for motion accuracy of machine tools. Proc. Workshop on Advanced Motion Control (AMC) 98, Coimbra, pg. 305-309

[C6]   Case, M.J., Kulentic, P.K., A microprocessor controller for the cyclocon-verter. Microelectronics in Power Electronics and Electrical Drives, Darmstadt 1982, ETG-Fachberichte 11, pg. 171

[C7]   Case, M. J., Van Wyk, Transputer based induction motor flux-control method. Proc. EPE 91, Firenze, pg. 2-192

[C8]   Cataliotti, A.,Tecniche di controllo del motore asincrono - Un nuovo schema di controllo vettoriale encoderless. Diss. Univ. Palermo 1998

[C9]   Cecati, C., Rotondale, N., On-line identification of electrical parameters of the induction motor using RLS estimation. Proc. IECON 98, Aachen, pg. 2263-2268

[C10]   Chapuis, Y.A., Girerd, C., Aubépart, F., Blondé, J.P., Braun, F., Quantization problem analysis on ASIC- based direct torque control of an induction motor. Proc. IECON 98, Aachen, pg. 1527-1532

[C11]   Chauprade, R, Abhondanti, A., Variable speed drives: Modern concepts and approaches. Proc. Int. Semiconductor Power Conv. Conf. 1982, Ind. Appl., pg. 20

[C12]   Chen, H.H., Microprocessor control of a three-pulse cycloconverter. Proc. IECI 1977, pg. 226

[C13]   Chiasson, J., Dynamic feedback linearization of the induction motor. IEEE Trans. Aut. Contr. 1993, pg. 1588

[C14]   Chiasson, J., Nonlinear controllers for induction motors. IEEE Trans. Aut. Contr. 1995

[C15]   Chiba, A., Fukao, T., A closed loop operation of super high speed reluctance motor for quick torque response. IEEE Trans. Ind. Electr. 1992, pg. 600

[C16]   Choi, J.S., Kim, S.U., Lee, E.J., Kim, Y.S., A sensorless vector control of the saturated induction motor using third harmonic voltages. Proc. IECON 98, Aachen, pg. 977-980

[C17]   Chollet, G., Commande par microprocesseur d'un commutateur de courant pour moteur asynchrone de traction. Conumel 83, Toulouse pg. IV-1

[C18]   Claussen, U., Adaptive zeitoptimale Lageregelung eines linearen Stellantriebes mit synchronem Linearmotor. Diss. TU Braunschweig 1979

[C19]   Claussen, U., Fromme, G., Motorregelung mit Mikrorechner. Regelungstechnische Praxis 1978, pg. 355

[C20]   Claussen, U., Leonhard, W., Microprocessor-controlled linear synchronous motor as positioning drive. Proc. Int. Conf. on Electrical Machines, Brussel 1978, L 5/2

[C21]   Clénet, S., Vinassa, J.M., Astier, S., Lefèvre, Y., Lajoie-Mazenc, M., Influence of the brushless DC motor torque compensation by acting on the current waveshapes on the torque speed characteristics. Proc. EPE 95, Sevilla, pg. 3.898-3.902

[C22]   Colak, I., Garvey, S., Wright, M.T., Mixed-frequency testing of induction machines using inverters. Proc. EPE 93, Brighton, pg. 5-317

[C23]   Collings, T.D., Wilson, W.J., A fast response current controller for microprocessor based SCR-DC motor drives. IEEE Trans. Ind. Appl. 1991, pg. 921

[C24]   Conrath, C., Commande á microprocesseur d'une machine asynchrone autopilotèe. Application à un système de levage. Conumel 83, Toulouse, pg. IV-7

[C25] Conroy, B.P., Sumner, M., Alexander, T., Application of encoderless vector control techniques in a medium performance induction motor drive. Proc. EPE 95, Sevilla, pg. 3.469-3.474

[C26] Cordes, S., Nestler, J., Matrixconverter zur Speisung von drehzahlveränderlichen Drehfeldmaschinen. Proc. SPS/IPC/DRIVES 99, Nürnberg, pg. 531-540

[C27] Cordeschi, G., Parasiliti, F., A variable structure approach for speed control of field-oriented induction motor. Proc. EPE 91, Firenze, pg. 2-222

[C28] Cornell E.P., Lipo, T.A., Modelling and design of controlled current induction motor drive systems. IEEE Trans. Ind. Appl. 1977, pg. 321

[C29] Coussens, P.J. Van den Bossche, A.P., Melkebeek, J.A., Nonlinear field oriented control in a rotor reference frame. Proc. EPE 95, Sevilla, pg. 1.820-1.825

[D1]   Da Costa Branco, P.J., Stephan, R.M., A simple adaptive scheme for indirect field orientation of an induction motor. Proc. EPE 91, Firenze, pg. 2-208

[D2]   Dakhouche, K., Roye, D., Digital vector control of induction machine using a PWM inverter. Proc. EPE 91, Firenze, pg. 2-227

[D3]   Dallago, E., Sassone, G., Study and application of a digital speed control in a DC brushless drive. Europ. Trans. on Electr. Power Engng. 1991, pg. 169

[D4]   Dallhammer, R., Gotthardt, K.-J., The operation of microcomputers in railways with the example of the H-Bahn. Microelectronics in Power Electronics and Electrical Drives, Darmstadt 1982, ETG-Fachberichte 11, pg. 409

[D5]   Dannhauer, D., Prüfstand für Werkzeugmaschinenantriebe mit einer permanenterregten Synchronmaschine als Belastungseinheit. Diss. TU Braunschweig 1993

[D6]   Daum, D., Unterdrückung von Oberschwingungen durch Pulsbreitensteuerung. ETZ-A, 1972, pg. 528

[D7]   Daum, D., Digitale Steuereinrichtung für Stromrichteranlagen. ETZ-A, 1973, pg. 299

[D8]   Daum, D., Untersuchung eines Einphasenstromrichters mit nahezu sinusförmigem Netzstrom und gut geglätteten Gleichgroßen. Diss. Univ. Bochum 1974

[D9]   Davis, R.M., Melling, J., Quantitative comparison of commutation circuits for bridge inverters. Proc. IEE 1977, pg. 237

[D10] Davis, R.M., A variable speed drive that took a century and a half to develop. Electrical Times, 1983

[D11] De Carli, A., Marola, G., Optimal pulse-width modulation for three-phase voltage supply. Microelectronics in Power Electronics and Electrical Drives, Darmstadt 1982, ETG-Fachberichte 11, pg. 197

[D12] De Carli, A, Steady state behaviour of converter fed systems. 3rd IFAC Symp. Control in Power Electronics and Electrical Drives. Lausanne, 1983, Survey

[D13] De Carli, A., Qualification of drives for motion control. Proc. EPE 91, Firenze, pg. 2-589

[D14] De Carli, A., Onofri, I., Design and implementation of a robust control strategy for mechanical systems. Proc. Workshop on Advanced Motion Control (AMC) 98, Coimbra, pg. 47-52

[D15] De Doncker, R.W.A.A., Synthesis and digital implementation of adaptive field orientation controllers for induction machines with air gap flux control and deep bar compensation. Diss. Kath. U. Leuven, 1986

[D16] De Doncker, R.W.A.A., Field-oriented controllers with rotor deep bar compensation circuits. IEEE Trans. Ind. Appl. 1992, pg. 1062

[D17]   de Fornel, B., Bach, J.L., Pietrzak-David, M., Control law of an AC variable speed drive without mechanical sensors. Microelectronics in Power Electronics and Electrical Drives Darmstadt 1982, ETG-Fachberichte 11, pg. 329

[D18]   Degner, M.W., Lorenz, R.D., Wide bandwidth flux, position and velocity estimation in AC machines at any speed (including zero) using multiple saliencies. Proc. EPE 97, Trondheim, pg. 1.536-1.541

[D19]   de Jager, W.A.G., Tubbing, G.H., A vector oriented control strategy for a 4-quadrant line side converter. Proc. EPE 93, Brighton, pg. 5-213

[D20]   de la Vallée, H., Grenier, D., Legat, J.-D., Labrique, F., A dedicated multiprocessor architecture for AC motor control. Proc. IECON 98, Aachen, pg. 1632-1637

[D21]   Delécluse, Ch., Grenier, D., A measurement method of the exact variation of the self and mutual inductances of a buried magnet synchronous motor and its application to the reduction of torque ripples. Proc. Workshop on Advanced Motion Control (AMC) 98, Coimbra, pg. 191-197

[D22]   Dell'Aquila, A., Cupertino, F., Salvatore, L., Stasi, S., Kalman filter estimators applied to robust control of induction motor drives. Proc. IECON 98, Aachen, pg. 2257-2262

[D23]   Deleroi, W., Winterling, M.W., Commutation analysis of synchronous machine drives. Proc. EPE 91, Firenze, pg. 4-572

[D24]   Depenbrock, M., Einphasenstromrichter mit sinusförmigem Netzstrom und gut geglätteten Gleichgrößen. ETZ-A, 1973, pg. 466

[D25]   Depenbrock, M.; Einphasenstromrichter mit optimiertem Leistungsfaktor. ETZ-A, 1974, pg. 360

[D26]   Depenbrock, M., Direkte Selbstregelung (DSR) für hochdynamische Drehfeldantriebe mit Umrichterspeisung, ETZ-Archiv 1985, pg. 211

[D27]   Depenbrock, M., Direct self control (DSC) of inverter-fed induction machine. IEEE Trans. Power Electr. 1988, pg.420

[D28]   Depenbrock, M., Hoffmann, F., Koch, S., Speed sensorless high performance control for traction drives. Proc. EPE 97, Trondheim, pg. 1.418-1.423

[D29]   Depenbrock, M., Foerth, Ch., Koch, S., Speed sensorless control of induction motor at very low stator frequencies. Proc. EPE 99, Lausanne

[D30]   Depping, F., Voits, M., Automatic selection of control algorithms for an electrical drive with microcomputer-based speed control. 3rd IFAC Symp. Control in Power Electronics and Electrical Drives, Lausanne 1983, pg. 507

[D31]   Dessaint, L.A., Saad, M., Hebert, B., Al-Haddad, K., An adaptive controller for a direct-drive scara robot. IEEE Trans. Ind. Electr. 1992, pg. 105

[D32]   Dietrich, D., Naunin, D., Closed loop speed control of an induction motor by microcomputers. Microelectronics in Power Electronics and Electrical Drives, Darmstadt 1982, ETG-Fachberichte 11, pg. 335

[D33]   Dirr, R, Neuffer, I., Schlüter, W., Waldmann, H., Neuartige elektronische Regeleinrichtungen für doppeltgespeiste Asynchronmotoren großer Leistung. Siemens Zeitschrift 1971, pg. 362

[D34]   Dittrich, A., Hofmann, W., Stoev, A., Thieme, A., Design and control of a wind power station with double fed induction generator. Proc. EPE 97, Trondheim, pg. 2.723-2.728

[D35]   Dobrovski, I.A., Beiträge zum Entwurf und zur Dimensionierung von Thyristor-Gleichstrom-Umrichtern. Diss. TU Braunschweig 1976

[D36]   Dodds, J.S., Vittek, J., Mienkina, M., Implementation of a sensorless induction motor drive control system with prescribed closed-loop rotor magnetic flux and speed dynamics. Proc. EPE 97, Trondheim, pg. 4.492-4.497

[D37] Dote, Y., Stabilization of controlled current induction motor drive system via new nonlinear state observer. IEEE Trans. Ind. El. Contr. Instr. 1980, pg. 77

[D38] Dote, Y., Anbo, K., Combined parameter and state estimation of controlled current induction motor drive system via stochastic nonlinear filtering techniques. IEEE Ind. Appl. Ann. Meet. Conf. Rec. 1979, pg. 838

[D39] Dote, Y., Manabe, T., Murakami, S., Microprocessor-based force control for manipulator using variable structure with sliding mode. 3rd IFAC Symp. Control in Power Electronics and Electrical Drives, Lausanne 1983, pg. 145

[D40] Drozdowski, P., Field oriented control of the induction motor fed by a 3x3 matrix converter. Proc. EPE 99, Lausanne

[D41] Du, T., Brdys, M.A., Taufiq, J.A., Analytical parameter sensitivity expressions for a field oriented induction motor drive. Proc. EPE 91, Firenze, pg. 3-623

[D42] Dunford, W.G., Dewan, S.G., The design of a control circuit for a two-quadrant chopper based on the motorola 6800 microprocessor. IEEE Trans. Ind. Appl. 1980, pg. 495

[E1]   Eckhardt, H., Berechnung des quasistationären Betriebs einer pulswechselrichtergespeisten Synchronmaschine kleiner Leistung. ETZ-Archiv 1982, pg. 321

[E2]   Eder, E., Die Ungleichförmigkeit der von einem dekadischen Frequenzteiler erzeugten Pulsfrequenzen. Regelungstechnik 1976, pg. 53

[E3]   Edvardsen, P.A., Nestli, T.F., Nilsen, R., Kolstad, H., Steady state power flow and efficiency optimizing analysis of a variable speed constant frequency generating system. Proc. EPE 97, Trondheim, pg. 2.691-2.694

[E4]   Eggert, B., 1.5 MW wind power station with low AC line distortion using a standard doubly-fed generator with field orientation control. Proc. EPE 97, Trondheim, pg. 2.739-2.742

[E5]   Eibel, J., Jötten, R., Control of a 3-level switching inverter, feeding a three-phase machine, by a microprocessor. Microelectronics in Power Electronics and Electrical Drives, Darmstadt 1982, ETG-Fachberichte 11, pg. 217

[E6]   Eisenack, H., Hofmeister, H., Digitale Nachbildung von elektrischen Netzwerken mit Dioden und Thyristoren. AfE 1972, pg. 32

[E7]   Elger, H., Weiß , M., Untersynchrone Stromrichterkaskade als drehzahlregelbarer Antrieb für Kesselspeisepumpen. Siemens Zeitschrift 1968, pg. 308

[E8]   Empringham, L., Wheeler, P., Clare, J., Bi-directional switch commutation for matrix converters. Proc. EPE 99, Lausanne

[E9]   Endo, T., Tajima, F., Okuda, H., Iizuka, K., Kawaguchi, Y., Uzuhashi, H., Okada, Y., Microcomputer-controlled brushless motor without a shaft-mounted position sensor. Proc. Int. Power Electr. Conf., Tokyo 1983, pg. 938

[E10] Engelhardt, W., Schlingenheber und deren Antriebe an Fertigstraßen von Breitbandwalzwerken. Siemens Zeitschrift 1973, Beiheft Antriebs-technik und Prozeßautomatisierung in Hütten-und Walzwerken, pg. 89

[E11] Enjeti, P.N., Ziogas, P.D., Lindsay, J.F., A current source PWM inverter with instantaneous current control capability. IEEE Trans. Ind. Appl. 1991, pg. 582

[E12] Enjeti, P.N., Ziogas, P.D., Lindsay, J.F., Rashid, M.H., A new current control scheme for AC motor drives. IEEE Trans. Ind. Appl. 1992, pg. 842

[E13] Ernst-Cathor, J., Drehzahlvariable Windenergieanlage mit Gleichstromzwischenkreis- Umrichter und optimum-suchendem Regler. Diss TU Braunschweig 1986

[E14]   Ertugrul, N., Acarnley, P.P., French, C.D., Real-time estimation of rotor position in PM motors during transient operation. Proc. EPE 93, Brighton, pg. 5-311

[E15]   Eschenauer, H., Henning, G., Tröndle, H.-P., Digitale Simulation des dynamischen Verhaltens und der Antriebsregelung von Großantennen. Siemens Zeitschrift 1974, pg 833

[E16]   Espelage, P.M., Chiera, J.A.,Turnbull, F.G., A wide range static inverter suitable for AC induction motor drives. IEEE Trans. Ind. Gen. Appl. 1969, pg. 438

[E17]   Ettlinger, G., Leitgeb. W., Poppinger, H., Simotron-Antriebe mittlerer Leistung. Siemens Zeitschrift 1971, pg. 186

[F1]   Fadel, M., de Fornel, B., Control laws of a synchronous machine fed by a PWM voltage source inverter. Proc. EPE 89, Aachen, pg. 859

[F2]   Farrer, W., Miskin, J.D., Quasi-sine-wave fully regenerative inverter. Proc. IEE 1973, pg. 969

[F3]   Favre, J.-P., Microprocessor-based speed control of a SCR-DC-motor. Microelectronics in Power Electronics and Electrical Drives, Darmstadt 1982, ETG-Fachberichte 11, pg.257

[F4]   Fedor, P., Fedak, V., Bober, P., Timko, J., Field-oriented induction motor drive with indirect rotor flux sensing. Proc. EPE 91, Firenze, pg. 2-214

[F5]   Fenker, O., Schumacher, W., Control of an induction motor without shaft encoder using the VECON chip. Proc. EPE 97, Trondheim, pg. 1.430-1.433

[F6]   Fenker, O., Wirkungsgradoptimierte und drehgeberlose Regelung von Asynchronmaschinen. Diss. TU Braunschweig 1998

[F7]   Ferranti, M., Ferraris, L., Laner, E., Villata, F., Induction motor drives for traction application. Proc. EPE 93, Brighton, pg. 5-282

[F8]   Ferraris, P., Fratta, A., Vagati, A., Villata, F., About the vector control of induction motors for special applications without speed sensor. Proc. Evolution and modern Aspects of Induction machines, 1986, Torino, pg. 444

[F9]   Ferretti, G., Magnani, G., Rocco, P., Online identification and compensation of torque disturbances in permanent magnet AC motors. Proc. IECON 98, Aachen, pg. 1521 1526

[F10]   Fetz, J., Horstmann, D., Comparison of different field oriented control methods for an induction motor fed by a PWM controlled inverter. Proc. EPE 89, Aachen, pg. 1079

[F11]   Fieger, K., Über die Anwendung der Abtasttheorie auf nichtlineare Regelkreise mit nichtstetigen Stellgliedern. Diss. TU Braunschweig 1967

[F12]   Fieger, K.: Zum dynamischen Verhalten thyristorgespeister Gleichstrom-Regelantriebe. ETZ-A, 1969, pg. 311

[F13]   Fieger, K., Middel, J., Vereinfachte Behandlung linearer Abtastregelkreise. Regelungstechnik 1967, pg. 445

[F14]   Finlayson, P.T., Washburn, D.C., Cycloconverter controlled synchronous machine for load compensation on AC power systems. IEEE Trans. Ind. Appl. 1974, pg. 806

[F15]   Fischer, J., Digitale Fahrkurvenrechner für Positionieranlagen. AEG Mitteilungen 1976, pg. 269

[F16]   Fischer, R., Geissing, H., Halbleiterstromrichter in kreisstromfreier Gegenparallelschaltung für Umkehrantriebe. Siemens Zeitschrift 1964, pg. 240

[F17]   Flach, E., Hoffmann, R., Mutschler, P., Direct mean torque control of an induction motor. Proc. EPE 97, Trondheim, pg. 3.672-3.677

[F18]   Flöte, R., Velte, H., Die thyristorgespeiste Fördermaschine in Feldumkehrschaltung. AEG Mitteilungen 1972, pg. 254

[F19]   Flöter, W., Ripperger, H., Die TRANSVEKTOR-Regelung für den feldori-
        entierten Betrieb einer Asynchronmaschine. Siemens Zeitschrift 1971, pg.
        761

[F20]   Flügel, W., Drehzahlregelung der spannungsumrichtergespeisten Asynchron-
        maschine im Grund-und Feldschwächbereich. ETZ-Archiv 1982, pg. 143

[F21]   Foerst, R.: Der Löschwinkelregelkreis einer Hochspannungs-Gleichstrom-
        Übertragung mit symmetrischer Drehspannung. Diss. TH Darmstadt 1967

[F22]   Fossard, A.J., Clique, M., A general model for switched DC-DC converters
        including filters. 2nd IFAC-Symp. Control in Power Electronics and Electri-
        cal Drives, Düsseldorf 1977, pg. 117

[F23]   Frankl, G., Control modes for a variable speed drive with a converter-fed
        synchronous machine. Proc. EPE 91, Firenze, pg. 4-592

[F24]   Fratta, A., Vagati, A., Villata, F., Vector control of induction motors without
        shaft transducers. Proc. IEEE-PESC 88, Kyoto, pg.839

[F25]   Fratta, A., Grassi, R., Vagati, A., High performance spindle drives adopting
        IPM synchronous machines; theoretical goals and a practical solution. Proc.
        EPE 91, Firenze, pg. 3- 476

[F26]   Fratta, A., Vagati, A., Villata, F., Extending the voltage saturated perfor-
        mance of DC brushless drive. Proc. EPE 91, Firenze, pg. 4- 134

[F27]   Fratta, A., Vagati, A., Villata, F., On the evolution of AC machines for
        spindle drive applications. IEEE Trans. Ind. Appl. 1992. pg. 1081

[F28]   Freund, E., Hoyer, H., Das Prinzip nichtlinearer Systementkopplung mit der
        Anwendung auf Industrieroboter. Regelungstechnik 1980, pg. 80

[F29]   Frick, A., von Westerholt, E., de Fornel, B., Nonlinear control of induction
        motors via input- output decoupling. Europ. Trans. on Electr. Power Engng.
        1994, pg. 261

[F30]   Fritzsche, W., Genaue und schnelle Regelungen von Drehzahlen durch digi-
        tale Methoden. AEG Mitteilungen 1960, pg. 419

[F31]   Fujimaki, T., Ohniwa, K., Miyashita, O., Simulation of the chopper con-
        trolled DC series motor. 2nd IFAC-Symp. Control in Power Electronics and
        Electrical Drives, Düsseldorf 1977, pg. 609

[F32]   Fujimoto, H., Kawamura, A., Perfect tracking digital motion control based
        on two-degree-of-freedom multirate feedforward control. Proc. Workshop on
        Advanced Motion Control (AMC) 98, Coimbra, pg. 322-327

[F33]   Fujita, K., Sado, K., Instantaneous speed detection with parameter identifi-
        cation for AC servo systems. IEEE Trans. Ind. Appl. 1992, pg. 864

[F34]   Fujiyama K., Tomizuka, M., Katayama, R., Digital tracking controller design
        for CD player using disturbance observer. Proc. Workshop on Advanced
        Motion Control (AMC) 98, Coimbra, pg. 598-603

[F35]   Fukunda, S., Itoh, Y., Nii, A., A microprocessor-controlled speed regula-
        tor for commutator-less motor drives. Proc. Int. Power Electr. Conference,
        Tokyo 1983, pg. 938

[F36]   Furuhashi, T., Sangwongwanich, S., Okuma, S., A position- and velocity-
        sensorless control for brushless DC motors using an adaptive sliding mode
        observer. IEEE Trans. Ind. Electr. 1992, pg. 89

[G1]   Gabriel, R., Leonhard, W., Nordby, C., Microprocessor control of induction
        motors employing field coordinates. Proc. Int. Conf. on Electrical Variable
        Speed Drives, IEE 1979, London, pg. 146

[G2]   Gabriel, R., Leonhard, W., Nordby, C., Regelung der stromrichtergespeisten
        Drehstrom-Asynchronmaschine mit einem Mikrorechner. Regelungstechnik
        1979, pg. 379

[G3]   Gabriel, R., Leonhard, W., Nordby, C., Field orientated control of a standard
        AC motor using microprocessors. IEEE Trans. Ind. Appl. 1980, pg. 186

[G4]   Gabriel, R., Leonhard, W., Nordby, C., Microprocessor control of the converter-fed induction motor. Process Automation 1980, pg. 35

[G5]   Gabriel, R., Feldorientierte Regelung einer Asynchronmaschine mit einem Mikrorechner. Diss. TU Braunschweig 1982

[G6]   Gabriel, R., Leonhard, W., Microprocessor control of induction motor. Proc. Int. Semiconductor Power Converter Conf. 1982, Orlando, pg. 385

[G7]   Ganji, A.A., Lataire, Ph., Rotor time constant compensation of an induction motor in indirect vector controlled drives. Proc. EPE 95, Sevilla, pg. 1.431-1.436

[G8]   Garces, L.J., Ein Verfahren zur Parameteranpassung bei der Drehzahlregelung der umrichtergespeisten Käfigläufermaschine. Diss. TH Darmstadt 1978

[G9]   Garces, L.J., Parameter adaption for the speed controlled static AC drive with squirrel cage induction motor. IEEE Trans. Ind. Appl. 1980, pg. 173

[G10] Garcia-Cerrada, A., Taufiq, J.A., Convergence of rotor flux estimation in field oriented control. Proc. IEE-Power Electronics and Variable Speed Drives, London 1988, pg. 291

[G11] Garcia, G.O., Stephan, R.M., Watanabe, E.H. Comparing the indirect field-oriented control with a scalar method. Proc. IECON 91, Kobe, pg. 475

[G12] Gastli, A., Matsui, N., Stator flux controlled V/f PWM inverter with identification of IM parameters. IEEE Trans. Ind, Electr. 1992, pg. 334

[G13] Gebler, D., Holtz, J., Identification and compensation of gear backlash without output position sensor in high-precision servo systems. Proc. IECON 98, Aachen, pg. 662-666

[G14] Gekeler, M, Jötten, R., Control device for a pulsed 3-level inverter feeding a three-phase machine implemented by EPROMS and MSI digital circuitry. Microelectronics in Power Electronics and Electrical Drives, Darmstadt 1982, ETG-Fachberichte 11, pg. 207

[G15] Gelman, V., High performance current source inverter drive. Proc. Int. Semiconductor Power Conv. Conf. 1982, pg. 266

[G16] Gelot, N.S., Alsina, P.J., A discrete model of induction motors for real-time control applications. IEEE Trans. Ind. Electr. 1993, pg. 317

[G17] Georgiou, G., Le Pioufle, B., Nonlinear speed control of a synchronous servomotor with robustness. Proc. EPE 91, Firenze, pg. 3-42

[G18] Gerecke, E., New methods of power control with thyristors. Survey, 3rd IFAC Congress London 1966

[G19] Geschke, B., Haspelanlage einer Warmband-Kontistraße. Siemens Zeitschrift 1973, Beiheft Antriebstechnik und Prozeßautomatisierung in Hütten-und Walzwerken, pg. 97

[G20] Geyer, W., Regelungstechnische Probleme bei der Regelung eines Gleichstrommotors bei gleichzeitigem Eingriff in Anker-und Feldstromkreis. In Energieelektronik und geregelte elektrische Antriebe VDE-Buchreihe, Vol. 11 1962, pg. 472

[G21] Geyer, W., Ströle, D., Entwicklungstendenzen bei geregelten elektrischen Antrieben. Siemens Zeitschrift 1965, pg. 490

[G22] Geyer, W., Waller, S., Direktführung numerisch gesteuerter Werkzeugmaschinen mit einem Fertigungsleitrechner. Siemens Zeitschrift 1970, Beiheft Numerische Steuerungen, pg. 11

[G23] Gewecke, J., Untersynchrone Drehstrom-Bremsschaltungen für Hebezeuge. Siemens Zeitschrift 1929, pg. 9

[G24] Globevnik, M., Induction motor parameters measurements at standstill. Proc. IECON 98, Aachen, pg. 280-285

[G25] Gölz, G., Grumbrecht, P., Umrichtergespeiste Synchronmaschinen. AEG Mitteilungen 1973, pg. 141

[G26] Götz, F.-R., Synchronisierte Einzelantriebstechnik für Druckmaschinen. Proc. SPS/IPC/DRIVES 97, Nürnberg, pg. 390-399

[G27] Gosbell, V.J., Dalton, P.M., Current control of induction motors at low speed. IEEE Trans. Ind. Appl. 1992, pg. 482

[G28] Grenier, D., Yala, S., Akhrif, O., Dessaint, L.-A., Direct torque control of PM motor with nonsinusoidal flux distribution using state-feedback linearization techniques. Proc. IECON 98, Aachen, pg. 1515-1520

[G29] Greubel, F., Huth, G., Weiterentwicklung permanentmagnet-erregter Linearmotoren. Proc. SPS/IPC/DRIVES 99, Nürnberg, pg. 745-754

[G30] Grieve, D.W., Cross, D M., Recent developments in the design and application of large, high performance cycloconverter drive systems. Proc. IEE-Power Electronics and Variable Speed Drives, London 1988, pg. 309

[G31] Grötzbach, M., Analyse von Gleichstromstellerschaltungen im lückenden und nichtlückenden Betrieb. ETZ-Archiv, 1979, pg. 29

[G32] Groß H., Riedel, H., Angepaßte Vorschubantriebe für Werkzeugmaschinen. Siemens Zeitschrift 1976, pg. 30

[G33] Grotstollen, H., Zusammenhänge und Kennwerte bei der Bemessung beschleunigungsoptimaler Antriebe. ETZ-A, 1974, pg. 663

[G34] Grotstollen, H., Comparison of speed controlled DC drives with and without subordinate current control. 2nd IFAC Symp., Control in Power Electronics and Electrical Drives, Dusseldorf 1977, pg. 583

[G35] Grotstollen, H., Pfaff, G., Bürstenloser Drehstrom-Servoantrieb mit Erregung durch Dauermagnete. ETZ 1979, pg. 1382

[G36] Gruber, M., Möller-Nehring, P., Digitaler Geschwindigkeitsregler auf Mikroprozessorbasis im System MODULPAC C. Siemens Energietechnik 1979, pg. 117

[G37] Güthlein, H., Die neue elektrische Lokomotive 120 der Deutschen Bundesbahn in Drehstrom-Antriebstechnik. Elektrische Bahnen, 1979

[G38] Guinee, R.A., Lyden, C., A novel and accurate estimation method of the steady state electromagnetic torque reduction in a brushless motor drive system with inverter dead time. Proc. IECON 98, Aachen, pg. 485-490

[G39] Guinee, R.A., Lyden, C., Accurate modelling and simulation of a high performance permanent magnet adjustable speed drive system for embedded industrial applications. Proc. EPE 99, Lausanne

[G40] Gumaste, V., Slemon, G.R., Steady state analysis of a permanent magnet synchronous motor drive with voltage-fed inverter. IEEE Ind. Appl. Ann. Meet. Conf. Rec. 1980, pg. 618

[G41] Gutt, H.J., Wanderfeldmotoren im Vergleich zu Drehfeldmaschinen herkömmlicher Bauart. ETZ-A, 1971, pg. 342

[G42] Gutt, H.J., Scholl, F.D., Investigations on a new electrical drive concept for the vertical axis of a selective compliance arm robot. Proc. IEE-Power Electronics and Variable Speed Drives, London 1988, pg. 257

[G43] Gutzwiller, R., Methode der digitalen Simulation von Stromrichterschaltungen am Beispiel des Gleichstromstellers. ETZ-A, 1978, pg. 8

[H1]   Habetler, T.G., A space vector-based rectifier regulator for AC/DC/AC converters. Proc. EPE 91, Firenze, pg. 2-101

[H2]   Habetler, T.G., Divan, D.M., Control strategies for direct torque control using discrete pulse modulation. IEEE Trans. Ind. Appl. 1991, pg. 893

[H3]   Habetler, T.G., Profumo, F., Pastorelli, M., Tolbert, L.M., Direct torque control of induction machines using space vector modulation. IEEE Trans. Ind. Appl. 1992, pg. 1045

[H4]   Haböck, A., Speisung von Synchronmaschinen über Umrichter mit einge-
        prägtem Strom im Zwischenkreis. VDE Fachtagung Elektronik 1969, pg.
        163

[H5]   Haböck, A., Köllensperger, D., Stand und Entwicklung des Stromrichtermo-
        tors. Siemens Zeitschrift 1971, pg. 177

[H6]   Hackstein, D., Time optimal 3-dimensional trajectory control. 3rd IFAC
        Symp. Control in Power Electronics and Electrical Drives, Lausanne 1983,
        pg. 553

[H7]   Hadzimejlic, N., Sabanovic, A., Behlilovic, N., Electrical vehicle main drive
        control system. Proc. EPE 99, Lausanne

[H8]   Härer, H., Horn, U., Investigation of a doubly-fed induction generator using
        a simple cycloconverter. Proc. EPE 91, Firenze, pg. 2-254

[H9]   Hagiwara, A., Tsuchiya, J., Shimizu, T., Kimura, G., Watanabe, I., Naniwa,
        K., Speed sensorless field-oriented control based on phase difference. Proc.
        IECON 98, Aachen, pg. 1014-1017

[H10]   Halász, S., Optimal control of voltage source inverters supplying induction
        motors. 2nd IFAC Symp., Control in Power Electronics and Electrical Drives,
        Düsseldorf 1977, pg. 379

[H11]   Halász, S., Zaharov. A., Analysis of two-phase PWM technique in inverter
        AC drives. Proc. EPE 99, Lausanne

[H12]   Hannakam, L., Übergangsverhalten des Drehstrom-Schleifringläufermotors.
        Regelungstechnik 1959, pg. 393, 421

[H13]   Hannakam, L., Digitales Universalprogramm für Schaltvorgänge an Dreh-
        strom-Asynchronmaschinen. ETZ-A, 1961, pg. 493

[H14]   Hapiot, J.C., Pietrzak-David, M., Bach, J.L., Regulation de vitesse d'une
        machine asynchrone alimentée en courant sans capteur de vitesse. Conumel
        83, Toulouse, pg. IV-12

[H15]   Harashima, F., Yanase, T., State space analysis of AC motors fed by thyris-
        tor inverters. 1st IFAC Symp., Control in Power Electronics and Electrical
        Drives, Düsseldorf 1974, Vol. 1, pg. 149

[H16]   Harashima, F., Naitoh, H., Dynamic performance of converter-fed syn-
        chronous motors. 2nd IFAC-Symp. Control in Power Electronics and Elec-
        trical Drives, Düsseldorf 1977, pg. 369

[H17]   Harashima, F., Hayashi, H., Dynamic performance of current source inverter-
        fed induction motors. IEEE Ind. Appl. Ann. Mect. Conf. Rec. 1978, pg. 904

[H18]   Harashima, F., Naitoh, H., Haneyoshi, T., Dynamic performance of a self
        controlled synchronous motor fed by current source inverters. IEEE Trans.
        Ind. Appl. 1979, pg. 36

[H19]   Harley, R.G., Wishart, M.T., Diana, G., Teaching field oriented control of
        VSI and CSI fed asynchronous and permanent magnet synchronous machines
        using a programmable high speed controller. Proc. EPE 91, Firenze, pg. 2-
        556

[H20]   Harms, K, Beiträge zur Regelung der Drehstrom-Asynchronmaschine unter
        Berücksichtigung veränderlicher Maschinenparameter. Diss. TU Braun-
        schweig 1988

[H21]   Harnefors, L., Taube, P., Nee, H-P., An improved method for sensorless
        adaptive control of permanent-magnet synchronous motors. Proc. EPE 97,
        Trondheim, pg. 4.541-4.546

[H22]   Harnefors, L., Instability phenomena in sensorless control of induction mo-
        tors. Proc. EPE 99, Lausanne

[H23]   Hars, W., Siebenphasiger Synchronmotor mit Pulswechelrichter in hochin-
        tegrierter Bauweise. ETG Fachberichte Nr. 22, 1988, VDE Verlag Berlin,
        Offenbach

422   Bibliography

[H24]   Hartramph, R., Schwinköthe, W., Elektrodynamische Direktantriebe mit integriertem Wegmeßsystem. Proc. SPS/IPC/DRIVES 98, Nürnberg, pg. 700-701

[H25]   Hashimoto, S., Tanaka, K., Miki, I., An autotuning method for vector controlled induction motor drives considering stator core losses. Proc. EPE 99, Lausanne

[H26]   Hasse, K., Zur Dynamik drehzahlgeregelter Antriebe mit stromrichtergespeisten Asynchron-Kurzschlußläufermaschinen. Diss. TH Darmstadt 1969

[H27]   Hasse, K., Drehzahlregelverfahren für schnelle Umrichterantriebe mit stromrichtergespeisten Asynchron-Kurzschlußläufermotoren. Regelungstechnik 1972, pg. 60

[H28]   Hattori, M., Ishikawa,T., Taniguchi, M., Ohnishi, T., Kurata, Y., Kamiyama, K., Imagawa, K., Takahashi, J., Sukegawa, T., Saito, K.: An experience with cycloconverter-fed IM drives for a modern tandem cold mill. Proc. EPE 91, Firenze, pg. 2-260

[H29]   Healey, R.C., The implementation of a vector control scheme based on an advanced motor model. Proc. EPE 95, Sevilla, pg. 3.017-3.022

[H30]   Heinemann, G., Comparison of several control schemes for AC induction motors under steady state and dynamic conditions. Proc. EPE 89, Aachen, pg. 843

[H31]   Heinemann, G., Selbsteinstellende, feldorientierte Regelung für einen asynchronen Drehstromantrieb. Diss. TU Braunschweig 1992

[H32]   Heinemann, G., Leonhard, W., Self-tuning field-orientated control of an induction motor drive. Proc. Int. Power Electr. Conf. 90, Tokyo, pg. 465

[H33]   Heinemann, G., Papiernik, W., Hochdynamische Vorschubantriebe mit Linearmotoren. VDI-Z Special Antriebstechnik, pg. 26-33, Springer Verlag, April 1998

[H34]   Heinze, K., Tappeiner, H., Weibelzahl, M., Pulswechselrichter zur Drehzahlsteuerung von Asynchronmaschimen. Siemens Zeitschrift 1971, pg. 154

[H35]   Heintze, K., Wagner, R., Elektronische Gleichstromsteller zur Geschwindigkeitssteuerung von aus Fahrleitungen gespeisten Gleichstrom-Triebfahrzeugen. ETZ-A, 1966, pg. 165

[H36]   Heller, M., Die doppelt-gespeiste Drehstrommmaschine für drehzahlvariable Pumpspeicherkraftwerke. Diss. TU Braunschweig 1998

[H37]   Heller, M., Schumacher, W., Stability analysis of doubly-fed induction machines in stator flux reference frame. Proc. EPE 97, Trondheim, pg. 2.707-2.710

[H38]   Hemmingsson, M., Simonsson, B., Alaküla, M., Olsson, G., Control and design interaction in hybrid vehicles - Benefits of using power electronics. Proc. EPE 97, Trondheim, pg. 1.566-1.569

[H39]   Hengsberger, A., Putz, U., Vetters, L., Thyristor-Stromrichter für Bahnmotoren. AEG Mitteilungen 1964, pg. 435

[H40]   Henneberger, G., Dynamic behaviour and current control methods of brushless DC motors. Proc. EPE 89, Aachen, pg. 1531

[H41]   Henneberger, G., Brunsbach, B. J., Klepsch, Th., Field oriented control of synchronous and asynchronous drives without mechanical sensors using a Kalman-filter. Proc. EPE 91, Firenze, pg. 3-664

[H42]   Henneberger, G., Lutter, T., Brushless DC-motor with digital state controller. Proc. EPE 91, Firenze, pg. 4-104

[H43]   Hertlein, K., Entwicklungstendenzen bei Banddickenregelungen für Kaltwalzwerke. AEG Mitteilungen 1979, pg. 38

[H44] Heumann, K., Jordan, K.G., Das Verhalten des Käfigläufermotors bei veränderlicher Speisefrequenz und Stromregelung. AEG Mitteilungen 1964, pg. 107

[H45] Hill, W.A., Ho, E.Y.Y., Neuzil, I.J., Dynamic behavior of cycloconverter systems. IEEE Trans. Ind. Appl. 1991, pg. 750

[H46] Hillenbrand, F., A method for determining the speed and rotor flux of the asynchronous machine by measuring the terminal quantities only. 3rd IFAC Symp. Control in Power Electronics and Electrical Drives Lausanne 1983, pg. 55

[H47] Höscheler, B., Erhöhung der Genauigkeit bei Wegmeßsystemen durch selbstlernende Komponenten systematischer Fehler. Proc. SPS/IPC/DRIVES 99, Nürnberg, pg. 617-626

[H48] Hövermann, M., Orlik, B., Schumacher, U., Field oriented control of induction motor without speed sensor. Proc. EPE 97, Trondheim, pg. 4.512-4.517

[H49] Hövermann, M., Orlik, B., Schumacher, U., Schümann, U., Operation of speed sensorless induction motors using open loop control at low frequency. Proc. EPE 99, Lausanne

[H50] Hofer, K., Digital speed control by adaptive computation of firing angle using microprocessors. Microelectronics in Power Electronics and Electrical Drives, Darmstadt 1982, ETG-Fachberichte 11, pg. 295

[H51] Hoffmann, A H., Brushless synchronous motors for large industrial drives. IEEE Trans. Ind. Gen. Appl. 1969, pg. 158

[H52] Hoffmann, E., Schäufele, R., Einsatz einer regelbaren Speicherpumpe in der Netzregelung, Elektrizitätswirtschaft, Heft 25, 1993

[H53] Hoffmann, F., Koch, S., Steady state analysis of speed sensorless control of induction machines. Proc. IECON 98, Aachen, pg. 1626-1631

[H54] Hofmann, W., A suboptimal predictive current controller for VSI-PWM inverters. Proc. EPE 91, Firenze, pg. 3-316

[H55] Hofstetter, M., Umrichter mit quasiresonant gepulstem Gleichspannungszwischenkreis für ein Drehstromantriebssystem. Archiv für Elektrotechnik, 1993, pg. 309

[H56] Hoft, R.G., Casteel, J.B., Power electronic circuit analysis techniques, survey. 2nd IFAC Symp., Control in Power Electronics and Electrical Drives, Düsseldorf 1977, pg. 987

[H57] Hoft, R.G., Singh, D., Microcomputer-controlled single phase cycloconverter. IEEE Trans. Ind. Electr. and Contr. Instr. 1977, pg. 226

[H58] Holl, E., Neumann, G., Piepenbreier, B., Tölle, H.-J., Water-cooled inverter for synchronous and asynchronous electric vehicle drives. Proc. EPE 93, Brighton, pg. 5-289

[H59] Holliday, D., Green, T.C., Williams, D.W., On-line parameter measurement of induction machines. Proc. IECON 93, Maui, pg. 992

[H60] Holliday, D., Fletcher, J.E., Williams, B.W., Non-invasive rotor position and speed sensing of asynchronous motors. Proc. EPE 95, Sevilla, pg. 1.333-1.337

[H61] Holtz, J., Ein neues Zündsteuerverfahren für Stromrichter am schwachen Netz. ETZ-A, 1970, pg. 345

[H62] Holtz, J., Kraftkomponenten und deren betriebliche Steuerung beim eisenlosen Synchronlinearmotor. ETZ-A, 1975, pg. 396

[H63] Holtz, J., Schwellenberg, U., A new fast-response current control scheme for line controlled converters. Proc. Int. Semiconductor Power Converter Conf. 1982, pg. 175

[H64] Holtz, J., Stadtfeld, S., A predictive controller for the stator current vector of AC-machines fed from a switched voltage source. Proc. Int. Power Electric Conference, Tokyo 1983, pg. 1665

[H65] Holtz, J., Bube, E., Field-oriented asynchronous pulse-width modulation for high- performance AC machine drives operating at low switching frequency. IEEE Trans. Ind. Appl. 1991, pg. 574

[H66] Holtz, J., Thimm, T., Identification of the machine parameters in a vector-controlled induction motor drive. IEEE Trans. Ind. Appl. 1991, pg. 1111

[H67] Holtz, J., Pulsewidth modulation– a survey. IEEE Trans. Ind. Electr. 1992, pg. 410

[H68] Holtz, J., Beyer, B., Optimal synchronous pulse width modulation with a trajectory- tracking scheme for high dynamic performance. IEEE Trans. Ind. Appl. 1993, pg. 1098

[H69] Holtz, J., Speed estimation and sensorless control of AC drives. Proc. IECON 93, Maui, pg. 649

[H70] Holtz, J., Beyer, B., Optimal pulsewidth modulation for AC servos and low-cost industrial drives. IEEE Trans. Ind. Appl. 1994, pg. 1039

[H71] Holtz, J., Beyer, B., The trajectory tracking approach– a new method for minimum distortion PWM in dynamic high-power drives. IEEE Trans. Ind. Appl. 1994, pg. 1048

[H72] Holtz, J., Sensorless position control of induction motors - an emerging technology. Proc. AMC 98, Coimbra, pg. 1-14

[H73] Holtz, J., Sensorless position control - an emerging technology. Proc. IECON 98, Aachen, pg. 1.1-1.12

[H74] Hombu, M., Ueda, S., Ueda, A., A current source inverter with sinusoidal inputs and outputs. IEEE Trans. Ind. Appl. 1987, pg. 247

[H75] Hombu, M., Futami, M., Miyazaki, S., Kubota, Y., Bando, A., Azusawa, N., Harmonics analysis on a slip-power recovery system fed by a DC link GTO converter. Proc. EPE 95, Sevilla, pg. 3-239

[H76] Hopfensberger, B., Atkinson, D.J., Lakin, R.A., Application of vector control to the cascaded induction machine for wind power generation schemes. Proc. EPE 97 Trondheim, pg. 2.700-2.706

[H77] Hori, Y., Umeno, T., Uchida, T., Konno, Y., An instantaneous speed observer for high performance control of DC servomotor using DSP and low resolution shaft encoder. Proc. EPE 91, Firenze, pg. 3-647

[H78] Hosono, I., Katsuki, K., Yano, M., Static converter starting of large synchronous motors. IEEE Ind. Appl. Ann. Meet. Conf. Rec. 1976, pg. 536

[H79] Huang, H., Davat, B., Lajoie-Mazenc, M., Study of control strategies and regulation in cycloconverter-fed synchronous machines. Proc. EPE 89, Aachen, pg. 871

[H80] Huang, L., Tadokoro, Y., Matsuse, K., Deadbeat flux level control of direct-field- oriented high-horsepower induction motor using adaptive rotor flux observer. IEEE Trans. Ind. Appl. 1994, pg. 954

[H81] Hugel, J., Die Berechnung von Stromrichter-Regelkreisen. AfE 1970, pg. 224

[H82] Huisman, H., Zero ripple torque control in brushless DC motors: a straightforward approach. Proc. EPE 95, Sevilla, pg. 3.851-3.856

[H83] Huppunen, J., Pyrhönen, J., Sine wave filter for PWM-inverter driven medium-speed (¡ 30 000 rpm) solid-rotor induction motors. Proc. EPE 99, Lausanne

[H84] Hussels, P., Mehne, M., Hentschel, F., Synthesis of cycloconverter and current source inverter–Presentation of a new control strategy. Proc. EPE 93, Brighton, pg. 5–33

[I1]   Inaba, H., Hirasawa, K., Ando, T., Hombu, M., Nakazato, M., Development of a high- speed elevator controlled by current source inverter system with sinusoidal input and output. IEEE Trans. Ind. Appl. 1992, pg. 893

[I2]   Irisa, T., Takata, S., Ueda, R., Sonoda, T., Mochizuki, T., A novel approach on parameter self-tuning method in AC servo systems. 3rd IFAC Symp. Control in Power Electronics and Electrical Drives, Lausanne 1983, pg. 41

[I3]   Itako, K., Internal Report, Inst. für Regelungstechnik, TU Braunschweig, 1999

[I4]   Iwakane, I., Inokuchi, H., Kai, T., Hirai, J., AC servo motor drive for precise positioning control. Proc. Int. Power Electr. Conf. Tokyo 1983, pg. 1453

[I5]   Iwasaki, M., Matsui, N., Robust speed control of IM with torque feedforward control. IEEE Trans. Ind. Electr. 1993, pg. 553

[I6]   Iwasaki, M., Shibata, T., Matsui, N., Disturbance observer based nonlinear friction compensation in table drive systems. Proc. Workshop on Advanced Motion Control (AMC) 98, Coimbra, pg. 299-304

[I7]   Iwens, R.P., Lee, F.C., Triner, J.E., Discrete time domain modelling and analysis of DC-DC converters with continuous and discontinuous inductor current. 2nd IFAC Symp., Control in Power Electronics and Electrical Drives, Düsseldorf 1977, pg. 87

[J1]   Jacovides, L.J., Matuoka, M.F., Shimer, D.W., A cycloconverter synchronous motor drive for traction applications. IEEE Ind. Appl. Ann. Meet. Conf. Rec. 1980, pg. 549

[J2]   Jahns, T.M. Flux weakening regime of an interior permanent-magnet synchronous motor drive. Trans. IEEE-Ind. Appl. 1987, pg. 681-689

[J3]   Jakubowicz, A., Nougaret, M., Perret, R., Simplified model and closed loop control of a commutatorless DC motor. IEEE Ind. Appl. Ann. Meet. Conf. Rec. 1979, pg. 857

[J4]   Jansen, P.L., Lorenz, R.D., Novotny, D.W., Observer-based direct field orientation: Analysis and comparison of alternative methods.IEEE Trans. Ind. Appl. 1994, pg. 945

[J5]   Jansen, P.L., Thompson, C.O., Lorenz, R.D., High-quality torque control at zero and very high speed operation. IEEE IAS Magazine, 1995, No.4, pg. 7

[J6]   Jansen, P.L., Lorenz, R.D., Transducerless position and velocity estimation in induction and salient AC machines. Trans. IEEE Ind. Appl. 1995, pg. 240-247

[J7]   Jansen, P.L., Corley, M.J., Lorenz, R.D., Flux, position and velocity estimation in AC machines at zero and low speed via tracking of high frequenciy saliencies. Proc. EPE 95, Sevilla, pg. 3.154-3.159

[J8]   Jayne, M.G., Bowes, S.R., Bird, B.M., Developments in sinusoidal PWM-inverters. 2nd IFAC Symp., Control in Power Electronics and Electrical Drives, Düsseldorf 1977, pg. 145

[J9]   Jenni, F., Zwicky ,R., Speed and flux control of a pulse width modulated current source inverter drive. Proc. IEE Power Electronics and Variable Speed Drives, London 1988, pg. 374

[J10]   Jentsch, W., Mathematisches Modell der Asynchronmaschine für die digitale Simulation des dynamischen Verhaltens. Simposio sulle macchine elettriche ed i convertitori statici

[J11]   Jezernik, K., Curk, B., Harnik, J., Variable structure field oriented control of an induction motor drive. Proc. EPE 91, Firenze, pg. 2-161

[J12]   Jezernik, K., Hren, A., Drevensek, D., Robust continuous VSS tracking control of an IM drive. Proc. EPE 95, Sevilla, pg. 3.857-3.862

[J13]   Jezernik, K., Drevensek, D., Robust torque control of induction motor for electric vehicle. Proc. Workshop on Advanced Motion Control (AMC) 98, Coimbra, pg. 36-41

[J14]   Ji, J.K., Lee, D.C., Sul, S.K., LQG based speed controller for torsional vibration suppression in two mass motor drive system. Proc. IECON 93, Maui, pg. 1157

[J15]   Jönsson, R., Method and apparatus for controlling an AC induction motor by indirect measurement of the airgap voltage, PCT Patent Application No. PCT/SE 91/00086, International filing date 8.2.91. European Patent 0515469, Issued 1994. US-Patent 5 294 876, Issued 1994

[J16]   Jönsson, R., Leonhard, W., Control of an induction motor without a mechanical sensor, based on the principle of "natural field orientation" (NFO). Proc. IPEC Yokohama, 1995, pg. 303

[J17]   Jötten, R., French Patent 1.331.612, 1961

[J18]   Jötten, R, Regelungstechnische Probleme bei elastischer Verbindung zwischen Motor und Arbeitsmaschine. In Energieelektronik und geregelte elektrische Antriebe, VDE- Buchreihe, Vol. 11, 1962, pg. 446

[J19]   Jötten, R., Die Berechnung einfach und mehrfach integrierender Regelkreise der Antriebstechnik. AEG Mitteilungen 1962, pg. 219

[J20]   Jötten, R., Signalverarbeitung für die Regelung umrichtergespeister Drehfeldmaschinen. 1st IFAC Symp., Control in Power Electronics and Electrical Drives, Düsseldorf 1974, Vol. 1, pg. 681

[J21]   Jötten, R., Mäder, G., Control methods for good dynamic performance induction motor drives based on current and voltage as measured quantities. Proc. Int. Semiconductor Power Converter Conf. 1982, pg. 397

[J22]   Jötten, R, Zimmermann, P., Sub-and super-synchronous cascade with current source inverter. 3rd IFAC-Symp. Control in Power Electronics and Electrical Drives, Lausanne 1983, pg. 337

[J23]   Jötten, R., Kehl, C., A fast space-vector control for a three level voltage source inverter. Proc. EPE 91, Firenze, pg. 2-70

[J24]   Jones, C.I., van Nice, R., Frequenzgeber und -teiler mit veränderlicher Frequenz. German Patent 1.100.084, 1958

[J25]   Jones, R., Wymeersch, W., Lataire, Ph., Vector controlled drives in a steel processing application. EPE 93, Brighton, pg. 5-508

[J26]   Jufer, M., Indirect sensors for electric drives. Proc. EPE 95, Sevilla, pg. 1.836–1.841

[K1]   Kaczmarek, B., Mayeur, X., Legros, W., Capieaux, Ph., Commande optimisée par microprocesseurs d'une machine synchrone autopilotée. Conumel 83, Toulouse, pg. IV-19

[K2]   Kahlen, H., Maggetto, G., Electric and hybrid vehicles. Proc. EPE 97 Trondheim, pg. 1.030-1.054

[K3]   Kamiyama, K., Azusawa, N., Miyahara, Y., Ohmae, T., Kivisawa, T., Microprocessor controlled fast response speed regulator for thyristorised reversible regenerative DC-motor drives. IEEE Ind. Electr. and Control Instr. Ann. Meet. Rec. 1978, pg. 216

[K4]   Kamiyama, K., Nagasaki, H., Saito, K., Matsuka, S., Shiraishi, H., Ohmae, T., Microprocessor-based high-speed internal controllers for industrial drives. Microelectronics in Power Electronics and Electrical Drives, Darmstadt 1982, ETG-Fachberichte 11

[K5]   Kanzaki, T., Yanagida, K., Yasouka, I., Inverter control system for driving induction-motors in rapid transit cars using high-powered GTO thyristors. Proc. Int. Semiconductor Power Converter Conf. 1982, pg. 145

[K6]   Kasa, N., Watanabe, H., A sensorless position control by salient-pole brushless DC motor. Proc. EPE 97, Trondheim, pg. 4.529-4.534

[K7]   Kao, Y.T., Liu, C.H., Analysis and design of microprocessor-based vector-controlled induction motor drives. IEEE Trans. Ind. Electr. 1992, pg.46

[K8]   Kappius, F., Zur Drehzahlsteuerung von Drehstrom-Kurzschlußläufermotoren mit einer gleichstromgespeisten Stromrichteranordnung in sechspulsiger Mittelpunktschaltung. Diss. TH Braunschweig 1965

[K9]   Kastner, G., Pfaff, G., Sack, L., Design and performance of an acceleration drive with permanent magnets. IEEE Trans. Ind. Appl. 1991, pg. 908

[K10]   Kavanagh, R.C., Improved analysis and design techniques for digital tachometry. Diss. Univ. College Cork 1998

[K11]   Kavanagh, R.C., Murphy, J.M.D., The effects of quantization noise and sensor non-ideality on digital-differentiator-based velocity measurement. IEEE Trans. Instrum. Meas. 1998, pg. 1457-1463

[K12]   Kawamura, A., Yokoyama, T., Comparison of five control methods for digital feedback controlled PWM inverters. Proc. EPE 91, Firenze, pg. 2-35

[K13]   Kazmierkowski, M.P., Köpke, H.-J., Comparison of dynamic behaviour of frequency converter fed induction machine drives. 3rd IFAC Symp. Control in Power Electronics and Electrical Drives Lausanne 1983, pg. 313

[K14]   Kazerani, M., Ooi, B.T., Direct AC/AC matrix converter based on three-phase voltage- source converter modules. Proc. IECON 93, Maui, pg.812

[K15]   Kazmierkowski, M.P., Kasprowicz, A., Improved torque and flux vector control of PWM inverter-fed induction motor drive. Proc. EPE 93, Brighton, pg.5-115

[K16]   Kazuno, H., A wide range speed control of an induction motor with static Scherbius and Kramer systems. J. Electr. Engg. Japan 1969, pg. 10

[K17]   Kelber, Ch.R., Aktive Dämpfung der doppelt-gespeisten Drehstrommaschine. Diss. TU Braunschweig, 2000

[K18]   Kennel, R., Schröder, D., Predictive control strategy for converters. 3rd IFAC Symp. Control in Power Electronics and Electrical Drives, Lausanne 1983, pg. 415

[K19]   Kennel, R., Kobs, G., Weber, R., Digital control of industrial servo drives for machine tools. Proc. EPE 99, Lausanne

[K20]   Kerkman, R.J., Rowan, T.M., Leggate, D., Indirect field-oriented control of an induction motor in the field-weakening region. IEEE Trans. Ind. Appl. 1992, pg. 850

[K21]   Kerkman, R.J., Leggate, D., Seibel, B.J., Rowan, T.M., An overmodulation strategy for PWM voltage inverters. Proc. IECON 93, Maui, pg. 1215

[K22]   Kessler, C., Über die Vorausberechnung optimal abgestimmter Regelkreise. Regelungstechnik 1954: pg. 274, 1955: pg. 11, 40

[K23]   Kessler, C., Das symmetrische Optimum. Regelungstechnik 1958, pg. 395, 432

[K24]   Kessler, C., Meinhardt, W., Neuffer, I., Rube, G., Die Gleichstrom-Fördermaschine mit Siemens-TRANSIDYN-Regelung. Siemens Zeitschrift 1958, pg. 555

[K25]   Kessler, C., Ein Beitrag zur Theorie mehrschleifiger Regelungen. Regelungstechnik 1960, pg. 261

[K26]   Kessler, G., Das zeitliche Verhalten einer kontinuierlichen elastischen Bahn zwischen aufeinanderfolgenden Walzenpaaren. Regelungstechnik 1960: pg. 436, 1961: pg. 154

[K27]   Kessler, G., Digitale Regelung der Relation zweier Drehzahlen. ETZ-A, 1961, pg. 574

[K28]   Kessler, G., Digitale Regelung der Drehzahl und der Relation zweier Drehzahlen. In Digitale Signalverarbeitung in der Regelungstechnik, VDE-Buchreihe, Vol. 8, 1962, pg. 152

[K29]   Kessler, G., Electrical drive systems for paper-, rubber- and plastics-industries, survey. IFAC Conference, Brussels 1976

[K30]   Khaburi, D.A., Tabar, F.M., Comsa, A., Sargos, F.M., A DSP-based con-
        troller for sensorless control of synchronous machines. Proc. EPE 97, Trond-
        heim, pg. 4.547-4.552
[K31]   Khambadkone, A., Holtz, J., Low switching frequency high power inverter
        drive based on field oriented pulsewidth modulation. Proc. EPE 91, Firenze,
        pg. 4-672
[K32]   Khanniche, M.S., Belaroussi, M., Sethuraman, S.K., An algorithm for gen-
        erating optimised PWM for real time micro control applications. Proc. IEE-
        Power Electronics and Variable Speed Drives, London 1988, pg. 386
[K33]   Kiel, E., Schumacher, W., Gabriel, R., PWM Gate-array for AC drives. Proc.
        EPE 87, Grenoble, pg. 653
[K34]   Kiel, E., Schumacher, W., Ehrenberg, J., Letas, H. H., Schrader-Hausmann,
        U., High performance digital control of uninterruptible power supply (UPS)
        using an application specific integrated circuit (ASIC). Proc. EPE 91,
        Firenze, pg. 3-174
[K35]   Kiel, E., Anwendungsspezifische Schaltkreise in der Drehstrom-Antriebs-
        technik. Diss. TU Braunschweig, 1994
[K36]   Kiel, E., Schumacher, W., Der VeCon-Prozessor, Signalverarbeitung für
        Antriebe auf einem Chip. Proc. SPS/IPC/DRIVES 94, Stuttgart, pg. 533-
        539
[K37]   Kiel, E., Schumacher, Vecon; A high-performance single-chip-servovontroller
        for AC drives. IPEC 95 Yokohama, 1995, pg. 1289
[K38]   Kiel, E., Schumacher, W., VECON: High-performance digital control of AC
        drives by one-chip servo controller. Proc. EPE 95, Sevilla, pg. 3.005-3.010
[K39]   Kiel, E., Loy, T., Der digitale Antriebsregler auf dem Weg vom ein-
        fachen Aktor zur dezentralen intelligenten Bewegungssteuerung. Proceedings
        SPS/IPC/DRIVES 98, Nürnberg, pg. 476-487
[K40]   Kielgas, H., Nill, R, Converter propulsion systems with three phase induction
        motors for electrical traction. Proc. Int. Semiconductor Power Converter
        Conf. Rec. 1977, pg. 304
[K41]   Kim, G.S., Ha, I.J., Ko, M.S., Control of induction motors for both high
        dynamic performance and high power efficiency. IEEE Trans. Ind. Electr.
        1992, pg. 323
[K42]   Kim, G.S., Kim, J.Y., Choi, J.Y., Choy, I., Recursive rotor resistance adap-
        tation algorithm for induction motor control. Proc. IECON 98, Aachen, pg.
        883-887
[K43]   Kim, H.Y., Shin, M.H., Hyun, D.S., Improved vector control of an induction
        motor with on-line tuning of its parameters. Proc. IECON 98, Aachen, pg.
        854-858
[K44]   Kim, J.M.S., Dewan, S.B., Closed loop voltage regulation of dual convert-
        ers with circulating-current mode, used in four-quadrant DC magnet power
        supplies. IEEE Trans. Ind. Appl. 1991, pg. 1026
[K45]   Kim, T.W., Kawamura, A., Slip frequency estimation for sensorless low
        speed control of induction motor. Proc. Workshop on Advanced Motion Con-
        trol (AMC) 96, Tsu City, pg. 156-161
[K46]   Kim, Y.H., Kim, S.S., Hong, I.P., Speed sensorless vector control of high-
        speed induction motors using intelligent control algorithm. Proc. IECON 98,
        Aachen, pg. 888-892
[K47]   Kim, Y.R., Sul, S.K., Park, M.H., Speed sensorless vector control of induc-
        tion motor using extended Kalman filter. IEEE Trans. Ind. Appl. 1994, pg.
        1225
[K48]   Kinniment, D.J., Acarnley, P.P., Jack, A.G., An integrated circuit controller
        for brushless DC drives. Proc. EPE 91, Firenze, pg. 4-111

[K49] Kinniment, D.J., Kappos, E., Acarnley, P.P., Experience of the use of ASIC methods in a motor control application. Proc. EPE 93, Brighton, pg. 458

[K50] Kirlin, R.L., Trzynadlowski, A.M., Bech, M.M., Blaabjerg, F., Pedersen, J.K., Analysis of spectral effects of random PWM strategies for voltage-source inverters. Proc. EPE 97 Trondheim, pg. 1.146-1.151

[K51] Kisck, D.O., Bucurenciu, S., Bucurenciu, M., Sirbu, G., Kisck, M., Real-time parameter estimation of vector-control induction motor fed by CR-PWM extended Kalman filter. Proc. EPE 99, Lausanne

[K52] Klaassen, H., Selbsteinstellende, feldorientierte Regelung einer Asynchronmaschine und geberlose Drehzahlregelung, Diss. TU Braunschweig, 1996

[K53] Klaes, N.R.,Identifikationsverfahren für die betriebspunktabhängigen Parameter einer wechselrichtergespeispen Induktionsmaschine. Diss. Univ. Bochum, 1992

[K54] Klaes, N.R., On-line tuning of the rotor resistance in an inverter-fed induction machine with direct-self-control. Europ. Trans. on Electr. Power Engineering. 1994, pg. 5

[K55] Klautschek, H., Das Verhalten der Induktionsmaschine bei Speisung über Strom-Zwischenkreisumrichter. ETZ 1974, pg. 283

[K56] Klautschek, H., Asynchronmaschinenantriebe mit Strom-Zwischenkreisumrichtern. Siemens Zeitschrift 1976, pg. 23

[K57] Klein, G, Stebel, H., Steuerung und Regelung von Hochlaufscheren und -sägen. BBC Nachrichten 1964, pg. 200

[K58] Kleinsorge, N., Putz, U., Stemmler, H., Large adjustable speed AC drives. Proc. EPE 87, Grenoble, pg. 491

[K59] Kliman, G.B., Plunkett, A.B., Development of a modulation strategy for a PWM inverter drive. IEEE Trans. Ind. Appl. 1979, pg. 72

[K60] Klug, R.- D., Effects and correction of switching dead times in 3-phase PWM inverter drives. Proc. EPE 89, Aachen, pg. 1261

[K61] Kobayashi, H., Endo, S., Kobayashi, S., Kempf, C.J., Robust digital tracking controller for high speed positioning systems - a new design approach and implementation techniques. Proc. Workshop on Advanced Motion Control (AMC) 96, Tsu City, pg. 65-70

[K62] Koch, W.H., Thyristor controlled pulsating field reluctance motor system. Electric Machines and Electromechanics 1977, pg. 201

[K63] Köllensperger, D., Die Synchronmaschine als selbstgesteuerter Stromrichtermotor. Siemens Zeitschrift 1967, pg. 830

[K64] Körber, J., Teich, W., Lokomotiven mit Drehstromantriebstechnik zur IVA 79, Beispiele eines breiten Anwendungsgebietes. Elektrische Bahnen 1979, pg. 151

[K65] Kolar, J.W., Ertl, H., Zach, F.C., Influence of the modulation method on the conduction and switching losses of a PWM converter system. IEEE Trans. Ind. Appl. 1991, pg. 1063

[K66] Konhäuser, W., Naunin, D., Digital control of a subsynchronous converter cascade by microcomputers. Microelectronics in Power Electronics and Electrical Drives, Darmstadt 1982, ETG-Fachberichte 11, pg. 369

[K67] Konishi, T., Kamiyama, K., Miyahara, Y., Ohmae, T., Sugimoto, N., An application technology of microprocessor to adjustable speed motor drives. IEEE Ind. Appl., Ann. Meet. Conf. Rec. 1978, pg. 669

[K68] Konishi, T., Kamiyama, K., Ohmae, T., A performance analysis of microprocessor-based control systems applied to adjustable speed motor drives. IEEE Trans. Ind. Appl. 1980, pg. 378

[K69] Koopmann, K., Meersmann, H., Restendloses Aufteilen von Walzgut durch rotierende Scheren. Siemens Zeitschrift 1964, pg. 218

[K70] Korb, F., Die Gefäßfolgesteuerung, ein Mittel zur Verminderung der Blindleistung von Stromrichteranlagen. ETZ-A, 1958, pg. 273

[K71] Korb, F., Drehzahlgeregelter Positionierantrieb. BBC Nachrichten 1968, pg. 102

[K72] Koseki, T., Morizane, T., Ohsaki, H., Masada, E., Novel linear induction drive: control scheme and converters. Proc. EPE 91, Firenze, pg. 1-481

[K73] Koyama, M., Yano, M., Two degrees of freedom speed controller using reference system model for motor drives. Proc. EPE 91, Firenze, pg. 3-60

[K74] Krafka, P., Bünte, A., Grotstollen, H., Comparison of induction machine control with orientation on rotor flux or stator flux in a very wide field weakening region. Proc. EPE 95, Sevilla, pg. 3.486-3.491

[K75] Kranenburg, R., Klaassen, H., Torri, G., Gatti, E., High performance induction machine drive for a flying shear, Proc. EPE 95, Sevilla, pg. 3-481

[K76] Krause, J., Zur Regelgüte von Thyristorstromrichter-Umkehrantrieben. Messen, Steuern, Regeln 1969, pg. 430

[K77] Krause, P.C., Method of multiple reference frames applied to the analysis of symmetrical machinery. IEEE-PAS, 1968, pg. 218

[K78] Krishnan, R., Stefanovic, V.R., Control principles in current source induction motor drives. IEEE Ind. Appl. Ann. Meet. Conf. Rec. 1980, pg. 605

[K79] Krishnan, R., Control and operation of PM synchronous motor drives in the field- weakening region. Proc. IECON 93, Maui, pg. 745

[K80] Krüger, L., Naunin, D., Dynamic modelling and model based control of an induction machine. Proc. EPE 97 Trondheim, pg. 3.819-3.824

[K81] Krüger, M., Bordstromversorgung mit schnellaufendem Synchrongenerator, Diss. TU Braunschweig, 1992

[K82] Krzeminski, Z., Speed and rotor resistance estimation in observer system of induction motor. Proc. EPE 91, Firenze, pg. 3-538

[K83] Kubo, K., Watanabae, M., Ohmae, I., Kamiyama, K., A software-based speed regulator for motor drives. Proc. Int. Power Electr. Conf., Tokyo 1983, pg. 1500

[K84] Kubota, H., Matsuse, K., Disturbance torque estimation for field oriented induction motor drives without rotational transducers. Proc. IECON 92, San Diego, pg. 336

[K85] Kubota, H., Speed sensorless field-oriented control of induction motor with rotor resistance adaptation. IEEE Trans. Ind. Appl. 1994, pg. 1219

[K86] Kübler, E., Der Stromrichtermotor. ETZ-A, 1958, pg. 15

[K87] Kühne, S., Berechnen statt messen - das Drehmoment elektrischer Antriebe. Proc. SPS/IPC/DRIVES 98, Nürnberg, pg. 140-149

[K88] Kühnert, S., Antriebe mit untersynchroner Stromrichterkaskade fur Kesselspeisepumpen. BBC Nachrichten 1975, pg. 588

[K89] Künzel, S., Papiernik, W., Permanenterregte Hauptspindelantriebe für Werkzeugmaschinen mit großem Feldschwächbereich. SPS/IPC/DRIVES 98, Nürnberg, pg. 99-109

[K90] Kulski, Z.H., Method of the air-gap electromagnetic torque identification for a current-fed AC drive system. Europ. Trans. on Electr. Power Engng. 1994, pg. 287

[K91] Kume, T., Iwakane, T., Sawa, T., Yoshida, T., Nagai, I., A wide constant power range vector-controlled AC motor drive using winding changeover technique. IEEE Trans. Ind. Appl. 1991, pg. 934

[K92] Kume, T., Sonoda, S., Ryu, H., Watanabe, E., A novel PWM technique to decrease lower harmonics. Proc. EPE 91, Firenze, pg. 3-223

[K93] Kume, T., Sawa, T., Yoshida, T., Sawamura, M., Sakamoto, M., A high speed vector controlled spindle motor drive with closed transition between with and without encoder control. IEEE Trans. Ind. Appl. 1992, pg. 421

[K94] Kunz, U., Eine digitale zeitoptimale Lageregelung. Diss. TH Darmstadt 1971

[K95] Kurz, R., Hochgenaue Drehzahlregelung mit Hilfe des Phase locked-loop. Elektrie 1978, pg. 658

[K96] Kusko, A., Application of microprocessors to AC and DC electric motor drive systems. IEEE Ind. Appl. Ann. Meet. Conf. Rec. 1977, pg. 1079

[L1]   Labahn, D., Untersuchungen an einem Stromrichtermotor in sechs-und zwölfpulsiger Schaltung. Diss. TU Braunschweig, 1961

[L2]   Läuger, A., The commutatorless DC motor with three-phase current excitation. 2nd IFAC Symp., Control in Power Electronics and Electrical Drives, Düsseldorf 1977, pg. 619

[L3]   Lagasse, J., Prajoux, R. Behaviour of control systems including controlled converters, especially rectifiers: A review of existing theories, survey. 1st IFAC Symp., Control in Power Electronics and Electrical Drives, Düsseldorf 1974, pg. 1

[L4]   Laithwaite, E.R., Kuznetsov, S.B., Development of an induction machine commutated thyristor inverter for traction drives. IEEE Ind. Appl. Ann. Meet. Conf. Rec. 1980, pg. 580

[L5]   Lampersberger, M., Weiss, H., Ebner, A., Schmidhofer, A., Improving the performance of a rotating tie-frequency converter control. Proc. EPE 99, Lausanne

[L6]   Langer, J., Umrichterspeisung von Synchronmotoren fur Rohrmühlen. BBC Mitteilungen 1970, pg. 112

[L7]   Langweiler, F., Richter, M., Flußerfassung in Asynchronmaschinen. Siemens Zeitschrift 1971, pg. 768

[L8]   Larabi, A., Grenier, D., Labrique, F., A simplified vector control scheme for an induction motor revisited. Proc. Workshop on Advanced Motion Control (AMC) 98, Coimbra, pg. 24-29

[L9]   Lataire, P., Induction motor drives with power transistor inverters. IEEE Ind. Appl. Ann. Meet. Conf. Rec. 1978, pg. 763

[L10] Lataire, Ph., Maggetto, G., Modern teaching on variable speed induction motor drives. Proc. EPE 89, Aachen, pg. 1445

[L11] Lataire, Ph., Van Muylem, H., Maggetto, G., Experiments for teaching on current- source inverter-fed induction motors with indirect flux control. Proc. EPE 91, Firenze, pg. 3-612

[L12] Latzel, W., Berechnung der Laststörung von drehzahlgeregelten Gleichstromantrieben mit geschachteltem Stromregelkreis. Regelungstechnik 1965, pg. 375

[L13] Lauffer, H., Die Drehstrommaschine mit polradwinkelabhängig eingeprägten Läuferströmen. AfE 1968, pg. 289

[L14] Lawrenson, P.J., Ray, W.F., Davis, RM., Stephenson, J.M., Fulton, N.N., Blake, R.J., Controlled-speed switched-reluctance motors: present status and future potential. Drives/Motors/Controls 82, Leeds

[L15] Lecocq, D., Lataire, Ph., Study of a variable speed, double fed induction motor drive system with both stator- and rotor- voltages controllable. Proc. EPE 91, Firenze, pg. 2-337

[L16] Lee, D.C., Sul, S.K., Park, M.H., High performance current regulator for a field- oriented controlled induction motor drive. IEEE Trans. Ind. Appl. 1994, pg. 1247

[L17]   Le-Huy, H., Perret, R, Roye, D., Microprocessor control of a current fed synchronous motor drive. IEEE Ind. Appl. Ann. Meet. Conf. Rec. 1979, pg. 873

[L18]   Leidhold, R., Garcia, G., Losses minimization in a variable speed field oriented controlled induction motor. Proc. IECON 98, Aachen, pg. 865-870

[L19]   Leitgeb, W., Zur Bemessung drehzahlveränderlicher Antriebe konstanter Leistung mit stromrichtergespeisten Drehfeldmaschinen. ETZ 1973, pg. 584

[L20]   Leimgruber, J., Stationary and dynamic behaviour of a speed controlled synchronous motor with cos δ or commutation limit line control. 2nd IFAC Symp., Control in Power Electronics and Electrical Drives, Düsseldorf 1977, pg. 463

[L21]   Leonard, H.W., A new system of electric propulsion. Trans. AIEE 1892, pg. 566

[L22]   Leonard, H.W., Volts vs Ohms – Speed regulation of electric motors. Trans. AIEE 1896, pg. 377

[L23]   Leonhard, A., Asynchroner und synchroner Betrieb der der allgemeinen doppeltgespeisten Drehstrommaschine. AfE 1936, pg. 483-502

[L24]   Leonhard, A., Kennlinien von Gleichstrommotoren. Elektrotechnik und Maschinenbau 1942, pg. 261

[L25]   Leonhard, A., Anlassen von Asynchronmotoren über KUSA-Widerstand. Elektrotechnik und Maschinenbau 1943, pg. 122

[L26]   Leonhard, A., Die Asynchronmaschine bei Laständerungen. Elektrotechnik und Maschinenbau 1965, pg. 3

[L27]   Leonhard, A., Die Drehstrommaschine mit eingeprägten Sekundärströmen. ETZ-A, 1968, pg. 97

[L28]   Leonhard, R., Schröder, D., New precalculating current control strategy for DC drives. Proc. EPE 87, Grenoble, pg. 659

[L29]   Leonhard, W., Elements of reactor controlled reversible induction motor drives. AIEE Trans. 1959, Pt II, pg. 106

[L30]   Leonhard, W., Digitale Regelung und Programmregelung in Digitale Signalverarbeitung in der Regelungstechnik. VDE-Buchreihe, Vol. 8, 1962, pg. 101

[L31]   Leonhard, W., Müller, H., Ein stetig wirkender digitaler Drehzahlregler. ETZ-A, 1962, S.381

[L32]   Leonhard, W., Zählende Rechenschaltungen für Regelaufgaben. AfE 1964, pg. 215

[L33]   Leonhard, W., Regelkreise mit symmetrischer Übertragungsfunktion. Regelungstechnik 1965, pg. 4

[L34]   Leonhard, W., Dynamic behaviour of a nonlinear sampled-data control loop using pulse width controlled thyristors as power amplifiers. 3rd IFAC Congress, London 1965, Paper 34 C.

[L35]   Leonhard, W., Regelkreis mit gesteuertem Stromrichter als nichtlineares Abtastproblem. ETZ 1965, pg. 513

[L36]   Leonhard, W., Regelprobleme bei der stromrichtergespeisten Drehstrom-Asynchronmaschine. VDE-Fachberichte 1968, pg. 50

[L37]   Leonhard, W., Cascade control, a versatile structural principle for technical and cybernetical control systems. Control and Instruments Systems 1970, pg. 28

[L38]   Leonhard, W., Optimale Zielsteuerung eines kräftefrei beweglichen Massepunktes. AfE 1970, pg. 65

[L39]   Leonhard, W., Zeitoptimale Scherenregelung. AfE 1976, pg. 61

[L40] Leonhard, W., Field orientated control of a variable speed alternator connected to the constant frequency line. IEEE Conf.Rec. Control in Power Systems, College Station 1979

[L41] Leonhard, W., Elektrische Regelantriebe für den Maschinenbau. VDI-Zeitschrift 1981, pg. 419

[L42] Leonhard, W., Microcomputer Control of High Dynamic Performance AC-Drives – A Survey, Automatica 1986, pg. 1

[L43] Leonhard, W., Adjustable speed AC drives. Proc. IEEE 1988, pg. 455

[L44] Leonhard,W., 30 years space vectors, 20 years field orientation, 10 years digital signal processing with controlled AC drives. EPE-Journal 1991, pg. 13, pg. 89

[L45] Leonhard, W., Field orientation with controlled AC drives - the natural solution to a travelling wave problem. Automatika 1995, Zagreb, pg. 5-9

[L46] Leonhard, W., Electrical drives - universal sources of mechanical energy. Proc. Workshop on Advanced Motion Control (AMC) 96, Tsu City, pg. 7-14

[L47] Le Pioufle, B., Louis, J.P., Influence of the dynamics of the mechanical speed of a synchronous servomotor on its torque regulation. Proc. EPE 91, Firenze, pg. 3-412

[L48] Lessmeier, R., Schumacher, W., Leonhard, W., Microprocessor-controlled AC servo drives with synchronous or induction motors: Which is preferable? IEEE Trans. Ind. Appl. 1986, pg. 812

[L49] Lessmeier, R., Digitale Signalverarbeitung für mehrachsige Antriebssysteme mit Drehstrom-Stellmotoren. Diss. TU Braunschweig 1989

[L50] Letas, H.H., Leonhard, W., Dual axis servo drive in cylindrical coordinates using permanent magnet synchronous motors with microprocessor control. Conumel 83, Toulouse 1983, pg. II-23

[L51] Letas, H.-H., Mikrorechner-geregelter Synchron-Stellantrieb. Diss. TU Braunschweig 1985

[L52] Levi, E., Vuckovic, V., Effects of main flux saturation on behavior of a stator and airgap flux oriented induction machine. Proc. EPE 91, Firenze, pg. 3-618

[L53] Levi, E., Adaptive decoupling circuit for universal field oriented controller operating in the airgap flux frame. Proc. EPE 93, Brighton, pg. 5-57

[L54] Levi, E., Vuckovic, V., Sokola, M., Rotor flux estimation in vector controlled induction machines incorporating the iron loss compensation. Proc. EPE 95, Sevilla, pg. 3.997-3.1002

[L55] Levi, E., Sokola, M., A novel saturation adaptive rotor flux estimator for rotor flux orientated induction machines. Proc. EPE 97 Trondheim, pg. 1.518-1.523

[L56] Li, Y., Jiang, L., Sensorless control of PMSM with an adaptive observer. Proc. EPE 99, Lausanne

[L57] Li, Y.D., de Fornel, B., Pietrzak-David, M., New approach for flux control of PWM fed induction motor drive. Proc. IEE-Power Electronics and Variable Speed Drives, London 1988, pg. 301

[L58] Liao, F., Sheng, J., Lipo, T.A., A new energy recovery scheme for doubly fed, adjustable-speed induction motor drives. IEEE Trans. Ind. Appl. 1991, pg. 728

[L59] Lienau, W., Commutation modes of a current source inverter. 2nd IFAC Symp., Control in Power Electronics and Electrical Drives, Düsseldorf 1977, pg. 219

[L60] Lienau, W., Müller-Hellmann, A., Skudelny, H.Ch., Power converters for feeding asynchronous traction motors of single phase AC vehicles. Proc. Int. Semiconductor Power Converter Conf. Rec. 1977, pg. 295

[L61] Lin, A.K., Koepsel, W.W., A microprocessor speed control system. IEEE Trans. Ind. Electr. and Control Instrumentation 1977, pg. 241

[L62] Lin, F.J., Liaw, C.M., Control of indirect field-oriented induction motor drives considering the effects of dead-time and parameter variations. IEEE Trans. Ind. Electr. 1993, pg.486

[L63] Lipo, T.A., Krause, P.C., Stability analysis of a rectifier inverter induction motor drive. IEEE Trans. Power Appl. Syst. 1969, pg. 55

[L64] Lipo, T.A., State variable steady state analysis of a controlled current induction motor drive. IEEE Trans. Ind. Appl. 1975, pg. 704

[L65] Lipo, T.A., Analysis and control of torque pulsations in current-fed induction motor drives. Proc. Power Electronics Specialists Conf. 1978

[L66] Lipphardt, G., Huth, S., A new field supply concept for industrial adjustable speed DC drives. Proc. EPE 99, Lausanne

[L67] Liu, C., Li, C., Wei, L. Li, F., Improvement of speed sensorless control method of induction motor. Proc. IECON 98, Aachen, pg. 986-990

[L68] Liu, S., Stiebler, M., A continuous discrete time state estimator for a synchronous motor fed by a PWM voltage source inverter. Proc. EPE 89, Aachen, pg. 865

[L69] Li, Y., Lee, F.C., A comparative study of zero-current-transition (ZCT) schemes for three-phase inverter applications. Proc. CPES Seminar 1999, Blacksburg, pg. 143-151

[L70] Lockwood, M., Fox, A.M., A novel high power transistor inverter. Proc. Int. Power Electronics Conference, Tokyo 1983, pg. 637

[L71] Löser, F., Sattler, Ph.K., Optimizing of the efficiency by the control of inverter-fed induction machines especially regarding saturation and heat effects. 3rd IFAC Symp. Control in Power Electronics and Electrical Drives, Lausanne 1983, pg. 25

[L72] Lorenz, R.D., Van Patten, K.W., High resolution velocity estimation for all-digital, AC servo drives. IEEE Trans. Ind. Appl. 1991, pg. 701

[L73] Lorenz, R.D., Yang, S.M., Efficiency-optimised flux trajectories for closed-cycle operation of field-orientation induction machine drives. IEEE Trans. Ind. Appl. 1992, pg. 574

[L74] Loron, L., Stator parameters influence on the field-oriented control tuning. Proc. EPE 93, Brighton, pg. 5-79

[L75] Louis, J.P., Nonlinear and linearised models for control systems including static converters. 3rd IFAC Symp. Control in Power Electronics and Electrical Drives, Lausanne 1983, pg. 9

[L76] Low, K.S., Deng, Y.Z., Guo, X.L., Two-degree-of-freedom control of a PMSM drive without a mechanical sensor. Proc. IECON 98, Aachen, pg. 498-502

[L77] Low, T.S., Lee, T.H., Chang, K.T., A nonlinear speed observer for permanent-magnet synchronous motors. IEEE Trans. Ind. Electr. 1993, pg.307

[L78] Ludtke, I., Arias, A., Jayne, M.G., Improving direct torque control of induction motors. Proc. EPE 99, Lausanne

[L79] Ludwig, E., German Patent 1.072.704, 1960

[L80] Luenberger, D.G., Observing the state of a linear system. IEEE Trans. Mil. Electr. 1964, pg. 74

[M1]   Maag, R.B., Characteristics and application of current source slip regulated AC induction motor drives. IEEE Ind. Gen. Appl. Ann. Meet. Conf. Rec. 1971, pg. 44

[M2]   Maeda, A., Huai Chin, T., Tanaka, H., Koga, T., Ohtani, T., Today's AC drive for industrial application in Japan. Proc. EPE 91, Firenze, pg. 2-618

[M3]   Maes, J., Melkebeek, J., Adaptive flux observer for sensorless induction motor drives with enhanced dynamic performance. Proc. EPE 99, Lausanne

[M4]   Magyar, P., The dynamic behaviour of the current control loop of a microcomputer-controlled DC drive in discontinuous current mode operation. Microelectronics in Power Electronics and Electrical Drives, Darmstadt 1982, ETG-Fachberichte 11, pg. 183

[M5]   Magyar, P., Schnieder, E., Vollstedt, W., Digitale Regelung und Steuerung einer stromrichtergespeisten Gleichstrommaschine mit Mikrorechner. Regelungstechnik 1982, pg. 378

[M6]   Maikath, R., Putz, U., Saupe, R., An all digital control for the LCI feeding a synchronous machine. EPE 91, Firenze, pg. 4-586

[M7]   Malesani, L., Tenti, P., Gaio, E., Piovan, R., Improved current control technique of VSI PWM inverters with constant modulation frequency and extended voltage range. IEEE Trans. Ind. Appl. 1991, pg. 365

[M8]   Malesani, L., Tomasin, P., PWM current control techniques of voltage source inverters- a survey. Proc. IECON 93, Maui, pg. 670

[M9]   Manninen, V., Application of direct torque control modulation technology to line converter. Proc. EPE 95, Sevilla, pg. 1.292-1.296

[M10] Martins, C.A., Carvalho, A.S., Roboam, X., Meynard, T.A., Evolution of induction motor control methods and related voltage-source inverter topologies. Proc. Workshop on Advanced Motion Control (AMC) 98, Coimbra, pg. 15-20

[M11] Marwali, M.N., Keyhani, A., Tjanaka, W., An integrated digital signal processor based system for induction motor vector control using off-line parameter identification. Proc. EPE 97 Trondheim, pg. 3.782-3.787

[M12] Mathys, S., Microprocessor-based multimode synchronous pulse width modulation. Microelectronics in Power Electronics and Electrical Drives, Darmstadt 1982, ETG-Fachberichte 11, pg. 237

[M13] Matsui, N., Sensorless operation of brushless DC motor. Proc. IECON 93, Maui, pg. 739

[M14] Matuonto, M., Monti, A., High performance field oriented control for cycloconverter-fed synchronous machine: from theory to aplication. Proc. EPE 95, Sevilla, pg. 3.458-3.463

[M15] Mauch, K., Ito, M.R., A multi-microprocessor AC drive controller. IEEE Ind. Appl. Ann. Meet. Conf. Rec., 1980, pg. 634

[M16] Maurer, F., Vergleich verschiedener Zündsteuerverfahren für netzgeführte Stromrichter. ETZ-A, 1974, pg. 50

[M17] Maurer, F., Stromrichtergespeiste Synchronmaschine als Vierquadrant-Regelantrieb. Diss. TU Braunschweig 1975

[M18] Mc Laren, S.G., Direct digital control of thyristors. 1st IFAC Symp., Control in Power Electronics and Electrical Drives, Düsseldorf 1974, Vol. 2, pg. 147

[M19] McMurray, W., SCR inverter commutated by an auxiliary impulse. IEEE Trans. Communications and Electronics, 1964, pg. 824-829

[M20] McMurray, W., Thyristor commutation in DC choppers - a comparative study. IEEE Trans. Ind. Appl. 1978, pg. 547-558

[M21] Meiners, U., Beiträge zur adaptiven Regelung elektrischer Antriebe. Diss. TU Braunschweig, 1996

[M22] Mertens, A., Harmonic distortion in three-phase inverters controlled by synchronous sigma-delta-modulation. Europ. Trans. on Electr. Power Engng. 1992, pg. 351

[M23] Mestha, L.K., Evans, P. D., Optimisation of losses in PWM inverters. Proc. IEE-Power Electronics and Variable Speed Drives, London 1988, pg. 394

[M24] Meyer, D., Vergleich klassischer und moderner Regelverfahren am Beispiel einer Schachtförderanlage. Diss. TU Braunschweig 1995

[M25] Meyer, H.G., Static frequency changers for starting synchronous machines. BBC Review 1964, pg. 526

[M26] Meyer, M., Möltgen, G., Wesselak, F., Der Quecksilberdampf-Stromrichter als Stellglied in Regelkreisen. Siemens Zeitschrift 1960, pg. 592

[M27] Meyer, M., Über die untersynchrone Stromrichterkaskade. ETZ 1961, pg. 589

[M28] Meyer, S., Sondermann, E., Mehrmotorenantrieb für Rotationsdruckmaschinen. Siemens Zeitschrift 1977, pg. 387

[M29] Michels, K., Fuzzy control for electrical drives ? Proc. EPE 97 Trondheim, pg. 1.102-1.105

[M30] Miki, I., Nakao, O., Nishiyama, S., A new simplified current control method for field- oriented induction motor drives. IEEE Trans. Ind. Appl. 1991, pg. 1081

[M31] Millet, C., Leroux, D., Li, Y., Feuvrie, B., Bergmann, C., Modern experimental tools for parameter identification and field-oriented control design of a low power induction machine. Proc. EPE 97, Trondheim, pg. 4.597-4.602

[M32] Miti, G.K., Renfrew, A.C., A microprocessor-based current profiling scheme for field-weakening applications in brushless DC motors. Proc. EPE 99, Lausanne

[M33] Miyashita, I., Ohmori, Y., Improvement of robustness on speed sensorless vector control of induction motor. Proc. EPE 91, Firenze, pg. 4-660

[M34] Miyashita, I., Ohmori, Y., A new speed observer for an induction motor using the speed estimation technique. Proc. EPE 93, Brighton, pg. 349

[M35] Mittenzwei, K., Probst, E., Thyristorgespeiste Fördermaschinen. BBC Nachrichten 1976, pg. 153

[M36] Möltgen, G., Neuffer, I., Stand der Technik der Antriebe mit netzgeführten Stromrichtern. In Energieelektronik und geregelte elektrische Antriebe, VDE-Buchreihe Vol. 11, 1960, pg. 239

[M37] Möltgen, G., Salzmann, Th., Power factor and current harmonics with cycloconverters fed by a three-phase supply. 2nd IFAC Symp., Control in Power Electronics and Electrical Drives, Düsseldorf 1977, pg. 181

[M38] Moghbelli, H., Adams, G.E., Hoft, R.G., Performance of a 10 HP switched reluctance motor and comparison with induction motors. IEEE Trans. Ind. Appl. 1991, pg. 531

[M39] Moerschell, J., Signal processor based field oriented vector control for an induction motor drive. Proc. EPE 91, Firenze, pg. 2-145

[M40] Moerschell, J., Tursini, M., A new vector control scheme for inverter-fed permenent magnet synchronous motor using DSP. Proc. EPE 91, Firenze, pg. 4-683

[M41] Mok, H.S., Kim, J.S., Kim, R.Y., Park, M.H., Sul, S.K., A stator flux orientation speed control of induction motor without speed sensor. Proc. EPE 91, Firenze, pg. 4-678

[M42] Mokrytzki, B., Pulse width modulated inverters for AC motor drives. IEEE Trans. Ind. Gen. Appl. 1967, pg. 493

[M43] Mokrytzki, B., The controlled slip static inverter drive, IEEE Trans. Ind. Appl. 1968, pg. 312

[M44] Monse, M., Müller,V., Cerv, H., Schulze, S., Modelling and dynamic simulation of electromechanical systems of equipment in the rolling mill field. Proc. Workshop on Advanced Motion Control (AMC) 98, Coimbra, pg. 469-474

[M45] Moreira, J.C., Lipo, T.A., Blasko, V., Simple efficiency maximiser for an adjustable frequency induction motor drive. IEEE Trans. Ind. Appl. 1991, pg.940

[M46] Moreira, J.C., Lipo, T.A., Modelling of saturated AC machines including air gap flux harmonic components. IEEE Trans. Ind. Appl. 1992, pg. 343

[M47] Moreira, J.C., Hung, K.T., Lipo, T.A., Lorenz, R.D., A simple and robust adaptive controller for detuning correction in field-oriented induction machines. IEEE Trans. Ind. Appl. 1992, pg. 1359

[M48] Moreno-Eguilaz, J.M., Peracaula, J., Real-time power measuring and monitoring for an efficient vector-controlled induction motor drive. Proc. IECON 98, Aachen, pg. 871-876

[M49] Morimoto, S., Sanada, M., Takeda, Y., Efficiency improvement of permanent magnet synchronous motor in constant power speed region. Proc. EPE 99, Lausanne

[M50] Moschetti, A., Microprocessor-based PWM-system for inverter-fed squirrel-cage motor traction drives. Microelectronics in Power Electronics and Electrical Drives, Darmstadt 1982, ETG-Fachberichte 11, pg. 425

[M51] Moynihan, J., Kavanagh, R.C., Egan, M.G., Murphy, J.M.D., Indirect phase current detection for field oriented control of a permanent magnet synchronous motor drive. Proc. EPE 91, Firenze, pg. 3-641

[M52] Mozdzer, A., Bose, B.K., Three phase AC power control using power transistors, IEEE Trans. Ind. Gen. Appl. 1976, pg. 499

[M53] Müller, M., Roboter mit Tastsinn, Diss. TU Braunschweig, Vieweg Verlag Braunschweig-Wiesbaden, 1993

[M54] Mühlöcker, H., Elektrische Ausrüstung einer großen Windkraftanlage. Siemens-Energietechnik 1979, pg. 443

[M55] Murakami, T., Yu, F., Ohnishi, K., Torque sensorless control in multidegree-of- freedom manipulator. IEEE Trans. Ind. Electr.1993, pg. 259

[M56] Murphy, J.M.D., Hoft, R.G., Howard, L.S., Controlled slip operation of an induction motor with optimum waveforms. 2nd Int. Conf. on Electr. Variable Speed Drives, IEE London 1979, pg. 157

[M57] Murphy, J.M.D., Egan, M.D., An analysis of induction motor performance with optimum PWM waveforms. Proc. Int. Conf. Electrical Machines, 1980, pg. 642

[M58] Mutoh, N., Sakai, K., Ueda, A., Tomita, H., Nandoh, K., Stabilizing methods at high frequencies for an induction motor driven by a PWM inverter. Proc. EPE 91, Firenze, pg. 2 352

[M59] Mutschler, P., Schulze, H., Control of current source inverter by single board computer for squirrel-cage motor drive of urban transport system. Microelectronics in Power Electronics and Electrical Drives, Darmstadt 1982, pg. 417

[N1]   Nabae, A., Otsuka, K., Uchino, H., Kurosawa, R., An approach to flux control of induction motors operated with variable frequency power supply. IEEE Trans. Ind. Appl. 1980, pg. 342

[N2]   Nabae, A., Akagi, H., Osanai, T., Unity power factor commutator-less motors based on a new commutation principle. Proc. Int. Semiconductor Power Converter Conf. 1982, pg. 342

[N3]   Nagase, H., Matsuda, Y., Ohnishi, K., Ninomiya, H., Koike, T., High performance induction motor drive system using a PWM inverter. IEEE-IAS Ann. Meeting, 1983

[N4]   Nakano, T., Oksawa, H., Endoh, K., A high performance cycloconverter-fed synchronous machine drive system. Proc. Int. Semiconductor Power Converter Conf. 1982, pg. 334

[N5]   Nashiki, M., Dote, Y., High performance current-controlled PWM transistor inverter-fed brushless servomotor. Proc. Int. Semiconductor Power Converter Conf. 1982, pg. 349

[N6]   Naunin, D., Ein Beitrag zum dynamischen Verhalten der frequenzges-
        teuerten Asynchronmaschine. Diss. TU Berlin 1968
[N7]   Naunin, D., The calculation of the dynamic behaviour of electric machines
        by space-phasors. Electric Machines and Electromechanics, 1979, pg. 33
[N8]   Neft, C.L., Schauder, C.D., Theory and design of a 30 HP matrix converter.
        IEEE Trans. Ind. Appl. 1992, pg. 546
[N9]   Neuffer, I., Stromrichterantriebe mit Momentumkehr. Siemens Zeitschrift
        1965, pg. 1079
[N10]   Nguyen, M.T., Tu Xuan, M., Simond, J.J., Digital transient torque or force
        measurement for rotating or linear AC machines. Proc. EPE 95, Sevilla, pg.
        1.319-1.321
[N11]   Nielsen, P.E., Thomsen, E.Chr., Nielsen, M.T., Digital voltage vector control
        with adaptive parameter tuning. Proc. EPE 89, Aachen, pg. 313
[N12]   Nielsen, P.E., Mueller, K., Vammen, S., Speed and position AC servo drive
        based on current vector control. Proc. EPE 91, Firenze, pg. 3-635
[N13]   Nilsen, R., Modeling, identification and control of an induction machine.
        Diss. NTH Trondheim, 1987
[N14]   Nishio, T., Minoguti, K., Uno, S., Hombu, M., Futami, M., Higuchi, M., Con-
        trol characteristics of an adjustable speed generating system with a flywheel
        excited by a DC link converter. Proc. EPE 97 Trondheim, pg. 2.695-2.700
[N15]   Nitschke, H.J., Der bürstenlose Motor, ein neuer universeller drehzahlregel-
        barer Drehstromantrieb. AEG Mitteilungen 1973, pg. 73
[N16]   Nobuyuki, K., Kamatani, M., Watanabe, H., Current source inverter drive –
        Speed sensorless vector controlled induction motor. Proc. IECON 93, Maui,
        pg. 983
[N17]   Nonaka, S., Neba Y., A PWM GTO current source converter-inverter system
        with sinusoidal inputs and outputs. IEEE Trans. Ind. Appl. 1989, pg. 76
[N18]   Nonaka, S., Neba Y., Quick regulation of sinusoidal output current in PWM
        converter- inverter system. IEEE Trans. Ind. Appl. 1991, pg. 1055
[N19]   Nonaka, S., Yamasaki, K., Hirohata, S., PWM current source converter-
        inverter system for railways application. Proc. EPE 91, Firenze, pg. 4-72
[N20]   Nonaka, S., Kawaguchi, T., A new variable speed AC generator system using
        a brushless self-excited type synchronous machine. IEEE Trans. Ind. Appl.
        1992, pg. 490
[N21]   Nonaka, S., Neba Y., Current regulated PWM-CSI induction motor drive
        system without a speed sensor. IEEE Trans. Ind. Appl. 1994, pg. 116
[N22]   Numeroli, R., Gatti, E., Torri, G., Kranenburg, R., Four quadrant, large
        power, IGBT vector controlled adjustable speed drive. Design and test. Proc.
        EPE 95, Sevilla, pg. 1.508-1.512
[N23]   Nuß , U., Huck, S., Systemdynamischer Zusammenhang zwischen Strom-
        modell, Spannungsmodell und Beobachter bei der Asynchronmaschine. etz-
        Archiv, 1990, pg. 227
[N24]   Nystrom, A., Hylander, J., Thorberg, K., Harmonic currents and torque
        pulsations with pulse width modulation methods in AC motor drives. Proc.
        IEE-Power Electronics and Variable Speed Drives, London 1988, pg. 378
[O1]   Obermeier, C., Kellermann, H., Brandenburg, G., Heinzl, J., Sensorless
        field-orientated speed control of a hybrid stepper motor using an extended
        Kalman filter. Proc. EPE 97 Trondheim, pg.1.238-1.243
[O2]   Obermeier, Ch., Realisierung eines positions-und drehzahlgeregelten Ser-
        voantriebs mittels eine Scheibenläufer-Schrittmotors. Proc. SPS/IPC/
        DRIVES 98, Nürnberg, pg. 171-180

[O3]   Odai, M., Hori, Y., Speed control of a 2-inertia system with gear backlash us-
        ing gear torque compensator. Proc. Workshop on Advanced Motion Control
        (AMC) 98, Coimbra, pg. 234-239
[O4]   Ogasawara, S., Akagi H., Nabae, A., The generalised theory of indirect vector
        control for AC machines. IEEE Trans. Ind. Appl. 1988, pg. 470
[O5]   Ogasawara, S., Akagi H., An approach to position sensorless drive for brush-
        less DC motors. IEEE Trans. Ind. Appl. 1991, pg. 928
[O6]   Ogino, Y., Murakami, Y., Ohishi, K., Nakaoka, M., High performance ultra-
        low speed control of inverter-fed linear induction motor using vector control
        scheme. Proc. EPE 93, Brighton, pg. 5-464
[O7]   Ohm, H.J., Oppermann, K., Raatz, E., Steuerung einer durchlaufenden Säge
        mit einem Kleinstrechner. AEG Mitteilungen 1976, pg. 266
[O8]   Ohmae, T., Matsuda, T., Suzuki, T., Azuwada, N., Kamiyama, K., Kon-
        ishi, T., A microprocessor-controlled fast-response speed regulator with dual
        mode current loop for DCM drives. IEEE Trans. Ind. Appl. 1980, pg. 388
[O9]   Ohnuki, T., Miyashita, O., Lataire, Ph., Maggetto, G., A three-phase PWM
        rectifier without voltage sensors. Proc. EPE 97 Trondheim, pg. 2.881-2.886
[O10]   Ohtani, T., Takada, N., Tanaka, K., Vector control of induction motor with-
        out shaft encoder. IEEE Trans. Ind. Appl. 1992, pg. 157
[O11]   Ohyama, K., Asher, G.M., Sumner,M., Comparison of the practical per-
        formance and operating limits of sensorless induction motor drives using
        a closed loop flux observer and a full order flux observer. Proc. EPE 99,
        Lausanne
[O12]   Oldenkamp, J.L., Peak, S.C., Selection and design of an inverter driven in-
        duction motor for a traction drive system. GE-Report 83 CRD 191, 1983
[O13]   Olomski, J., Bahnplanung und Bahnführung von Industrierobotern. Diss.
        TU Braunschweig, Vieweg Verlag Braunschweig-Wiesbaden 1989
[O14]   Omori, Y., Nakanishi, T., Kobayashi, H., A speed sensorless spatial vector
        controlled inverter added an auto-measuring function. Proc. EPE 95, Sevilla,
        pg. 3.452-3.457
[O15]   Orlik, B., Zum Betriebsverhalten mikrorechner-geregelter und pulswechsel-
        richter- gespeister Synchronmaschinen mit Permanenterregung. Diss. TU
        Braunschweig 1987
[O16]   Orlik, B., Economical inverter-fed AC drives with fully digital control by
        means of an 8 bit microcontroller. Proc. EPE 91 Firenze, pg. 4-578
[O17]   Orlik, B., Feldorientierte sensorlose Drehzahlregelung von Drehstrom-Asyn-
        chronmaschinen. Proc. SPS/IPC/DRIVES 94, Stuttgart, pg. 541-552
[O18]   Oumamar, A., Louis, J.P., El-Hefnawy, A., Design of an optimal, auto-
        adaptive current loop for DC motor. Realization with a hybrid device in-
        cluding a microprocessor. 2nd IFAC Symp., Control in Power Electronics
        and Electrical Drives, Düsseldorf 1977, pg. 593
[O19]   Oyama, J., Mamo, M., Abe, T., Higuchi, T., Yamada, E., High frequency
        method of sensor elimination in interior permanent magnet motors. Proc.
        EPE 97, Trondheim, pg. 4.535-4.540
[P1]   Pachter, M., Speed control of a field controlled d.c. traction motor. Auto-
        matica 1981, pg. 627
[P2]   Pagano, E., Perfetto, A., A comparison between operating conditions of
        inverter-fed asynchronous motors. 2nd IFAC Symp., Control in Power Elec-
        tronics and Electrical Drives, Düsseldorf 1977, pg. 399
[P3]   Paice, D.A., Induction motor speed control by stator voltage control. IEEE
        Trans. Power Appl. Syst. 1968, pg. 585
[P4]   Panaitescu, R.C., Norum, L.E., Kalman-filter based sensorless control of an
        IPM motor drive. Proc. EPE 97, Trondheim, pg. 4.567-4.572

[P5]   Papiernik, W., Teckentrup, M., Kartesische Bahnregelung für Servoantriebe. Proc. SPS/IPC/DRIVES 98, Nürnberg, pg. 353-362

[P6]   Parasuram, M.K., Ramaswami, B., Analysis and design of a current-fed inverter. 2nd IFAC Symp., Control in Power Electronics and Electrical Drives, Düsseldorf 1977, pg. 235

[P7]   Patel, H.S., Hoft, R.G., Generalised techniques of harmonic elimination and voltage control in thyristor inverters. IEEE Trans. Ind. Appl., Part I, Harmonic elimination 1973, pg. 310; Part II, Voltage control techniques 1974, pg. 666

[P8]   Pedersen, J.K., Blaabjerg, F., Jensen, J.W., Thoegersen, P., An ideal PWM-VSI inverter with feedforward and feedback compensation. Proc. EPE 93, Brighton, pg. 5-501

[P9]   Peixoto, Z.M.A., Seixas, P.F., Parameter identification for induction machines at standstill. Proc. EPE 99, Lausanne

[P10]   Pena, R.S., Clare, J.C., Asher, G.M., Implementation of vector control strategies for a variable speed doubly-fed induction machine for wind generation system. Proc. EPE 95, Sevilla, pg. 3.075-3.080

[P11]   Peng, F.Z., Fukao, T., Robust speed identification for speed-sensorless vector control of induction motors. IEEE Trans. Ind. Appl. 1994, pg. 1234

[P12]   Pereira, L.F.A., Haffner, J.F., Hemerly, E.M., Gründling, H.A., A simulation framework for flux estimation and vector control of induction machines. Proc. IECON 98, Aachen, pg. 1587-1591

[P13]   Pereira, L.F.A., Haffner, J.F., Hemerly, E.M., Gründling, H.A., Direct vector control for a servopositioner using an alternative rotor flux estimation algorithm. Proc. IECON 98, Aachen, pg. 1603-1608

[P14]   Peresada, S., Tilli, A., Tonielli, A., Indirect field-oriented control of induction motor: New design leads to improved performance and efficiency. Proc. IECON 98, Aachen, pg. 1609-1614

[P15]   Peresada, S., Tilli, A., Tonielli, A., Robust active-reactive power control of a doubly-fed induction generator. Proc. IECON 98, Aachen, pg. 1621-1625

[P16]   Perin, A.J., da Cunha, A.A., de Oliveira, M.A., Variable optimal PWM for voltage source rectifiers and current source inverters. Proc. EPE 89, Aachen, pg. 1279

[P17]   Perng, S.S., Lai, Y.S., Liu, C.H., Sensorless vector controller for induction motor drives with parameter identification. Proc. IECON 98, Aachen, pg. 1008-1013

[P18]   Pesse, G., Pagá, T., A permanent magnet synchronous motor flux control scheme without position sensor. Proc. EPE 97, Trondheim, pg. 4.553-4.560

[P19]   Peters, K.W., Theoretische und praktische Untersuchungen an einer reihenschlußerregten, stromrichtergespeisten Reluktanzmaschine. Diss. TU Braunschweig 1979

[P20]   Peters, W., SYNAX, ein Motion-Control-System für wellenlose Druck- und Papierverarbeitungsmaschinen. Proc. SPS/IPC/DRIVES 97, Nürnberg, pg. 377-389

[P21]   Petersen, I.R., Pulle, D.W.J., Robust Kalman filtering in direct torque control. Proc. EPE 97, Trondheim, pg. 4.585-4.590

[P22]   Peterson, B., Valis, J., Oscillations in inverter fed induction motor drives. Proc. EPE 91 Firenze, pg. 2-369

[P23]   Peterson, B., Induction machine speed estimation; observations on observers. Diss. TH Lund, 1996

[P24]   Peter-Contesse, L.-O., Pietrzak-David, M., Ben Ammar, F., de Fornel, B., High-performance control for high-power induction machine without speed sensor: choice and comparison of two methods. Proc. EPE 95, Sevilla, pg. 3.430-3.435

[P25]   Pfaff, G., Zur Dynamik des Asynchronmotors bei Drehzahlsteuerung mit veränderlicher Speisefrequenz. ETZ-A 1964, pg. 719

[P26]   Pfaff, G., Time-response analysis of inverter-fed induction motors with controlled stator current. 2nd IFAC Symp., Control in Power Electronics and Electrical Drives, Düsseldorf 1977, pg. 493

[P27]   Pfaff, G., Scheurer, H.G., Parameterunempfindliche Regelung von Gleichstrommotoren im Feldschwächbereich. ETZ-A, 1969, pg. 400

[P28]   Pfaff, G., Weschta, A., Wick, A., Design and experimental results of a brushless AC servo motor. IEEE Ind. Appl. Ann. Meet. Conf. Rec. 1982, pg. 692

[P29]   Pfaff, G., Wick, A., Direkte Stromregelung bei Drehstromantrieben mit Pulswechselrichter. Regelungstechn. Praxis 1983, pg. 472

[P30]   Pfaff, G., Segerer, H., Resistance corrected and time discrete calculation of rotor flux in induction motors. Proc. EPE Aachen 89, pg. 499

[P31]   Pfitscher, G.H., A microprocessor control scheme for naturally commutated thyristor converter with variable frequency supply. 3rd IFAC Symp. Control in Power Electronics and Electrical Drives, Lausanne 1983, pg. 431

[P32]   Phillips, K.P., Current source converter for AC motor drives. IEEE Trans. Ind. Appl. 1972, pg. 679

[P33]   Pienaar, A.J., A contribution to the solution of instability problems in supply commutated power electronic converter-fed systems. Diss. Randse Africaanse Universiteit Johannesburg 1980

[P34]   Pietrzak-David, M., de Fornel, B., Universal PWM inverter fed asynchronous drive with a dynamic flux control. Proc. EPE 89, Aachen, pg. 1329

[P35]   Plaetrick, H., Digitale Lageregelungen. AEG Mitteilungen 1976, pg. 291

[P36]   Plunkett, A.B., Direct flux and torque regulation in a PWM inverter-induction motor drive. IEEE Trans. Ind. Appl. 1977, pg. 139

[P37]   Plunkett, A.B., A current-controlled PWM transistor inverter drive. IEEE Ind. Appl. Ann. Meet. Conf. Rec. 1979, pg. 785

[P38]   Plunkett, A.B., Lipo, T.A., New methods of induction motor torque regulation. IEEE Trans. Ind. Appl. 1976, pg. 47

[P39]   Plunkett, A.B., Turnbull, F.G., Load commutated inverter synchronous motor drive without a shaft position sensor. IEEE Trans. Ind. Appl. 1979, pg. 63

[P40]   Pohl, K., Rumold, G., Digitaler Regler in TTL-Technik für genaue Drehzahl- und Drehzahlverhältnisregelung. Siemens Zeitschrift 1972, pg. 335

[P41]   Pollmann, A., Gabriel, R, Zündsteuerung eines Pulswechselrichters mit Mikrorechner. Regelungstechnische Praxis 1980, pg. 145

[P42]   Pollmann, A., A digital pulse-width modulator employing advanced modulation techniques. IEEE Trans. Ind. Appl. 1983, pg. 409

[P43]   Pollmann, A., Comparison of PWM techniques. Microelectronics in Power Electronics and Electrical Drives, Darmstadt 1982, ETG-Fachberichte, pg. 231

[P44]   Pollmann, A., Ein Beitrag zur digitalen Pulsbreitenmodulation bei wechselrichtergespeisten Asynchronmotoren. Diss. TU Braunschweig, 1984

[P45]   Pollmann, A., Tosetto, A., Brea, P., Microprocessor based control system for rolling mills. Proc. EPE 91 Firenze, pg. 1-64

[P46]   Prajoux, R., Jalade, J., Marpinard, J.C., Mazankine, J., A modelling method for the behaviour of converters operating in control loops. 2nd IFAC Symp., Control in Power Electronics and Electrical Drives, Düsseldorf 1977, pg. 67

[P47]   Preuß, H.-P., Störungsunterdrückung durch Zustandsregelung. Regelungstechnik 1980, pg. 227, 266

[P48] Probst, U., Comparison of disturbance suppression strategies for servo-
       drives. Proc. EPE 93, Brighton, pg. 5-442
[P49] Probst, E., Antriebstechnik in Konti-Walzwerken. BBC Nachrichten 1977,
       pg. 485
[P50] Profumo, F., Pastorelli, M., Ferraris, P., De Doncker, R.W., Comparison
       of Universal Field Oriented (UFO) controller in different reference frames.
       Proc. EPE 91 Firenze, pg. 4-689
[P51] Putz, U., Thyristor industrial drives. Proc. Int. Semiconductor Power Con-
       verter Conf. Rec. 1977, pg. 340
[P52] Pyrhönen, J., Niemelä, M., Kaukonen, J., Luukko, J., Pyrhönen, O., Tiiti-
       nen, P., Väänänen, J., Synchronous motor drives based on direct flux linkage
       control. Proc. EPE 97 Trondheim, pg.1.434-1.439
[P53] Pyrhönen, O., Analysis and control of excitation, field weakening and stabil-
       ity in direct torque controlled electrically excited synchronous motor drives.
       Diss. Lappeenrantta 1998
[P54] Pyrhönen, J., Pyrhönen, O., Niemelä, M., Luukko, J., A direct torque con-
       trolled synchronous motor drive concept for dynamically demanding appli-
       cations. Proc. EPE 99, Lausanne
[Q1]   Quang, Ng.Ph., Schnelle Drehmomenteinprägung in Drehstromstellantrie-
       ben. Diss. TU Dresden, 1991
[Q02] Quang, Ng.Ph., Schönfeld, R., Dynamische Stromregelung zur Drehmo-
       menteinprägung in Drehstromantrieben mit Pulswechselrichter. Archiv für
       Elektrotechnik, 1993, pg. 317
[R1]   Raatz, E., Der Einsatz von adaptiven Drehzahlreglern in der Antriebstech-
       nik. AEG Mitteilungen 1970, pg. 375
[R2]   Raatz, E., Der Einfluß von elastischen Übertragungselementen auf die Dy-
       namik geregelter Antriebe. AEG Mitteilungen 1973, pg. 205
[R3]   Raducanu, D.C., Hochdynamisches Positionsregelsystem Axumerik. BBC
       Nachrichten 1977, pg. 301
[R4]   Rapp, H., Examination of transient phenomena in induction machines,
       caused by an incorrectly adjusted rotor time constant in a field-orientated
       control scheme. Europ. Trans. on Electr. Power Engng. 1993, pg. 397
[R5]   Rasmussen, H., Knudsen, M., Toennes, M., Inverter and motor model adap-
       tation at stand-still using reference voltages and measured currents. Proc.
       EPE 95, Sevilla, pg. 1.367-1.372
[R6]   Rasmussen, K.S., Thoegersen, P., Model based energy optimiser for vector
       controlled induction motor drives. Proc. EPE 97 Trondheim, pg. 3.711-3.716
[R7]   Rasmussen, K.S., Field magnitude adaptation in induction motors. Proc.
       EPE 97 Trondheim, pg. 3.922-3.926
[R8]   Rasmussen, H., Vadstrup, P., Boersting H., Nonlinear control of induction
       motors: a performance study. Proc. Workshop on Advanced Motion Control
       (AMC) 98, Coimbra, pg. 109-116
[R9]   Ray, W.F., Al-Bahadly, I.H., Sensorless methods for determining the rotor
       position of switched reluctance motors. Proc. EPE 93, Brighton, pg. 6-7
[R10] Rebhan, M., Modulare Echtzeitsimulation in der elektrischen Energietech-
       nik. Proc. EPE Firenze 91, pg. 1-74
[R11] Retif, J.M., Allard, B., Jorda, X., Perez, A., Use of ASIC's in PWM tech-
       niques for power converters. Proc. IECON 93, Maui, pg. 683
[R12] Ribbeck, G., Gudakowski, M., Digitale Mischungsregelung System BLEN-
       DOMAT. AEG Mitteilungen 1963, pg. 217
[R13] Ribeiro, L.A.S., Jacobina, C.B., Parameter and speed estimation for induc-
       tion machines based on dynamic models. Proc. EPE 95, Sevilla, pg. 1.496-
       1.501

[R14]   Ribickis, L.S., Ivbuls, U.V., Greivulis, J.P., Petrov, S.S., Static Krämer drive with specially designed voltage inverter. Proc. EPE 91 Firenze, pg. 2-301

[R15]   Richardson, J. Bezanov, G., A PWM 3-phase on-line controller for AC drives. Proc. EPE Firenze 91, pg. 3-299

[R16]   Richter, W., Microprocessor-controlled inverter-fed synchronous motor drive. Proc. IEE Conf. on Variable Speed Drives, London 1979, pg. 161

[R17]   Richter, W., Einsatz eines Mikrorechners zur Regelung eines Stromrichter-motors bei Betrieb mit maximalem Leistungsfaktor. Diss. TU Braunschweig 1981

[R18]   Richter, W., Microprocessor control of the inverter-fed synchronous motor. Process Automation 1981, pg. 100

[R19]   Riedo, P.-J., Cascade digital control by state variable feedback method applied to DC-motor. Microelectronics in Power Electronics and Electrical Drives, Darmstadt 1982 ETG-Fachberichte 11, pg. 249

[R20]   Riedo, P.-J., Cascade control by state variable feedback method applied to a synchronous motor. 3rd IFAC Symp. Control in Power Electronics and Electrical Drives, Lausanne 1983, pg. 111

[R21]   Riefenstahl, U., Einsatz von Mikrorechnern zur Steuerung und Regelung elektrischer Antriebssysteme. Elektrie 1978, pg. 245

[R22]   Riefenstahl, U., Nguyen H., Bannack, A., Tensile force and angular synchro-nism control for flexibly coupled drives by the example of a cage-stranding machine. Proc. EPE 99, Lausanne

[R23]   Riehlein, D., Getriebeloser Antrieb für eine Zementmahlanlage. Siemens Zeitschrift 1971,S. 189

[R24]   Riese, M. A microcontroller implementation of speed sensorless field oriented induction machine. Proc. EPE 97, Trondheim, pg. 4.476-4.479

[R25]   Robert, P.Ph., Gautier, M., Bergmann, C., Millet, C., Feuvrie, B., Experi-mental identification of asynchronous machine. Proc. EPE 97, Trondheim, pg. 4.620-4.625

[R26]   Robyns, B., Buyse, H., Labrique, F., Comparison of the sensitivity of the flux control to parameter uncertainties in two induction actuator indirect field oriented control schemes. Proc. EPE 93, Brighton, pg. 5-402

[R27]   Robyns, B., Buyse, H., Sensitivity of the flux control to parameter uncer-tainties in induction motor direct F.O.C. strategies using a simple flux esti-mation. Proc. EPE 95, Sevilla, pg. 3.674-3.679

[R28]   Robyns, B., Meuret, R., Sente, P., Current digital control influence on per-formance of induction motor indirect field orientation control. Proc. EPE 99, Lausanne

[R29]   Rohner, R., Lineare Servomotoren im Maschinenbau. Proceedings SPS/IPC/DRIVES 98, Nürnberg, pg. 533–539

[R30]   Rojek, P., Bahnführung eines Industrieroboters mit Multiprozessorsystem. Diss. TU Braunschweig, 1987

[R31]   Rowan, T.M., Kerkman, R.J., Leggate, D., A simple on-line adaptation for indirect field orientation of an induction machine. IEEE Trans. Ind. Appl. 1991, pg. 720

[R32]   Rufer, A.-Ch., Schibli, N., Briguet, Ch., A direct coupled 4-quadrant multi-level converter for 16 2/3 Hz traction systems. Proc. Conf. Power electronics and variable speed drives, IEE 1996, London, pg. 448-453

[R33]   Ruff, M., Grotstollen, H., Identification of the saturated mutual inductance of an asynchronous motor at standstill by recursive least squares algorithm. Proc. EPE 93, Brighton, pg. 5-103

[S1]   Salama, S., Lennon, S., Overshoot and limit cycle free current control method for PWM inverters. Proc. EPE 91 Firenze, pg. 3-247

[S2]   Salo, M., Tuusa, H., A vector controlled PWM current source inverter-fed induction motor drive with a novel rotor time constant estimation method. Proc. EPE 99, Lausanne

[S3]   Salvatore, L., Stasi, S., Tarchoni, L., A new EKF-based algorithm for flux estimation in induction machines. IEEE Trans. Ind. Electr. 1993, pg. 496

[S4]   Salzmann, Th., Wokusch, H., Direktumrichterantrieb für große Leistungen und hohe dynamische Anforderungen. Siemens Energietechnik 1980,pg. 409

[S5]   Salzmann, Th., Kratz, G., Däubler, Chr., High power drive system with advanced power circuitry and improved digital control. IEEE Trans. Ind. Appl. 1993, pg.168

[S6]   Santisteban, J.A., Stephan, R.M., Comparative study of vector control methods for induction machines. Proc. EPE 93, Brighton, pg. 5-390

[S7]   Sathiakumar, S., Nguyen, M.T., Shrivastava, Y., Sensorless-speed control of induction motor. Proc. EPE 99, Lausanne

[S8]   Sato, N., A study of commutatorless motor. Electr. Engin. in Japan 1964, pg. 42

[S9]   Sato, N., A brushless DC motor with armature induced voltage commutation. IEEE Trans. Ind. Gen. Appl. 1971, pg. 1485

[S10]   Sato, N., Semenov, V.V., Adjustable speed drive with a brushless DC motor. IEEE Trans. Ind. Gen. Appl. 1971, pg. 539

[S11]   Sato, N., Ishida, M., Sawaki, N., Some characteristics of induction motor driven by current source inverter. Proc. Int. Semiconductor Power Converter Conf. Rec. 1977, pg. 477

[S12]   Sattler, P.K., Balsliemke, G., Experimental and theoretical investigations concerning the design of induction machines with high power density with respect to the inverter supply and speed regulation. 2nd IFAC Symp., Control in Power Electronics and Electrical Drives, Düsseldorf 1977, pg. 387

[S13]   Sattler, Ph.K., Stärker, K., Estimation of speed and pole position of an inverter fed permanent excited synchronous machine. Proc. EPE 89, Aachen, pg. 1207

[S14]   Sawaki, N., Sato, N., Steady-state and stability analysis of induction motor driven by current source inverter. IEEE Trans. Ind. Gen. Appl. 1977, pg. 244

[S15]   Schäfer, U., Brandenburg, G., Compensation of Coulomb friction in industrial elastic two-mass systems through model reference adaptive control. Proc. EPE 89, Aachen, pg. 1409

[S16]   Schäfer, U., Comparative study of inverter-fed drives with different motor types. Proc. EPE 95, Sevilla, pg. 1.502-1.507

[S17]   Schauder, C., Adaptive speed identification for vector control of induction motors without rotational transducers. IEEE Trans. Ind. Appl. 1992, pg. 1054

[S18]   Schierling, H., Self-commisioning, a novel feature of inverter-fed motor drives. Proc. IEE-Power Electronics and Variable Speed Drives, London 1988, pg. 287

[S19]   Schlabach, L.A., Analysis of discontinuous current in a 12-pulse thyristor DC motor drive. IEEE Trans. Ind. Appl. 1991, pg. 1048

[S20]   Schnieder, E., Control of DC-drives by microprocessors. 2nd IFAC Symp., Control in Power Electronics and Electrical Drives, Düsseldorf 1977, pg. 603

[S21]   Schnieder, E., Digitale Nachbildung stromrichtergespeister Maschinen unter besonderer Berücksichtigung der wechselrichtergespeisten reihenschlußerregten Reluktanzmaschine. Diss. TU Braunschweig 1978

[S22]   Schnieder, E., Modelling and simulation of numerically controlled thyristorfed-drives. Proc. CONUMEL 80, p. III/12-23

[S23]   Schönfeld, R., Einfluß der Getriebelose auf Stabilität und Genauigkeit von Lageregelkreisen. Messen, Steuern, Regeln 1965, pg. 40

[S24]   Schönfeld, R., Krug, H., Näherungsverfahren zur dynamischen Beschreibung der Stromrichterstellglieder. Messen, Steuern, Regeln 1972, pg. 246

[S25]   Schönfeld, R, Das dynamische Verhalten des Stromrichterstellgliedes im Lückbereich. Messen, Steuern, Regeln 1977, pg. 79

[S26]   Schönfeld, R., Entwicklungstendenzen der elektrischen Antriebstechnik. Elektrie 1981, pg. 451

[S27]   Schönfeld, R., Franke, M., Hasan, H., Müller, F., Intelligent drives in systems with decentralised intelligence. Proc. EPE 93, Brighton, pg. 5-489

[S28]   Schönfeld, R., Control of nonlinear motion processes in processing machines by individual electric drives. Proc. Workshop on Advanced Motion Control (AMC) 96, Tsu City, pg. 447-451

[S29]   Schönung, A., Stemmler, H., Geregelter Drehstrom-Umkehrantrieb mit gesteuertem Umrichter nach dem Unterschwingungsverfahren. BBC Nachrichten 1964, pg. 555

[S30]   Schönung, A., Stemmler, H., Static frequency changers with subharmonic control in conjunction with reversible variable speed drives. BBC Review 1964, pg. 555

[S31]   Schräder, A., Eine neue Schaltung zur Kreisstromregelung in Stromrichteranlagen. ETZ-A, 1969, pg. 331

[S32]   Schröder, D., Untersuchung der dynamischen Eigenschaften von Stromrichterstellgliedern mit natürlicher Kommutierung. Diss. TH Darmstadt 1969

[S33]   Schröder, D., Die dynamischen Eigenschaften von Stromrichterstellgliedern mit natürlicher Kommutierung. ETZ-A, 1970, pg. 242

[S34]   Schröder, D., Grenzen der Regeldynamik von Regelkreisen mit Stromrichterstellgliedern. Regelungstechnik 1973, pg. 322

[S35]   Schröder, D., Einsatz adaptiver Regelverfahren bei Regelkreisen mit Stromrichterstellgliedern. VDE-Aussprachetag Freiburg 1973, pg. 81

[S36]   Schröder, D., Selbstgeführter Stromrichter mit Phasenfolgelöschung und eingeprägtem Strom. ETZ-A, 1975, pg. 520

[S37]   Schroedl, M., Sensorless control of permanent magnet synchronous machines at arbitrary operating points using a modified "inform" flux model. Europ. Trans. on Electr. Power Engng. 1993, pg. 277

[S38]   Schroedl, M., Hennerbichler, D., Wolbank, T.M., Induction motor drive for electric vehicles without speed and position sensors. Proc. EPE 93, Brighton, pg. 5-271

[S39]   Schülting, L., Skudelny, H.-Ch., A control method for permanent magnet synchronous motors with trapezoidal electromotive force. Proc. EPE 91, Firenze, pg. 4-117

[S40]   Schütze, T., Strönisch, V., Low floor trams with IGBT- 3- level inverter. Proc. EPE 93, Brighton, pg. 6-92

[S41]   Schumacher, W., Microprocessor controlled AC servo drive. Microelectronics in Power Electronics and Electrical Drives, Darmstadt 1982, ETG-Fachberichte, pg. 311

[S42]   Schumacher, W., Leonhard, W., Transistor-fed AC-servo drive with microprocessor control. Proc. Int. Power Electronic Conf., Tokyo 1983, pg. 1465

[S43]   Schumacher, W., Letas, H.-H., Leonhard, W., Microprocessor-controlled AC-servo-drives with synchronous and asynchronous motors. Proc. IEE Conf. on Power Electronics and Variable-Speed-Drives, London 1984, pg. 233

[S44]   Schumacher, W., Mikrorechner-geregelter Asynchron-Stellantrieb. Diss. TU Braunschweig 1985

[S45]   Schumacher, W., Rojek, P., Letas, H.H., Hochauflösende Lage- und Dreh-
        zahlerfassung optischer Geber für schnelle Stellantriebe. Elektronik, 1985,
        S.65
[S46]   Schumacher, W., Kiel, E., Elektrische Antriebe mit moderner Mikroelek-
        tronik: Der Schlüssel zur Mechatronik. Proc. SPS/IPC/DRIVES 94, Stutt-
        gart, pg. 607-614
[S47]   Schuster, A., A drive-system with digitally controlled matrix-converter feed-
        ing an AC-induction machine. Proc. Conf. Power electronics and variable
        speed drives, IEE 1996, London, pg. 378-382
[S48]   Seefried, E., Stromregelung im Lückbereich von Stromrichter-Gleichstrom-
        antrieben. Elektrie 1976, pg. 308
[S49]   Seeger, G., Selbsteinstellende, modellgestützte Regelung eines Roboters,
        Diss. TU Braunschweig, Vieweg Verlag Braunschweig-Wiesbaden, 1991
[S50]   Serrano-Iribarnegaray, L., The modern space-phasor theory. Part I: Its co-
        herent formulation and its advantages for transient analysis of converter-fed
        AC machines. Europ. Trans. on Electr. Power Engng. 1993, pg. 171. Part
        II: Comparison with the generalised machine theory and the space vector
        theory. Europ. Trans. on Electr. Power Engng. 1993, pg. 213
[S51]   Shepherd, W., On the analysis of three-phase induction motor with voltage
        control by thyristor switching. IEEE Trans. Ind. Gen. Appl. 1968, pg. 304
[S52]   Shepherd, W., Stanway, J., Slip power recovery in an induction motor by
        the use of a thyristor inverter. IEEE Trans. Ind. Appl. 1969, pg. 74
[S53]   Shinohara, K., Mieno, Y., Current control and magnet configurations for
        high torque to current ratio in surface mounted permanent magnet syn-
        chronous motors. Proc. EPE 97 Trondheim, pg. 3.865-3.869
[S54]   Shrivastava, Y., Nguyen, M.T., Sathiakumar, S., Speed estimator for induc-
        tion motor. Proc. EPE 99, Lausanne
[S55]   Simons, S., Robuste kartesische Bahnregelung eines sensorgeführten Indus-
        trieroboters mit digital geregelten Drehstromantrieben. Diss. TU Braun-
        schweig 1995
[S56]   Singh, D., Hoft, R.G., Microcomputer-controlled single-phase cyclocon-
        verter. Proc. IECI 1978, pg. 233
[S57]   Skudelny, H.-Ch., Stromrichterschaltungen für Wechselstrom-Triebfahrzeu-
        ge. ETZ-A 1966, pg. 249
[S58]   Sng, E.K.K., Liew, A.C., On-line tuning of rotor flux observers for field
        oriented drives using improved stator-based flux estimator for low speeds.
        Proc. EPE 95, Sevilla, pg. 1.437-1.442
[S59]   Sobottka, U., Einfluß der Temperatur auf das Betriebsverhalten der dreh-
        zahlgeregelten Asynchronmaschine. Siemens Zeitschrift 1969, pg. 760
[S60]   Soran, I. F., Kisch, D. O., Improved scheme for speed control of an asyn-
        chronous machine by field oriented method. Proc. EPE 91, Firenze, pg. 2-218
[S61]   Speth, W., Selbstanpassende Regelsysteme in der Antriebstechnik. Diss. TU
        Braunschweig 1971
[S62]   Sprecher, J., David, A., Desdoude, M., High power adjustable speed drive for
        power plant: Current source structure with thyristors for induction machine.
        Proc. EPE 95, Sevilla, pg. 3.056-3.061
[S63]   Steimel, A.,. Control of the induction machine in traction. Proc. PEMC 98,
        Prag, pg. K4.1-4.6
[S64]   Steimer, P.K., Redundant, fault-tolerant control system for a 13 MW high
        speed drive. Proc. IEE-Power Electronics and Variable Speed Drives, London
        1988, pg. 343

[S65] Steiner, M., Deplazes, R., Stemmler, H., A new transformerless topology for AC-fed traction vehicles using multi-star induction motors. Proc. EPE 99, Lausanne

[S66] Steinke, J.K., Switching frequency optimal PWM control of a three-level inverter. Proc. EPE 89, Aachen, pg. 1267

[S67] Stemmler, H., Speisung einer langsam laufenden Synchronmaschine mit einem direkten Umrichter. VDE-Fachtagung Elektronik 1969, pg. 177

[S68] Stemmler, H., Ein- und mehrpulsige Unterschwingungswechselrichter, Steuerverfahren, Strom- und Spannungsverhältnisse. 1st IFAC Symp., Control in Power Electronics and Electrical Drives, Düsseldorf 1974, Vol. 1, pg. 457

[S69] Stemmler, H., Wirk- und Blindleistungsregelung von Netzkupplungsumformern 50 Hz–16 2/3 Hz mit Umrichterkaskade. BBC-Mitteilungen 1978, pg. 614

[S70] Stemmler, H., Guggenbach, P., Configurations of high power voltage source inverter drives. Proc. EPE 93, Brighton, pg. 5-7

[S71] Stemmler, H., Power electronics in electric traction applications. Proc. IECON 93, Maui, pg. 707

[S72] Stephan, R.M., Americo, M., Lima, A.G.G., Rodrigues, A., Electric energy conservation in pumping processes at a refinery. Proc. EPE 95, Sevilla, pg. 3.633-3.637

[S73] Stepina, J., Fundamental equations of the space vector analysis of electrical machines. Acta Technica CSAV 13, 1968, pg. 184

[S74] Stephan, R.M. Field oriented and field acceleration control for induction motors: is there a difference? Proc. IECON 91, Kobe, pg. 567

[S75] Stoschek, J., 5 MVA-Umrichter für einen Linearmotorantrieb. BBC Nachrichten 1975, pg. 192

[S76] Ströle, D., Vogel, H., TRANSIDYN-Regelungen für Walzwerkantriebe. Regelungstechnik 1960, pg. 194

[S77] Ströle, D., Adaptivsysteme der elektrischen Antriebstechnik. ETZ-A, 1967, pg. 182

[S78] Sütö, Z., Nagy, I., Jákó, Z., Periodic responses of a nonlinear current controlled IM drive. Proc. EPE 97 Trondheim, pg. 3.847-3.852

[S79] Sugi, K., Naito, Y., Kurosawa, R., Kano, Y., Katayama, S., Yoshida, T., A microcomputer based high capacity cycloconverter drive for main rolling mill. Proc. Int. Power Electr. Conf., Tokyo 1983, pg. 744

[S80] Sugiura, M., Yamamoto, S., Sawaki, J., Matsuse, K., The basic characteristics of two-degrees-of-freedom PID position controller using a simple design method for linear servo drives. Proc. Workshop on Advanced Motion Control (AMC) 96, Tsu City, pg. 59-64

[S81] Sukegawa, T., Kamiyama, K., Mizuno, K., Matsui, T., Okuyama, T., Fully digital, vector-controlled PWM VSI-fed AC drives with an inverter dead-time compensation strategy. IEEE Trans. Ind. Appl. 1991, pg. 552

[S82] Sukegawa, T., Kamiyama, K., Takahashi, J., Imiki, T., Matsutake, M., A multiple PWM GTO line-side converter for unity power factor and reduced harmonics. IEEE Trans. Ind. Appl. 1992, pg. 1302

[S83] Sumner, M., Asher, G.M., Pena, R., The experimental investigation of rotor time constant identification for vector controlled induction motor drives during transient operating condition. Proc. EPE 93, Brighton, pg. 5-51

[S84] Svensson, T., Alaküla, M., The modulation and control of a matrix converter - synchronous machine drive, Proc. EPE 91, Firenze, pg. 469-476

[S85] Syrbe, M., Vermaschte Regelkreise, eine Möglichkeit zur Vereinfachung von Regelaufgaben. Tagung Heidelberg 1956, Oldenbourg 1957, pg. 78

[T1]   Tadakuma, S., Ehara, M., Historical and predicted trends of industrial AC drives. Proc. IECON 93, Maui, pg. 655

[T2]   Tadros, Y., Junge, G., Salama, S., Design aspects of high power PWM inverter with IGBT. Proc. EPE 91, Firenze, pg. 2-83

[T3]   Taguchi, T., Aida, K., Mukai, K., Yanagisawa, T., Kanai, T., Variable speed pumped storage system fed by large-scale cycloconverter. Proc. EPE 91, Firenze, pg. 2-237

[T4]   Tajima, H., Hori, Y., Speed sensor-less field-orientation control of the induction machine. IEEE Trans. Ind. Appl. 1993, pg. 175

[T5]   Takahashi, I., Noguchi, T., A new quick-response and high efficiency control strategy of an induction motor. IEEE-Trans. Ind. Appl. 1986, pg.820

[T6]   Takahashi, I., Kanmachi, Ultra-wide speed control with a quick torque response AC servo by a DSP. Proc. EPE 91. Firenze, pg. 3-572

[T7]   Takeshita, T., Matsui, N., Sensorless control and initial position estimation of salient-pole brushless DC motors. Proc. Workshop on Advanced Motion Control (AMC) 96, Tsu City, pg. 18-23

[T8]   Tamura, Y., Tanaka, S., Tadakuma, S., Control method and upper limit of output frequency in circulating current type cycloconverter. Proc. Int. Semiconductor Power Converter Conf. 1982, pg. 313

[T9]   Taniguchi, K., Tomiyama, Y., Iwatani, T., Irie, H., Acoustic noise reduced PWM converter and inverter using BJT's and IGBT's. Proc. EPE 91. Firenze, pg. 3-252

[T10]   Taufiq, J.A., Evaluating the step response of a traction VSI drive. IEE-Power Electronics and Variable Speed Drives, London 1988, pg. 340

[T11]   Tez, E.S., MOTONIC-Chip set: High performance intelligent controller for industrial variable speed AC drives. Proc. IEE-Power Electronics and Variable Speed Drives, London 1988, pg. 382

[T12]   Theocharis, J., Petridis, V., Harmonic insertion in PWM inverter drive schemes. Europ. Trans. on Electr. Power Engng. 1992, pg 143

[T13]   Theuerkauf, H., Digitale Nachbildung drehzahlgeregelter Drehstrom-Antriebe mit Stromrichterspeisung. IFAC Symp., Control in Power Electronics and Electrical Drives, Düsseldorf 1974, Vol. 1, pg. 709

[T14]   Theuerkauf, H., Zur digitalen Nachbildung von Antriebsschaltungen mit umrichtergespeisten Asynchronmaschinen. Diss. TU Braunschweig 1975

[T15]   Thompson, K.R., Acarnley, P.P., Parameter estimation from step excitation tests on a stationary induction motor. Proc. EPE 97, Trondheim, pg. 4.591-4.596

[T16]   Tiitinen, P., Pohjalainen, P., Lalu, J., The next generation motor control method: direct torque control (DTC). EPE Journal 1995, no. 1, pg. 14-18

[T17]   Timmer, R.A., PWM frequency inverters in the metal industry. Proc. EPE 91. Firenze, pg. 1-48

[T18]   Timpe, W., Cycloconverter drives for rolling mills. IEEE Trans. Ind. Appl. 1982, pg. 400

[T19]   Thoegersen, P., Toennes. M., Jaeger, U., Nielsen, S.E., New high performance vector controlled AC drive with automatic energy optimizer. Proc. EPE 95, Sevilla, pg. 3.381-3.386

[T20]   Thomas, J.L., Boidin, M., An internal model control structure in field oriented controlled VSI induction motors. Proc. EPE 91. Firenze, pg. 2-202

[T21]   Török, V., Valis, J., High accuracy and fast response digital speed measurement for control of industrial motor drives. 2nd IFAC Symp., Control in Power Electronics and Electrical Drives, Düsseldorf 1977, pg. 721

[T22] Török, V., Near optimum on-line modulation of PWM inverters. 3rd IFAC Symp. Control in Power Electronics and Electrical Drives, Lausanne 1983, pg. 247

[T23] Török, V., Loreth, K., The world's simplest motor for variable speed control? The Cyrano motor, a PM-biased SR-motor of high torque density. Proc. EPE 93, Brighton, pg. 6-44

[T24] Tomita, M., Sato, M., Yamaguchi, S., Doki, S., Okuma, S., Sensorless estimation of rotor position of cylindrical brushless DC motors using eddy currents. Proc. Workshop on Advanced Motion Control (AMC) 96, Tsu City, pg. 24-28

[T25] Tomizuka, M., Model based prediction, preview and robust controls in motion control systems. Proc. Workshop on Advanced Motion Control (AMC) 96, Tsu City, pg. 1-6

[T26] Tondos, M.S., Minimizing electromechanical oscillations in the drives with resilient couplings by means of state and disturbance observers. Proc. EPE 93, Brighton, pg. 5-360

[T27] Tondos, M., Mysinski, W., A microprocessor-based speed regulator using state and disturbance observer. Proc. EPE 97, Trondheim, pg. 4.626-4.631

[T28] Toennes, M., Rasmussen, H., Robust self tuning of AC servo drive. Proc. EPE 91. Firenze, pg. 3-49

[T29] Tripathi, A., Sen, P.C., Comparative Analysis of fixed and sinusoidal band hysteresis current controllers for voltage source inverters. IEEE Trans. Ind. Electr. 1992, pg. 63

[T30] Tröndle, H.-P., Regelung elastisch gekoppelter Vielmassensysteme. Siemens Forschungs- und Entwicklungs-Berichte 1975, pg. 45

[T31] Tröndle, H.-P., Regelung einer Spiegelantenne. Siemens Forschungs- und Entwicklungs- Berichte 1975, pg, 75

[T32] Trzynadlowski, A.M., Kirlin, R.L., Legowski, S., Space vector PWM technique with minimum switching losses and a variable pulse rate. Proc. IECON 93, Maui, pg. 689

[T33] Tso, S.K., Ho, P.T., Dedicated microprocessor scheme for thyristor phase control of multiphase converters. Proc. IEE 1981, pg. 128

[T34] Tsuji, M., Yamada, E., Parasiliti, F., Tursini, M., A digital parameter identification for a vector controlled induction motor. Proc. EPE 97, Trondheim, pg. 4.603-4.608

[T35] Tsuji, T., Sakakibara, J., Naka, S., CSI drive induction motor by vector approximation. IEEE Trans. Ind. Appl. 1991, pg.715

[T36] Turnbull, F.G., Selected harmonic reduction in static DC-AC inverters. IEEE Trans. Comm. Electr. 1964, pg. 374

[T37] Tuttas, C., Pseudo torque control (PTC) - a new control concept for induction motors. Proc. EPE 97 Trondheim, pg. 3.788-3.793

[U1]   Uchijima, T., Takigawa, M., Saijo, T., Speed control over a wide operating range of brushless DC motor applying optimal feedback control. Proc. EPE 91. Firenze, pg. 3-54

[U2]   Urbanke, C., Microcomputer control systems for rapid transit vehicles with AC-drives. Microelectronics in Power Electronics and Electrical Drives, Darmstadt 1982, ETG-Fachberichte 11, pg. 401

[U3]   Utkin, V.I., Variable structure systems with sliding mode. IEEE Trans. Autom. Contr. 1977, pg. 212

[V1]   van der Broeck, H., Analysis of harmonics in voltage fed inverter drives caused by PWM schemes with discontinuous switching operation. Proc. EPE 91. Firenze, pg. 3-261

[V2]   van der Broeck, H., Analysis of the voltage harmonics of PWM voltage fed inverters using high switching frequencies and different modulation functions. Europ. Trans. on Electr. Power Engng. 1992, pg. 341

[V3]   van der Broeck, H., Gerling, D., Bolte, E., Switched reluctance drive and PWM induction motor drive compared for low cost applications. Proc. EPE 93, Brighton, pg. 6-71

[V4]   van Haute, S., Henneberger, S., Hameyer, K., Belmans, R., de Temmerman, J., de Clerq, J., Design and control of a permanent magnet synchronous motor drive for a hybrid vehicle. Proc. EPE 97 Trondheim, pg. 1.470-1.575

[V5]   van Wyk, J.D., Holtz, H.R., A simple and reliable four quadrant variable frequency AC drive for industrial application up to 50 kW. 2nd IEE Int. Conf. Electr. Variable Speed Drives, London 1979, pg. 34

[V6]   van Wyck, J.D., Ferreira, J.A., Evaluation and future prospects of an integration technology for hybrid multi-kilowatt power electronic converters. Proc. EPE 97 Trondheim, pg. 1.194-1.199

[V7]   Vas, P., Alaküla, M., Hallenius, K.E., Brown, J.E., Field-oriented control of saturated AC machines. Proc. IEE-Power Electronics and Variable Speed Drives, London 1988, pg. 283

[V8]   Vas, P., Stronach, A.F., Rashed, M., Zordan, M., Chew, B.C., DSP implementation of sensorless DTC induction motor and PM synchronous motor drives with minimized torque ripples. Proc. EPE 99, Lausanne

[V9]   Völker, C.F., German Patent 906 951, 1944

[V10]   Vollstedt, W., Magyar, P., Adaptive control of DC drives by microcomputer. Microelectronics in Power Electronics and Electrical Drives, Darmstadt 1982, ETG-Fachberichte 11, pg. 289

[V11]   von Musil, R., Schmatloch, W., Bemessung und Ausführung großer drehzahlveränderbarer Synchronmotoren. Siemens Energie und Automation 1987, Special: Drehzahlveränderbare elektrische Großantriebe, S. 32-41

[V12]   Voss, U., Gleichstromstellertechnik in Nahverkehrsfahrzeugen. Elektrische Bahnen 1977, pg. 162

[V13]   Vranka, P., Griva, G., Profumo, F., Practical improvement of a simple V-I flux estimator for sensorless F.O. controllers operating in the low speed region. Proc. IECON 98, Aachen, pg. 1615-1620

[V14]   Vukosavic, S.L., Stojic, M.R., On-line tuning of the rotor time constant for vector- controlled induction motor in position control applications. IEEE Trans. Ind. Electr. 1993, pg. 130

[W1]   Wade, S., Dunnigan, M.W., Williams, B.W., Improvements for induction machine vector control. Proc. EPE 95, Sevilla, pg. 1.542-1.546

[W2]   Wagner, R., Elektronischer Gleichstromsteller für die Geschwindigkeitssteuerung elektrischer Triebfahrzeuge. Siemens Zeitschrift 1964, pg. 14

[W3]   Wagner, R, Beitrag zur Theorie des direkten Gleichstrom-Gleichstrom-Umrichters. Diss. TU Braunschweig 1968

[W4]   Wagner, R., Thyristortechnik für Gleichstrombahnen. Siemens Zeitschrift 1974, pg. 780

[W5]   Waidmann, M., Drehstromantrieb für Gleichstrombahnen. Siemens Zeitschrift 1976, pg. 493

[W6]   Walczyna, A.M., Koczara, W., Simulation study of current-controlled doubly-fed induction machine. Proc. EPE 89, Aachen, pg. 876

[W7]   Walcyna, A.M., Comparison of dynamics of doubly-fed induction machine controlled in field- and rotor- oriented axes. Proc. EPE 91. Firenze, pg. 2-231

[W8]   Waldmann, H., Koordinatentransformationen bei der Mehrgrößenregelung von Wechsel-und Drehstromsystemen. Diss. TU Braunschweig 1977

[W9]   Walker, L.H., Espelage, P.M., A high performance controlled-current inverter drive. IEEE Trans. Ind. Appl. 1980, pg. 193

[W10] Wang, Y., Grotstollen, H., Control strategies for the discontinuous current mode of AC drives with PWM inverters. Proc. EPE 91. Firenze, pg. 3-217

[W11] Warnecke, O., Einsatz einer doppeltgespeisten Asynchronmaschine in der großen Windenergieanlage Growian. Siemens Energietechnik 1983, pg. 364

[W12] Waschatz, U., Adaptive control of electrical drives. Microelectronics in Power Electronics and Electrical Drives, Darmstadt 1982, ETG-Fachberichte 11, pg. 135

[W13] Waschatz, U., Adaptive Regelung elektrischer Antriebe mit Hilfe von Mikrorechnern. Diss. TU Braunschweig 1984

[W14] Weh, H., Schröder, U., Static inverter concepts for multiphase machines with square-wave current-field distributions. Proc. EPE 1985, Brussels, pg. 1.147-1.152

[W15] Weibull, H., Magnusson, T., Valis, J., Standstill testing of properties of induction motors for inverter control. Proc. EPE 91. Firenze, pg. 2-363

[W16] Weidauer, J., Flexible machines using distributed intelligent drives. Proc. Workshop on Advanced Motion Control (AMC) 98, Coimbra, pg. 481-486

[W17] Weihrich, G., Drehzahlregelung von Gleichstromantrieben unter Verwendung eines Zustands- und Störgrößenbeobachters. Regelungstechnik 1978, pg. 349, 392

[W18] Weihrich, G., Wohld, D., Adaptive speed control of DC drives using adaptive observers. Siemens Forschungs- und Entwicklungs-Berichte 1980, pg. 283

[W19] Weiss, H.W., Adjustable speed AC drive systems for pump and compressor applications. IEEE Trans. Ind. Appl. 1975, pg. 162

[W20] Weiss, H., Rotor circuit GTO converter for slip-ring induction machine. Proc. EPE 97 Trondheim, pg. 2.707-2.710

[W21] Weiss, H., Lampersberger, M., Control system for the voltage source DC link converter in the rotor circuit of a slip-ring induction machine. Proc. EPE 97 Trondheim, pg. 3.853-3.858

[W22] Weiss, H., Ebner, A., Schmidhofer, A., Rotating system tie-frequency convertor 60 Hz – 25 Hz with rotor converter at a doubly-fed induction machine. Proc. EPE 99, Lausanne

[W23] Weninger, R, Einfluß der Maschinenparameter auf Zusatzverluste, Momentoberschwingungen und Kommutierung bei der Umrichterspeisung von Asynchronmaschinen. AfE 1981, pg. 19

[W24] Wheatley, C.T., Drives. Proc. EPE 93, Brighton, pg. 1-33

[W25] Wheeler, P.W., Grant, D.A., A low loss matrix converter for AC variable speed drives. Proc. EPE 93, Brighton, pg. pg. 5-27

[W26] Wieser, R., High dynamic torque calculator for inverter-fed induction machines. Proc. EPE 95, Sevilla, pg. 3.771-3.776

[W27] Wieser, R.S., Kral, Ch., Pirker, F., Schagginger, M., The Vienna induction machine monitoring method; on the impact of the field oriented control structure on real operating behaviour of a faulty machine. Proc. IECON 98, Aachen, pg. 1544-1549

[W28] Wilharm, H., Fick, H., Einsatz eines Beobachters zur genauen Bestimmung von Torsionsschwingungen. VDI-Berichte 1978, pg. 69

[W29] Willis, G.H., A study of the thyratron commutator motor. Gen. Electric Review 1933

[W30] Willner, L., Falk, M., Beitrag zur Regelung mechanischer Schwingungen beim Haspeln von Band. Regelungstechnik 1977, pg. 101

[W31] Wishart, M.T., Harley, R.G., Diana, G., The application of field oriented control to the brushless DC machine. Proc. EPE 91. Firenze, pg. 3-629

[W32] Wollenberg, G., Schwarz, W., Anwendungen von Mikroprozessoren in numerischen Steuerungen für Werkzeugmaschinen. Elektrie 1980, pg. 452

[W33] Wolters, P., Auslegung stetiger Vorschubantriebe für Werkzeugmaschinen. Regelungstechnik 1978, pg. 363

[W34] Wu, B., Slemon, G.S., Dewan, B.D., PWM-CSI induction motor drive with phase angle control. IEEE Trans. Ind. Appl. 1991, pg. 970

[W35] Wu, R., Slemon, G.S., A permanent magnet motor drive without a shaft sensor. IEEE Trans. Ind. Appl. 1991, pg. 1005

[W36] Wu, B., Dewan, S.B., Slemon, G.R., PWM-CSI inverter for induction motor drives. IEEE Trans. Ind. Appl. 1992, pg. 64

[W37] Wu, Y., Fujikawa, K., Kobayashi, H., A torque control method of a two-mass resonant system with PID-P controller. Proc. Workshop on Advanced Motion Control (AMC) 98, Coimbra, pg. 240-245

[W38] Würslin, R., Pulsumrichtergespeister Asynchronmaschinenantrieb mit hoher Taktfrequenz und sehr großem Feldschwächbereich. Diss. Univ. Stuttgart, 1984

[W39] Wyss, W., Gierse, H., Dupré- Latour, B., AC servo drives. Proc. EPE 89, Aachen, pg. 1061

[X1] Xue, Y., Xu, X., Habetler, T.G., Divan, D.M., A stator flux-oriented voltage source variable-speed drive based on DC link measurement. IEEE Trans. Ind. Appl. 1991, pg. 962

[X2] Xu, X., Novotny, D.W., Implementation of direct stator flux orientation control on a versatile DSP system. IEEE Trans. Ind. Appl. 1991, pg. 694

[X3] Xu, X., Novotny, D.W., Bus utilization of discrete CRPWM inverters for field-oriented drives. IEEE Trans. Ind. Appl. 1991, pg. 1128

[X4] Xu, X., Novotny, D.W., Selection of the flux reference for induction machine drives in the field weakening region. IEEE Trans. Ind. Appl. 1992, pg. 1353

[Y1] Yamamoto, H., Tomizuka, M., Torsional vibration control for steel mill drives via anti-vibration command generation. Proc. IPEC 95, Yokohama, pg. 447-452

[Y2] Yamada, K., Komada, S., Ishida, M., Hori, T., Analysis of servo system realized by disturbance observer. Proc. Workshop on Advanced Motion Control (AMC) 96, Tsu City, pg. 338-343

[Y3] Yang, G., Chin, T.-H., Adaptive speed identification scheme for a vector controlled speed sensor less inverter induction motor drive. IEEE Trans. Ind. Appl. 1993, pg. 820

[Y4] Yeomans, K.A., The controlled retardation of Ward Leonard drives. Proc. IEE, 1962, pg. 559

[Y5] Yeomans, K.A., Ward Leonard drives, 75 years of development. Electronics & Power, 1968 pg. 144

[Z1] Zach, F.C., Thiel, F.A., Pulse width modulated inverters for efficiency optimal control of AC-drives-Switching angles and efficiency/loss profiles. 3rd IFAC Symp. Control in Power Electronics and Electrical Drives, Lausanne 1983, pg. 231

[Z2] Zägelein, W., Drehzahlregelung des Asynchronmotors unter Verwendung eines Beobachters mit geringer Parameterempfindlichkeit. Diss. Univ. Erlangen-Nürnberg, 1984

[Z3] Zai, L.C., De Marco, Chr.L., Lipo, T.A., An extended Kalman filter approach to rotor time constant measurement in PWM induction motor drives. IEEE Trans. Ind. Appl. 1992, pg. 96

[Z4] Zamora, J.L., Garcia-Cerrada, A., Unified formulation for rotor-flux estimation in an induction motor. Proc. IECON 98, Aachen, pg. 309-314

[Z5]   Zdenkovic, J., Kuljic, Z., Pasalic, N., Speed sensorless drive with induction machine based on natural field orientation. Proc. EPE 95, Sevilla, pg. 1.752-1.757

[Z6]   Zeitz, M., Nichtlineare Beobachter. Regelungstechnik 1979, pg. 241

[Z7]   Ziegler, M., Hofmann, W., Der Matrix-Umrichter - Kommutierung in nur zwei Schritten. Proc. SPS/IPC/DRIVES 99, Nürnberg, pg. 521-530

[Z8]   Zimmermann, P., Super-synchronous static converter cascade. 2nd IFAC Symp., Control in Power Electronics and Electrical Drives, Düsseldorf 1977, pg. 559

[Z9]   Zolghadri, M.R., Diallo, D., Roye, D., Direct torque control system for synchronous machine. Proc. EPE 97 Trondheim, pg. 3.694-3.699

[Z10]   Zubek, J., Abbondanti, A., Nordby, C., Pulsewidth modulated inverter motor drives with improved modulation. IEEE Trans. Ind. Appl. 1975, pg. 695

[Z11]   Zürneck, H., Ein drehzahlgeregelter spannungsgesteuerter Stromrichter-Asynchronmotor. Diss. TH Darmstadt 1965

Index